Advances in Retinal Image Processing

Advances in Retinal Image Processing

Guest Editors

P. Jidesh

Vasudevan (Vengu) Lakshminarayanan

MDPI

Basel • Beijing • Wuhan • Barcelona • Belgrade • Novi Sad • Cluj • Manchester

Guest Editors
P. Jidesh
Department of Mathematical and Computational Sciences
National Institute of Technology Karnataka
Karnataka
India

Vasudevan (Vengu) Lakshminarayanan
Theoretical and Experimental Epistemology Laboratory
University of Waterloo
Waterloo, ON
Canada

Editorial Office
MDPI AG
Grosspeteranlage 5
4052 Basel, Switzerland

This is a reprint of the Special Issue, published open access by the journal *Journal of Imaging* (ISSN 2313-433X), freely accessible at: https://www.mdpi.com/journal/jimaging/special_issues/6OWLNIZNK8.

For citation purposes, cite each article independently as indicated on the article page online and as indicated below:

Lastname, A.A.; Lastname, B.B. Article Title. *Journal Name* **Year**, *Volume Number*, Page Range.

ISBN 978-3-7258-6189-7 (Hbk)
ISBN 978-3-7258-6190-3 (PDF)
https://doi.org/10.3390/books978-3-7258-6190-3

Contents

About the Editors

P. Jidesh

P. Jidesh is currently working as an Associate Professor in the Department of Mathematical and Computational Sciences at the National Institute of Technology Karnataka, India. He has more than 18 years of academic experience and 3 years of industry experience. He has published several papers in image processing, machine learning, deep learning, and computer vision. He has guided several Ph.D. theses. He has executed several research projects funded by various agencies. His areas of interest include computer vision, image processing and analysis, retinal image analysis, deep learning, and machine learning applications.

Vasudevan (Vengu) Lakshminarayanan

Vasudevan (Vengu) Lakshminarayanan is currently serving as a professor in the Department of Optometry and Vision Sciences at the University of Waterloo, Canada. He completed his Ph.D. from UC Berkeley. Lakshminarayana was a KITP Scholar at the Kavli Institute for Theoretical Physics at UC Santa Barbara and an associate of the Michigan Center for Theoretical Physics, and he has held research and teaching positions at UC Irvine, UC Berkeley, the University of Michigan, and the University of Missouri, among others. He has been a visiting professor of Physics at IIT Delhi, a Royal Society of Edinburgh visiting professor at Glasgow, an international visiting professor of biophotonics in the Department of Information Engineering at the University degli Studi di Brescia, Italy, and a Gian Professor of Physics at IIT Madras. He also served on UNESCO's International Year of Light planning/organizing committee, is a founding member of the UNESCO ALOP Program, and is on the International Day of Light committee. He is on the optics advisory board of the International Center for Theoretical Physics at Trieste, Italy, and a consultant to the medical devices group of the US FDA. He has represented the United States at two IUPAP general assemblies, Additionally, he served as the chair of the US advisory committee for the International Commission on Optics, chair of the committee on international scientific affairs of the APS, member of the Public Policy Committee of APS, member of executive committee of the Forum on International Physics and the Forum on Education, an AAAS Science and technology policy fellow, and a director of the OSA amongst other professional services. His areas of research interest include Clinical Imaging, Vision Rehabilitation, Optical Devices, Technology and Optics, and Optical Systems.

Preface

Every scientific domain has benefited from advances in AI and related fields. Retinal imaging and image processing also witnessed a tremendous progression in this direction. This Special Issue highlights some of those advancements in retinal imaging and image processing. The articles in this Special Issue provide an insightful introduction to technological and scientific advancements in retinal image processing. Nine contributory papers make up this Special Issue. These papers examine state-of-the-art developments in various retinal imaging applications. These cutting-edge research findings will serve as a reference for future research activities in the retinal imaging domain.

P. Jidesh and Vasudevan (Vengu) Lakshminarayanan
Guest Editors

Journal of *Imaging*

MDPI

Editorial

Editorial on the Special Issue: "Advances in Retinal Image Processing"

P. Jidesh [1,*] and Vasudevan Lakshminarayanan [2]

1 Department of Mathematical and Computational Sciences, National Institute of Technology Karnataka, Surathkal 575025, Karnataka, India
2 Theoretical and Experimental Epistemology Laboratory, University of Waterloo, Waterloo, ON N2J 4A8, Canada
* Correspondence: jidesh@nitk.edu.in

Retinal disorders are one of the major causes of visual impairment. The damage to the retina can lead to severe visual challenges that can eventually cause blindness [1]. The lack of available early detection facilities is one of the challenges in handling this problem. Prompt diagnosis of a retinal disorder can reduce the risk of visual impairment in many cases [2]. Two of the prominent retinal imaging modalities that are widely used for retinal examination are fundus images and optical coherence tomographs. Fluorescent angiography is yet another non-invasive imaging modality used for retinal analysis [3]. An ophthalmologist analyzes these images to arrive at a diagnosis. Due to the large patient populations, especially in developing countries, a proper diagnosis becomes a challenging task for the clinicians. Moreover, a lack of adequate facilities for acquiring images, especially in remote areas, is another challenge for the timely diagnosis of disorders. Automated retinal analysis can handle large amounts of data with minimal manual intervention. The advent of AI and machine/deep learning modalities in the last decade has changed the face of the scientific world. Retinal image analysis has also witnessed commendable progress due to the advancement of AI and ML in the present context [4], with AI and ML models becoming more efficient and accurate [5]. Convolutional Neural Networks (CNNs) and their variants are also employed in retinal imaging applications [6]. A further development is reported in [7], wherein the authors go one step further and utilize transformer networks for retinal image processing. The ability of these models to handle large amounts of input data in a short time has revolutionized the domain of retinal image processing. The domain of retinal image enhancement and registration, multi-modal analysis, and multiple disorder detection are the current focuses of retinal imaging, and AI models are employed extensively in their implementations.

In this regard, this Special Issue focuses on multiple topics related to the retinal image acquisition, analysis, and processing (https://www.mdpi.com/journal/jimaging/special_issues/6OWLNIZNK8, accessed on 10 October 2025). Nine contributory papers are included as a part of this Special Issue. Among them, eight are research articles and one is a comprehensive review article. Each of these papers discuss various cutting-edge technologies proposed for various retinal imaging and image processing applications. All of these papers have undergone several rounds of rigorous review before being considered for publication. A brief description of the papers accepted for the Special Issues is highlighted below.

The first contributory paper, titled "Robust PCA with $L_{w,*}$ and $L_{2,1}$ Norms: A Novel Method for Low-Quality Retinal Image Enhancement", is related to retinal image preprocessing and enhancement for efficient analysis and detection of disorders. The authors

 https://doi.org/10.3390/jimaging11100366

have studied the role of Principal Component Analysis (PCA) with different norm constraints for enhancing low-quality retinal images that are degraded in their row forms in many applications. As claimed by the authors, their method introduces a novel parameter update approach and significantly improves retinal image quality, detecting cataracts and diabetic retinopathy.

A retinal image quality assessment technique is studied in the second contributory article titled "Subjective Straylight Index: A Visual Test for Retinal Contrast Assessment as a Function of Veiling Glare". A contrast assessment of retinal images was performed as a function of Veiling Glare. Contrast assessments are vital for analyzing and making a diagnosis from the images. If the image quality hampers the diagnosis and detection process it can eventually result in spurious analysis. The authors propose a new computational methodology to generate visual acuity charts affected by ocular scattering effects. A retinal contrast assessment is analytically studied in this work.

In the third contributed paper titled "When Sex Matters: Differences in the Central Nervous System as Imaged by OCT through the Retina", the authors compute mean value fundus images for the neuroretina layers as imaged via OCT of healthy individuals. Texture metrics were obtained from these images to assess whether women and men have the same retina texture characteristics in both eyes. The authors arrive at the conclusion that the differences between the right and left eyes manifest differently in females and males.

The main objective of the fourth contribution, titled "Arteriovenous Length Ratio: A Novel Method for Evaluating Retinal Vasculature Morphology and Its Diagnostic Potential in Eye-Related Diseases", is to propose a new method for assessing one of the morphological changes in the fundus through morphometric analysis of retinal images. The proposed method in this paper introduces a novel approach called the arteriovenous length ratio (AVLR). Unlike commonly used measures such as the arteriovenous width ratio or tortuosity, the AVLR focuses on assessing the relative length of arteries and veins in the retinal vasculature.

The fifth contributory paper, titled "Retinal Microvasculature Image Analysis Using Optical Coherence Tomography Angiography in Patients with Post-COVID-19 Syndrome", highlights the analysis of retinal microvasculature changes in patients after COVID-19. In this study, the authors investigated the retinal microvasculature in PCS patients using OCT-angiography and analyzed the macular retinal nerve fiber layer (RNFL) and ganglion cell layer (GCL) thickness via spectral domain-OCT (SD-OCT).

In the sixth contributory paper titled "Evaluating Retinal Disease Diagnosis with an Interpretable Lightweight CNN Model Resistant to Adversarial Attacks", the authors propose a lightweight CNN model for retinal disease diagnosis evaluation. Given the computational requirement of a CNN model, a lightweight model is required to handle large volumes of data efficiently. The advent of deep learning frameworks has changed the face of retinal image analysis. This OCT-based model built on a CNN framework has been found to work efficiently in diagnosing retinal diseases.

An early screening and diagnosis of retinal diseases can be performed from optical coherence tomography (OCT) images using deep learning methods. In the seventh contributory paper titled "Multi-Fundus Diseases Classification Using Retinal Optical Coherence Tomography Images with Swin Transformer V2", the authors propose a shifting window-based transformer for multi-fundus disease classification using OCT images. The transformers efficiently attend to important features as they use a multi-headed attention network for feature enhancement. Furthermore, the Swin transformers improve their computational efficiency.

In the eighth contributory paper titled "Advancements in Cataract Detection The Systematic Development of LeNet-Convolutional Neural Network Models", the authors portray various advancements in cataract detection using a CNN-based model. Despite the fact that cataracts can be diagnosed and treated successfully, patients often delay seeking medical attention due to the relatively asymptomatic nature of the disease. To address this challenge, this research focuses on the identification of cataract abnormalities using image processing techniques and machine learning for preliminary assessment.

Finally, in the ninth and final paper titled "Denoising of Optical Coherence Tomography Images in Ophthalmology Using Deep Learning: A Systematic Review", a systematic survey of denoising OCT images under a deep learning framework is carried out. This is a review contribution where the authors have completed an elaborative analysis on various state-of-the-art denoising methods for OCT image restoration. The efficiency of deep learning frameworks in handling the denoising task make them a better choice compared to traditional restoration models.

Conflicts of Interest: The authors declare no conflict of interest.

List of Contributions:

1. Likassa, H.T.; Chen, D.-G.; Chen, K.; Wang, Y.; Zhu, W. Robust PCA with $L_{w,*}$ and $L_{2,1}$ Norms: A Novel Method for Low-Quality Retinal Image Enhancement. *J. Imaging* **2024**, *10*, 151. https://doi.org/10.3390/jimaging10070151.
2. Ávila, F.J.; Casado, P.; Marcellán, M.C.; Remón, L.; Ares, J.; Collados, M.V.; Otín, S. Subjective Straylight Index: A Visual Test for Retinal Contrast Assessment as a Function of Veiling Glare. *J. Imaging* **2024**, *10*, 89. https://doi.org/10.3390/jimaging10040089.
3. Nunes, A.; Serranho, P.; Guimarães, P.; Ferreira, J.; Castelo-Branco, M.; Bernardes, R. When Sex Matters: Differences in the Central Nervous System as Imaged by OCT through the Retina. *J. Imaging* **2024**, *10*, 6. https://doi.org/10.3390/jimaging10010006.
4. Badawi, S.A.; Takruri, M.; Al-Hattab, M.; Aldoboni, G.; Guessoum, D.; ElBadawi, I.; Aichouni, M.; Chaudhry, I.A.; Mahar, N.; Nileshwar, A.K. Arteriovenous Length Ratio: A Novel Method for Evaluating Retinal Vasculature Morphology and Its Diagnostic Potential in Eye-Related Diseases. *J. Imaging* **2023**, *9*, 253. https://doi.org/10.3390/jimaging9110253.
5. Noor, M.; McGrath, O.; Drira, I.; Aslam, T. Retinal Microvasculature Image Analysis Using Optical Coherence Tomography Angiography in Patients with Post-COVID-19 Syndrome. *J. Imaging* **2023**, *9*, 234. https://doi.org/10.3390/jimaging9110234.
6. Bhandari, M.; Shahi, T.B.; Neupane, A. Evaluating Retinal Disease Diagnosis with an Interpretable Lightweight CNN Model Resistant to Adversarial Attacks. *J. Imaging* **2023**, *9*, 219. https://doi.org/10.3390/jimaging9100219.
7. Li, Z.; Han, Y.; Yang, X. Multi-Fundus Diseases Classification Using Retinal Optical Coherence Tomography Images with Swin Transformer V2. *J. Imaging* **2023**, *9*, 203. https://doi.org/10.3390/jimaging9100203.
8. Ganokratanaa, T.; Ketcham, M.; Pramkeaw, P. Advancements in Cataract Detection: The Systematic Development of LeNet-Convolutional Neural Network Models. *J. Imaging* **2023**, *9*, 197. https://doi.org/10.3390/jimaging9100197.
9. Ahmed, H.; Zhang, Q.; Donnan, R.; Alomainy, A. Denoising of Optical Coherence Tomography Images in Ophthalmology Using Deep Learning: A Systematic Review. *J. Imaging* **2024**, *10*, 86. https://doi.org/10.3390/jimaging10040086.

References

1. Abràmoff, M.D.; Garvin, M.K.; Sonka, M. Retinal imaging and image analysis. *IEEE Rev. Biomed. Eng.* **2010**, *3*, 169–208. [CrossRef] [PubMed]
2. Early Treatment Diabetic Retinopathy Study Research Group. Early photocoagulation for diabetic retinopathy. ETDRS Rep. 9. *Ophthalmology* **1991**, *98* (Suppl. S5), 766–785.
3. Baudoin, C.E.; Lay, B.J.; Klein, J.C. Automatic detection of microaneurysms in diabetic fluorescein angiography. *Rev. Epidemiol. Sante Publique* **1984**, *32*, 254–261. [PubMed]

4. Bahr, T.; Vu, T.A.; Tuttle, J.J.; Iezzi, R. Deep Learning and Machine Learning Algorithms for Retinal Image Analysis in Neurodegenerative Disease: Systematic Review of Datasets and Models. *Transl. Vis. Sci. Technol.* **2024**, *13*, 16. [CrossRef] [PubMed]
5. Patil, A.D.; Biousse, V.; Newman, N.J. Artificial intelligence in ophthalmology: An insight into neurodegenerative disease. *Curr. Opin. Ophthalmol.* **2022**, *33*, 432–439. [CrossRef] [PubMed]
6. Bhandari, M.; Shahi, T.B.; Neupane, A. Evaluating Retinal Disease Diagnosis with an Interpretable Lightweight CNN Model Resistant to Adversarial Attacks. *J. Imaging* **2023**, *9*, 219. [CrossRef] [PubMed]
7. Li, Z.; Han, Y.; Yang, X. Multi-Fundus Diseases Classification Using Retinal Optical Coherence Tomography Images with Swin Transformer V2. *J. Imaging* **2023**, *9*, 203. [CrossRef] [PubMed]

Journal of
Imaging

MDPI

Article

Robust PCA with $L_{w,*}$ and $L_{2,1}$ Norms: A Novel Method for Low-Quality Retinal Image Enhancement

Habte Tadesse Likassa [1,*,†], Ding-Geng Chen [1,2,*,†], Kewei Chen [1,†], Yalin Wang [3,†] and Wenhui Zhu [3,†]

1 Department of Biostatistics, College of Health Solutions, Arizona State University, Phoenix, AZ 85004, USA
2 Department of Statistics, University of Pretoria, Pretoria 0028, South Africa
3 Computer Science and Engineering, School of Computing and Augmented Intelligence, Arizona State University, Phoenix, AZ 85287-8809, USA
* Correspondence: hlikassa@asu.edu (H.T.L.); ding-geng.chen@asu.edu (D.-G.C.)
† These authors contributed equally to this work.

Abstract: Nonmydriatic retinal fundus images often suffer from quality issues and artifacts due to ocular or systemic comorbidities, leading to potential inaccuracies in clinical diagnoses. In recent times, deep learning methods have been widely employed to improve retinal image quality. However, these methods often require large datasets and lack robustness in clinical settings. Conversely, the inherent stability and adaptability of traditional unsupervised learning methods, coupled with their reduced reliance on extensive data, render them more suitable for real-world clinical applications, particularly in the limited data context of high noise levels or a significant presence of artifacts. However, existing unsupervised learning methods encounter challenges such as sensitivity to noise and outliers, reliance on assumptions like cluster shapes, and difficulties with scalability and interpretability, particularly when utilized for retinal image enhancement. To tackle these challenges, we propose a novel robust PCA (RPCA) method with low-rank sparse decomposition that also integrates affine transformations τ_i, weighted nuclear norm, and the $L_{2,1}$ norms, aiming to overcome existing method limitations and to achieve image quality improvement unseen by these methods. We employ the weighted nuclear norm $(L_{w,*})$ to assign weights to singular values to each retinal images and utilize the $L_{2,1}$ norm to eliminate correlated samples and outliers in the retinal images. Moreover, τ_i is employed to enhance retinal image alignment, making the new method more robust to variations, outliers, noise, and image blurring. The Alternating Direction Method of Multipliers (ADMM) method is used to optimally determine parameters, including τ_i, by solving an optimization problem. Each parameter is addressed separately, harnessing the benefits of ADMM. Our method introduces a novel parameter update approach and significantly improves retinal image quality, detecting cataracts, and diabetic retinopathy. Simulation results confirm our method's superiority over existing state-of-the-art methods across various datasets.

Keywords: RPCA; τ_i; $L_{w,*}$ norm; $L_{2,1}$ norm; image enhancement

1. Introduction

Estimating genuine low-rank components from corrupted high dimensional images is a significant advancement in health applications, particularly in biomedical image processing. Leveraging these inherent low-rank characteristics in biomedical image processing is crucial for transforming a wide array of real-world health-related applications, fostering substantial progress and innovation in the field. Among these, image enhancement has garnered significant attention from researchers, particularly in fields such as cataract diagnosis [1–4], cancer detection [5], retinal enhancement [6–9], medical image segmentation [10–12], scene categorization [13], crime detection [14], sparse coding [15], image denoising [16], communications and computational imaging [17], computer vision [18], and retinal disease detection [19,20] such as glaucoma [21,22]. However, the existing methods encounter significant challenges such as image blurring, noise, and corruptions in biomedical imaging.

Nowadays, deep learning methods have been widely applied to biomedical image processing for medical diagnosis [23,24]. An example is the deep hybrid network low-light image enhancement approach via a unified network with two different streams to capture the global content and the salient structures of the clear image [25]. Yet, this method requires a large volume of training data and storage to process millions of parameters. A hybrid retinal image enhancement algorithm was proposed by [26] for detecting diabetic retinopathy and improving the low quality, using the deep learning model. However, this method is computationally expensive and lacks robustness when the percentage of noise level in images is high. To improve poor-quality retinal fundus images, a simple but effective end-to-end unsupervised learning framework was proposed by [7]. Moreover, Zhu et al. (2023) [27] introduced an unpaired image-to-image translation method for converting low-quality images into their high-quality counterparts. Similarly, Liu et al. (2022) [28] proposed the pyramid constraint to create a degradation-invariant supervised learning enhancement network (PCE-Net). This approach reduces the need for clinical data and effectively enhances the hidden intrinsic dataset. However, challenges still persist in clinical scenarios. To address the challenges of uneven illumination, blurring, and various anomalies, in enhancing retinal images, Liu and Huang [29] introduced a combined approach for improving low-quality retinal images and segmenting blood vessels, utilizing a diffusion model for both tasks [30]. Furthermore, Oh et al. (2023) [31] introduced a novel retina image enhancement framework using scattering transform. This framework entails training an enhancement model that relies on paired images to convert low-quality images into their high-quality counterparts. However, these methods lack generalization on data outside of the training set and encounter problems with mode collapse with the GAN-based unsupervised method, including difficulties in optimizing the parameters. Additionally, deep learning methods via a new frontier of machine learning require more training [32], which takes more computational time [33,34], and have poor real-world clinical generalizability, limiting their practicality in medical imaging [35,36]. Moreover, deep learning models are often considered black boxes, lacking the interpretability crucial for clinical acceptance [37,38]. Hence, it is essential to consider traditional unsupervised machine learning methods to enhance the quality of retinal images.

To overcome the drawbacks of the deep learning methods, unsupervised learning methods have been proposed for retinal image processing [39]. For instance, a contrast-limited adaptive histogram equalization (CLAHE) method was proposed by [40], and a histogram equalization method (HEM) incorporating a tunable parameter was proposed by [41]. However, these methods fail to maintain image quality, often resulting in excessive blurring of the edges. In an effort to enhance retinal image quality, researchers have proposed the low-light image enhancement method (LLIEM). This innovative approach incorporates multi-resolution branches to gain a deeper understanding of diverse levels of local and global context through distinct streams as outlined in reference [42]. However, this method suffers loss of information and semantic content. The machine learning technique was proposed by [43] for retinal image enhancement and glaucoma detection—review and perspective—and a hybrid image enhancement algorithm (HIEA) was developed by [44], which incorporates a median filter for image denoising and is also time consuming. Although these methods do not require training datasets, they lack robustness in high-dimensional medical images with noisy data. To enhance the low-quality of images, a spatial domain filtering method incorporated with the $L_{w,*}$ norm by [45] and a new approach proposed by [46] suggest detecting microaneurysms by considering grey-scale transformations that reduce spatial dependence between images as in [47]. Gao et al. (2019) [48] applied adaptive retinal mechanisms to enhance fundus images as demonstrated by [49,50]. To enhance the quality of retinal images, several methods [51] have been proposed. However, these approaches require explicit training data and exhibit more computational complexity.

Recently, Jiang et al. (2023) [52] proposed an event-based low-illumination image enhancement technique. However, these methods lack the ability to accurately estimate the true underlying objects and reduce nonexistent blurring when enhancing the low-quality color fundus images. This affects diagnostic accuracy and hinders their direct application to fundus images. Given the constraints of conventional unsupervised learning techniques, there is a compelling need to introduce a novel method capable of robust generalization and effective noise handling in high-dimensional and complex image datasets.

To address the limitations of unsupervised learning methods in terms of robustness, interpretability, and computational efficiency for various imaging tasks, numerous methods have been developed ever since the epoch-making emergence of Robust Principal Component Analysis (RPCA) by [53,54]. These methods, proposed by [53], and a myriad of other methods [55,56], proposed to enhance the quality of images through robust low-rank-sparse image representation. For instance, Wright et al. (2009) and Kopriva et al. (2016) [57,58] demonstrated the potential of low-rank matrix approximation to enhance the quality of images. However, the full benefits of these techniques have yet to be extensively explored in the realm of biomedical imaging. In this regard, the low-rank approximation model [59–61] has been explored with great success in natural image recovery. The ADMM approach is employed to iteratively update optimization variables, similar to the work of [61–63]. However, while the performance of these methods appears promising, the methods assign the same singular values to different images, which affects the performance in complex and highly correlated images. Similarly, tensor low-rank representation (TLLR) for image denoising was proposed by [64], while [8,65] proposed sparse rank-constrained learning and its application for medical image grading. However, they fail to denoise highly correlated images, as they do not consider the $L_{2,1}$ and $L_{w,*}$ norms, which undermines the method's performance.

Despite its theoretical foundation and practical efficacy, the RPCA methodology put forth by [9,53,61,62,64,66] fails to differentiate between singular values in image data, employing a uniform approach to regularization across all singular values. This approach leads to an inaccurate estimation of the low-rank component of image data [53,67,68]. To overcome this issue, this paper proposes a novel RPCA method by adding τ_i, $L_{w,*}$ and $L_{2,1}$ norms. To be robust against the adverse effects, the new method combines τ_i, $L_{w,*}$ and $L_{2,1}$ norms for a better biomedical image processing. To reduce the misalignment problem in retinal image recovery, affine transformations are incorporated to render more accurate robust image enhancement. Our method benefits from the weighted nuclear norm, a norm which assigns varying weights to different retinal images through singular value decomposition, enhancing its adaptability and effectiveness. In this paper, an alternative novel method is proposed which is robust to the selection of different EyeQ and cataract images taken from the Kaggale datasets. The ADMM technique is considered, and a new set of equations is formed to estimate the optimization parameters and affine transformations in an iterative process. The simulation results demonstrate that the proposed method outperforms state-of-the-art techniques for enhancing retinal images on certain popular available datasets.

The key contributions of this paper can be summarized as follows.

(1) In this paper, we proposed a novel RPCA that integrates affine transformations to iteratively and accurately estimate the low-rank component from highly complicated retinal images. This work incorporates affine transformations to rectify distorted or misaligned retinal images, aiming to achieve improved quality. As a result of incorporating affine transformation, a new updated parameter is achieved. To tackle the computational load, all parameters are individually solved using ADMM and then updated iteratively in a round-robin manner.

(2) The novel approach aims to enhance robustness against diverse adverse effects, such as measurement noise, image blurring, and artifacts by integrating the previously unexplored $L_{w,*}$ and $L_{2,1}$ norms in retinal imaging enhancement techniques. In this work, the $L_{w,*}$ norm is employed to assign weights to singular values for each retinal image, providing essential adaptability for scenarios where specific features or dimensions need emphasis during decomposition. Additionally, the $L_{2,1}$ norm is utilized to effectively

eliminate correlated samples and outliers within complex retinal images, and it enables denoising, feature highlighting, and artifact removal from retinal images, resulting in clearer and more informative images, beneficial for medical diagnosis and analysis.

(3) The developed method is efficiently solved using the ADMM approach iteratively, ensuring robustness and effectiveness in addressing the complexities of retinal image enhancement.

(4) The method's effectiveness is demonstrated through extensive simulations with multiple retinal images, showing improved image quality by addressing degradation factors such as cataract, glaucoma and diabetic retinopathy in human eyes.

(5) This work not only proposes a novel RPCA method but also aims to draw scholars' attention to the development of low-rank image representation techniques in retinal, cataract, and cancer imaging, with the goal of reducing anomalies for improved clinical diagnosis in biomedical image processing, an aspect that has not been extensively explored in previous methods, highlighting the potential for advancements in this field.

This paper is structured as follows. Section 2 discusses the novel method of RPCA with $L_{w,*}$ and $L_{2,1}$ norms. We further explain the parameter estimation in the optimization techniques. Section 3 describes the nature of the dataset considered for medical image analysis. Section 4 presents the results of medical image analysis using visualization and numerical analysis. Section 5 provides some discussions and concluding remarks to summarize the paper.

2. Methods

Within this section, we describe the development of the new method for retinal image enhancement.

2.1. RPCA with $L_{w,*}$ and $L_{2,1}$ Norms

One of the major drawbacks of the existing methods, such as [53,69], is their inefficiency in adequately eliminating outliers and noise, and detecting cataracts, glaucoma, and diseases during biomedical image enhancement in human eyes. To overcome this limitation, the subsequent section introduces a pioneering approach for enhancing retinal images and also detecting cataracts and glaucoma.

2.2. The $L_{w,*}$ and $L_{2,1}$ Norms Method

Consider n low-quality retinal images, $\{\boldsymbol{R}_i^0\} \in \Re^{w\times h\times c}$, $i = 1,\cdots,n$, denoting the width and height of the images as w and h, respectively, with c representing the number of channels (e.g., "c = 3" for an RGB image).

Each of these retinal images depicts identical objects and exhibits high correlation with one another. Often, these images are marred by issues such as image blurring due to various adverse annoying effects. Then, it is possible to stack these images into a matrix: $\boldsymbol{N} = [vec(\mathbf{R}_1^0) \mid vec(\mathbf{R}_2^0) \mid \ldots \mid vec(\mathbf{R}_n^0)] \in \Re^{m\times n}$, as such $vec(\cdot)$ used to denote the vectorization operators for the purpose of stacking images. As such, the original images $\boldsymbol{N}$ can be further decomposed $\boldsymbol{N} = \boldsymbol{L} + \boldsymbol{E}$ [70,71], where $\boldsymbol{L} \in \Re^{m\times n}$ is a clean low-rank or enhanced image, and $\boldsymbol{E} \in \Re^{m\times n}$ denotes a sparse error matrix incurred by outliers or corruptions. The RPCA by [53], which decomposes highly corrupted retinal images as low-rank, can name enhanced image and anomalies as sparse in the form of optimization techniques given by

$$\min_{\boldsymbol{L},\boldsymbol{E},\boldsymbol{\Gamma}} \|\boldsymbol{L}\|_* + \alpha\|\boldsymbol{E}\|_1 \quad \text{s.t } \boldsymbol{N} = \boldsymbol{L} + \boldsymbol{E} \tag{1}$$

where $\|\boldsymbol{L}\|_* = \sum_{i=1}^{min(m,n)} \sigma_i(\boldsymbol{L})$, denoting the nuclear norm of the low-rank component matrix $\boldsymbol{L}$, $\sigma_i(\boldsymbol{L})$ indicates the singular values of $\boldsymbol{L}$, $\|\boldsymbol{E}\|_1$ is the L_1 norm given by $\sum_{i=1}^{n}|\boldsymbol{E}_i|$, and $\boldsymbol{\Gamma}$ is the Lagrangian multiplier.

Typically, $\boldsymbol{R}_i^0$ are usually not well matched, leading to imprecision in the low-rank-sparse decomposition of retinal images after mitigating adverse effects.

To address this, drawing inspiration from [72], we consider $\boldsymbol{\tau}_i$ to substantially misaligned retinal images $\boldsymbol{R}_i^0$ to achieve well-transformed images $\boldsymbol{R}_i = \boldsymbol{R}_i^0 o \boldsymbol{\tau}_i$, where o indicates the transforming operator. Then, by stacking these transformed retinal images into a matrix, we achieve $\boldsymbol{N}_{o\tau} = [vec(\mathbf{R}_1) \mid vec(\mathbf{R}_2) \mid \ldots \mid vec(\mathbf{R}_n)] \in \Re^{m\times n} = \boldsymbol{L} + \boldsymbol{E}$. Since the solution of $\boldsymbol{N}_{o\tau}$ is intractable due to the nature of the nonlinearity issue, we have to further linearize $\boldsymbol{N}_{o\tau}$. Solving for the parameters related to the constraints $\boldsymbol{N}_{o\tau} = \boldsymbol{L} + \boldsymbol{E}$ is intractable as a result of the nonlinearity issue. To overcome this obstacle, we proceed under the assumption that the alterations induced by these affine transformations $\boldsymbol{\tau}_i$ are minor, and an initial $\boldsymbol{\tau}_i$ is already available.

Then, we make a linearization to $\boldsymbol{N}_{o\tau}$ by taking the first-order-Taylor-approximation as $\boldsymbol{N}_{o(\tau+\Delta\tau)} \approx \boldsymbol{N}_{o\tau} + \sum_{i=1}^{n} J_i \boldsymbol{\Delta\tau} \omega_i \omega_i^T$; as such, $\boldsymbol{N}_{o\tau} \in \Re^{m\times n}$ is denoting a transformed image, $\boldsymbol{\Delta\tau} \in \Re^{p\times n}$ with p being the number of variables, $\boldsymbol{J}_i = \frac{\partial vec(\boldsymbol{R}_i o \boldsymbol{\tau}_i)}{\partial \boldsymbol{\tau}_i} \in \Re^{m\times p}$ represents the Jacobian of the i-th retinal images with respect to $\boldsymbol{\tau}_i$, and $\boldsymbol{\omega}_i$ is the standard basis for $\Re^n$. Thus, by adding τ_i to $\boldsymbol{N}$, $\boldsymbol{N}$ is changed to $\boldsymbol{N}_{o\tau} + \sum_{i=1}^{n} \boldsymbol{J}_i \boldsymbol{\Delta\tau} \omega_i \omega_i^T$ as in [62,72,73].

To make the proposed method more resilient and robust to noise and outliers, occlusions, blurring and artifacts, the $L_{2,1}$ norms are incorporated by combining the L_1 into the L_2 norm, which is employed to manifest the sparsity and the low-rank properties that are regarded as the enhanced retinal images. Also, we transform images and consider the $L_{2,1}$ suggested by [74] to tackle the misalignment problem and highly correlated samples between images. Moreover, the $L_{2,1}$ regularizer is taken into account as the rotational invariant of the L_1 norm and it captures the collinearity between retinal images which is preferred to overcome and address the lack of robustness due to outliers and anomalies [75].

To boost the performance of the proposed method, and tackle the drawbacks of the nuclear norm [53,59,60,62], the $L_{w,*}$ norm is incorporated to assign weights to singular values in retinal images as demonstrated in [76–78]. Subsequently, the overall problem can be formulated as an optimization problem as follows

$$
\begin{aligned}
&\min_{\boldsymbol{L},\boldsymbol{E},\boldsymbol{\Delta\tau},\boldsymbol{\Gamma}} \|\boldsymbol{L}\|_{w,*} + \alpha\|\boldsymbol{E}\|_{2,1} \\
&\text{s.t } \boldsymbol{N}_{o\tau} + \sum_{i=1}^{n} \boldsymbol{J}_i \boldsymbol{\Delta\tau} \boldsymbol{\omega}_i \boldsymbol{\omega}_i^T = \boldsymbol{L} + \boldsymbol{E}
\end{aligned} \tag{2}
$$

where $\|\boldsymbol{L}\|_{w,*} = \sum_{i=1}^{n} |w_i \sigma_i(\boldsymbol{L})|$, w_i is the weight given by $\frac{b/\sqrt{n}}{\sigma i(\boldsymbol{L})+\varepsilon}$, where $b > 0$ is representing a constant, n is the number of similar retinal images in L_j, $\varepsilon = 10^{-16}$ is used to reduce the complexity of dividing by zero, and $\sigma_i(\boldsymbol{L})$ denotes the singular value of a matrix $\boldsymbol{L}$ [76], then the $L_{2,1}$ norm can be given by $\|\boldsymbol{E}\|_{2,1} = \sum_{i=1}^{n} (\sum_{j=1}^{m} E_{ji}^2)^{1/2}$ which denotes the $L_{2,1}$ norm of $\boldsymbol{E}$ which denotes the $L_{2,1}$ norm of $\boldsymbol{E}$ [60,62], and Γ and α denote the Lagrangian multiplier and regularization parameter, respectively.

2.3. Parameter Estimation

To solve (2), we delve into the augmented Lagrangian function, characterized by:

$$
\begin{aligned}
\mathcal{L}(\boldsymbol{L},\boldsymbol{E},\boldsymbol{\Delta\tau},\boldsymbol{\Gamma}) = &\|\boldsymbol{L}\|_{w,*} + \boldsymbol{\alpha}\|\boldsymbol{E}\|_{2,1} + \langle \boldsymbol{\Gamma}, \mathbb{P} - \boldsymbol{L} - \boldsymbol{E} \rangle \\
&+ \frac{\mu}{2}\|\mathbb{P} - \boldsymbol{L} - \boldsymbol{E}\|_F^2
\end{aligned} \tag{3}
$$

where $\boldsymbol{\Gamma} \in \Re^{m\times n}$ denotes the Lagrangian multiplier, μ denotes the penalty parameter, and $\mathbb{P} = \boldsymbol{N}_{o\tau} + \sum_{i=1}^{n} J_i \boldsymbol{\Delta\tau} \omega_i \omega_i^T$. By utilizing an augmented Lagrange multiplier alongside an adaptive penalty as proposed in [79,80], (3) can be reformulated as:

$$
\begin{aligned}
\mathcal{L}(\boldsymbol{L},\boldsymbol{E},\boldsymbol{\Delta\tau},\boldsymbol{\Gamma}) = &\|\boldsymbol{L}\|_{w,*} + \alpha\|\boldsymbol{E}\|_{2,1} \\
&+ \frac{\mu}{2}\left\|\mathbb{P} - \boldsymbol{L} - \boldsymbol{E} + \frac{\boldsymbol{\Gamma}}{\mu}\right\|_F^2
\end{aligned} \tag{4}
$$

Directly solving (4) poses significant computational challenges; thus, we opt for iteratively updating the parameters alternately using the ADMM method [62,81].

Firstly, to update $\boldsymbol{L}$, we fix $\boldsymbol{E}$ and $\boldsymbol{\Delta\tau}$ as constant, so $\boldsymbol{L}^{(k+1)}$, updated by

$$\boldsymbol{L}^{(k+1)} = \underset{\boldsymbol{L}}{\operatorname{argmin}} \mathcal{L}\left\{\boldsymbol{L}, \boldsymbol{E}^{(k)}, \boldsymbol{\Delta\tau}^{(k)}\right\} \tag{5}$$

k is an index representing an iteration. By ignoring all parameters as a constant $\boldsymbol{L}$, Equation (5) can be rewritten as

$$\begin{aligned}\boldsymbol{L}^{(k+1)} = \underset{\boldsymbol{L}}{\operatorname{argmin}} \mathcal{L}\Bigg\{ & \alpha\|\boldsymbol{L}\|_{w,*} + \frac{\mu^{(k)}}{2}\Bigg\|\mathbb{P}^{(k)} - \boldsymbol{L} \\ & - \boldsymbol{E}^{(k)} + \frac{\boldsymbol{\Gamma}^{(k)}}{\mu^{(k)}}\Bigg\|_F^2\Bigg\}\end{aligned} \tag{6}$$

Problem (6) is equivalent to the weighted nuclear norm minimization (WNNM) problem [76,81,82], and the closed-form solution of the WNNM operator is given by

$$\boldsymbol{L} = \boldsymbol{U}\hat{\boldsymbol{\Sigma}}\boldsymbol{V}^T \tag{7}$$

where $\boldsymbol{L} = \boldsymbol{U}\hat{\boldsymbol{\Sigma}}\boldsymbol{V}^T, \hat{\boldsymbol{\Sigma}}$ is given by $(\boldsymbol{diag}(\sigma_1(\boldsymbol{L}),...,\boldsymbol{diag}(\sigma_n(\boldsymbol{L})))$ and $\sigma_i(\boldsymbol{L}) = \begin{cases} 0, c_2 < 0 \\ \frac{c_1+\sqrt{c_2}}{2}, c_2 \geq 0 \end{cases}$ where $c_1 = \sigma_i(K) - \varepsilon, c_2 = (\sigma_i(\boldsymbol{K}) + \varepsilon)^2 - 4C$, ε and C are small constants and $\boldsymbol{K} = -\frac{1}{2}(\boldsymbol{E}^k - \mathbb{P}^k - \frac{\boldsymbol{\Gamma}^k}{\mu^k})$.

Secondly, to update $\boldsymbol{E}$, we keep $\boldsymbol{L}$ and $\boldsymbol{\Delta\tau}$ as constants, then $\boldsymbol{E}^{(k+1)}$ is updated by

$$\boldsymbol{E}^{(k+1)} = \underset{\boldsymbol{E}}{\operatorname{argmin}} \mathcal{L}\left\{\boldsymbol{L}^{(k+1)}, \boldsymbol{E}, \boldsymbol{\Delta\tau}^{(k)}\right\} \tag{8}$$

from which $\boldsymbol{E}$, Equation (8), is reduced as

$$\begin{aligned}\boldsymbol{E}^{(k+1)} = \underset{\boldsymbol{E}}{\operatorname{argmin}}\Bigg\{ & \alpha\|\boldsymbol{E}\|_{2,1} + \frac{\mu^{(k)}}{2}\Bigg\|\mathbb{P}^{(k)} \\ & - \boldsymbol{L}^{(k)} - \boldsymbol{E} + \frac{\boldsymbol{\Gamma}^{(k)}}{\mu^{(k)}}\Bigg\|_F^2\Bigg\}\end{aligned} \tag{9}$$

By considering the lemma as in [83] as constants, the optimal parameter of the i-th column of $\boldsymbol{E}^{(k+1)}$, $\boldsymbol{E}_j^{(k+1)}$ is given by

$$\boldsymbol{E}_j^{(k+1)} = \begin{cases} \dfrac{\left\|\boldsymbol{u}_j^{(k)}\right\|_2 - \frac{\alpha}{\mu^{(k)}}}{\left\|\boldsymbol{u}_j^{(k)}\right\|_2}\boldsymbol{U}_j^{(k)}, & \text{if } \left\|\boldsymbol{U}_j^{(k)}\right\|_2 \geq \frac{\alpha}{\mu^{(k)}} \\ 0, & \text{otherwise} \end{cases} \tag{10}$$

where $\boldsymbol{U}^{(k)} = (\mathbb{P}^{(k)} - \boldsymbol{L}^{(k)} + \frac{\boldsymbol{\Gamma}^{(k)}}{\mu^{(k)}})$, and $\|.\|_2$ is denoting the Euclidean norm. Next, to optimize $\boldsymbol{\Delta\tau}$, $\boldsymbol{L}$ and $\boldsymbol{E}$ are considered fixed, and then $\boldsymbol{\Delta\tau}^{(k+1)}$ is given by

$$\boldsymbol{\Delta\tau}^{(k+1)} = \underset{\boldsymbol{\Delta\tau}}{\operatorname{argmin}} \mathcal{L}\left\{\boldsymbol{L}^{(k+1)}, \boldsymbol{E}^{(k+1)}, \boldsymbol{\Delta\tau}\right\} \tag{11}$$

Consider all other parameters independent of $\boldsymbol{\Delta\tau}$ as constant, from which we can obtain

$$\boldsymbol{\Delta\tau}^{(k+1)} = \underset{\Delta\tau}{\text{argmin}}\left\{\frac{\mu^{(k)}}{2}\left\|\mathbb{P}^{(k)} - \boldsymbol{L}^{(k+1)} - \boldsymbol{E}^{(k+1)} + \frac{\boldsymbol{\Gamma}^{(k)}}{\mu^{(k)}}\right\|_F^2\right\} \tag{12}$$

Solving (12) by considering the thresholds operator as in [72], we can achieve an optimal parameter given by

$$\boldsymbol{\Delta\tau}^{(k+1)} = \sum_{i=1}^{n} J_i^{+}\left(\boldsymbol{L}^{(k+1)} + \boldsymbol{E}^{(k+1)} - - \boldsymbol{N}_{o\tau} - \frac{\boldsymbol{\Gamma}^{(k)}}{\mu^{(k)}}\right)\boldsymbol{\omega}_i\boldsymbol{\omega}_i^T \tag{13}$$

where J_i^{+} denotes the Moore–Penrose pseudoinverse of J_i [84]. Following the same procedure as above, the Lagrangian multiplier $\boldsymbol{\Gamma}$ is updated through

$$\boldsymbol{\Gamma}^{(k+1)} = \boldsymbol{\Gamma}^{(k)} + \mu^{(k+1)}\left\{\mathbb{P}^{(k)} - \boldsymbol{L}^{(k+1)} - \boldsymbol{E}^{(k+1)}\right\} \tag{14}$$

Similarly, the regularization parameter μ is updated through

$$\mu^{(k+1)} = min\left\{\mu_{max}, \rho\mu^{(k)}\right\} \tag{15}$$

where ρ is a carefully selected constant and μ_{max} is an adjustable parameter that influences the convergence of the proposed method. The remaining parameters are updated independently while keeping all other variables fixed.

As we invoked with affine transformation, we also achieved a new updating parameter $\boldsymbol{\Delta\tau}$. To make the new method easy to understand, the pseudocode is given in Algorithm 1.

Algorithm 1 ADMM for RPCA with $L_{w,*}$ and $L_{2,1}$ norms.

Output Data Matrix $\boldsymbol{N} \in \Re^{m\times n}$, $\boldsymbol{L}^0 \in \Re^{m\times n}$, $\boldsymbol{E}^0 \in \Re^{m\times n}$, $\boldsymbol{\Delta\tau}^0 \in \Re^{p\times n}$, α, ρ
While not converged **Do**
Update: $\boldsymbol{L}^{(k+1)}$ using (7)
Update: $\boldsymbol{E}^{(k+1)}$ using (10)
Update: $\boldsymbol{\Delta\tau}^{(k+1)}$ using (13)
Update: $\boldsymbol{\Gamma}^{(k+1)}$ using (14)
Update: $\mu^{(k+1)}$ using (15)
End while
Outputs: L,E, $\boldsymbol{\Delta\tau}$

In this paper, we evaluate the performance of the new method first using statistical measures through peak signal-to-noise ratio (PSNR) and structural similarity index measure (SSIM). We further verify the generalizability of the proposed method using the Pearson correlation coefficient (PCC) and Visual Information Fidelity (VIF) index based on on EyeQ, Kaggle, and High-Resolution Fundus (HRF) retinal image datasets, each containing its own degraded and ground truth images. In each experimental simulation, we first consider the degraded retinal images, then apply the new method to these images to achieve enhanced images. Next, we compare the enhanced images obtained through the proposed method with the ground truth. Finally, we compare the robustness of our method with existing methods. All experimental simulations are performed in MATLAB. Additionally, we consider regularization parameters, including $\rho = 3 \times 10^{-3}$, $\mu = 3 \times 10^{-10}$ and $\lambda = 4$. For an easy understanding of the procedure of the proposed method, we support with the diagrammatic image representation shown in Figure 1.

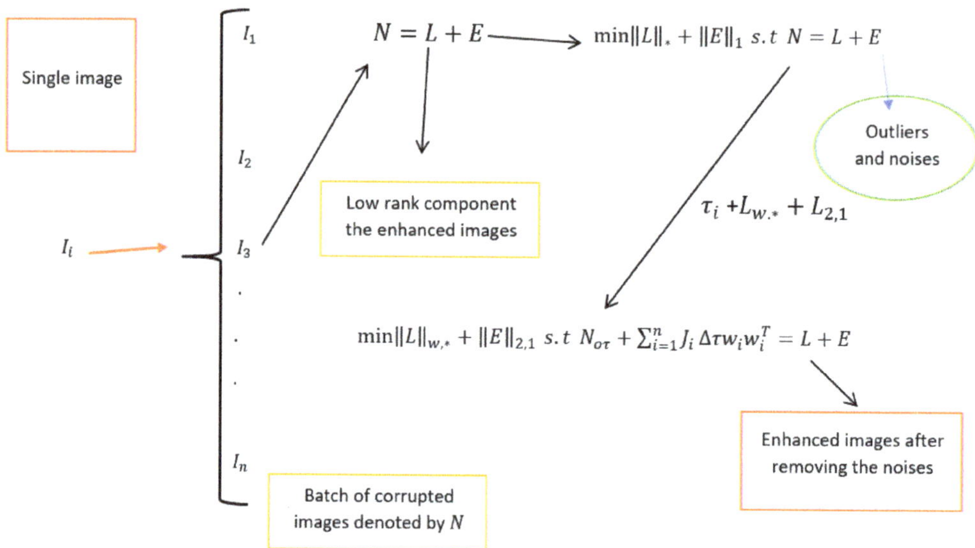

Figure 1. Flowchart of the robust PCA for retinal image decomposition.

2.4. Numerical Evaluation Criterion

The two popular criteria mainly used as quantitative evaluation indicators are the peak signal-to-noise ratio (PSNR) [85] and the structural similarity index measure (SSIM) [86,87].

The quality of retinal image enhancement by the proposed method is also validated using SSIM [86,87], which is given by

$$\mathrm{SSIM}(f, \hat{f}) = \frac{(2\mu_f \mu_{\hat{f}} + C_1)((2\sigma_f \sigma_{\hat{f}} + C_2))}{(\mu_f^2 + \mu_{\hat{f}}^2 + C_1)(\sigma_f^2 + \sigma_{\hat{f}}^2 + C_2)} \tag{16}$$

where $\mu_{\hat{f}}$, σ_f^2, $\sigma_{\hat{f}}^2$ are the mean and variance of the ground truth and enhanced retinal images.

The Pearson correlation coefficient PCC [88] between the ground truth image f and the enhanced image $\hat{f}$ is given by:

$$\mathrm{PCC} = \frac{\sum_{i=1}^{n}(f_i - \bar{f})(\hat{f}_i - \bar{\hat{f}})}{\sqrt{\sum_{i=1}^{n}(f_i - \bar{f})^2}\sqrt{\sum_{i=1}^{n}(\hat{f}_i - \bar{\hat{f}})^2}} \tag{17}$$

where:

$$\bar{f} = \frac{1}{n}\sum_{i=1}^{n} f_i \tag{18}$$

$$\bar{\hat{f}} = \frac{1}{n}\sum_{i=1}^{n} \hat{f}_i \tag{19}$$

The Visual Information Fidelity (VIF) [89] index between the ground truth image f and the enhanced image $\hat{f}$ is given by:

$$\mathrm{VIF} = \frac{\sum_{i=1}^{N} \log_2\left(1 + \frac{\sigma_{f_i}^2}{\sigma_{v_i}^2}\right)}{\sum_{i=1}^{N} \log_2\left(1 + \frac{\sigma_{f_i - \hat{f}_i}^2}{\sigma_{v_i}^2}\right)} \tag{20}$$

where f_i is the i-th block of the ground truth image f, $\hat{f}_i$ is the i-th block of the enhanced image $\hat{f}$, $\sigma_{f_i}^2$ is the variance of the ith block of the ground truth image f, $\sigma_{f_i - \hat{f}_i}^2$ is the variance of the difference between the i-th block of the ground truth and enhanced images, and $\sigma_{v_i}^2$ is the noise variance in the i-th block of the ground truth image.

3. Datasets

In this section, three various retinal image datasets are used to evaluate the performance of the new method described in Section 2.2. The first dataset is the Eye Quality (EyeQ) retinal images which is taken from https://github.com/HzFu/EyeQ (accessed on 16 June 2024). To see the effectiveness of the new method, we used two independent datasets (both the training and testing images). The second dataset consists of retinal images infected with cataracts and glaucoma taken from the https://www.kaggle.com/datasets/jr2ngb/cataractdataset Kaggle dataset (accessed on 16 June 2024). The third dataset is taken from the High-Resolution Fundus (HRF) Image Database https://www5.cs.fau.de/research/data/fundus-images/ (accessed on 16 June 2024). We conducted comprehensive simulations across scenarios where ground truth, considered clean images, were available (full-reference assessment) for three distinct datasets to validate the generalizability of our method.

To see the efficacy of the new approach, we compared it with several existing approaches. Specifically, we considered HEM by [41], HIEA by [44], LLIE by [42], and one low-rank sparse method, TLLR by [64]. The methods we compared our proposed approach with are HEM by [41], TLLR by [64], LLIE by [42], and HIEA by [44]. In the upcoming subsections, we will delve into the datasets utilized, the numerical evaluation criteria employed, and the findings from our medical image analyses.

3.1. EyeQ Retinal Image Data

The Eye Quality (EyeQ) Assessment Dataset is a re-annotated subset of the EyePACS dataset, created for fundus image quality evaluation. The EyeQ dataset [7,90,91] consists of 28,480 training and 15,128 testing retinal images. The original retinal image dataset is manually labeled into three quality levels: good, usable, and reject. First, we considered 10 in low-quality images from the training dataset and 10 from the testing images, which are independent of the training dataset, to see the effectiveness of a novel approach in enhancing the low-quality images. This dataset encompasses the ground truth and degraded images, where the ground truth refers to the normal high-quality retinal images, while the degraded images, also called the low-quality retinal images, are simulated from the ground truth using light disturbance, image blurring, and retinal artifacts as outlined in [7,90,91]. The first two experiments predominantly focus on retinal image enhancement considering 10 true color degraded retinal images with a size of 800 × 800 pixels from the training dataset, while the second has a size of 256 × 256 pixels from the testing set; we considered this to assess the performance of the proposed method through visualization and numerical analysis.

3.2. Kaggle Cataract Retinal Image Data

The Kaggle dataset is another commonly used resource for evaluating the performance of the proposed method described in Section 2.1. The EyePACS dataset on Kaggle contains high-resolution retinal images used for cataract and other eye disease research. It includes thousands of annotated images, supporting diagnostic algorithm development. This dataset aids advancements in automated disease detection. Researchers use EyePACS to improve cataract diagnosis and eye care. This dataset contains retinal images that show symptoms of cataracts, glaucoma, and other diseases affecting human eyes. The cataract retinal images used for this analysis are taken from https://www.kaggle.com/datasets/jr2ngb/cataractdataset dataset (accessed on 16 June 2024), which consists of approximately 3500. The main focus of the final simulation results relies on the cataract retinal images, each having a size of 2464 × 1632 pixels.

3.3. High-Resolution Fundus Retinal Image Data

To verify the performance of the proposed method, we also considered the High-Resolution Fundus (HRF) dataset, a meticulously curated collection of retinal images tailored for developing and evaluating algorithms in medical image analysis. This dataset has three subjects: the first is the normal retinal images, the second is the retinal images infected with glaucoma, and the last one is retinal images with diabetic retinopathy, each with dimensions of 3304 × 2336 pixels. Its objective is to advance ophthalmology by considering high-quality images. It is also essential in the implementation of the new development of methods for detecting retinal diseases like diabetic retinopathy, macular degeneration, and glaucoma. The dataset includes 45 images, with 15 healthy retinas, 15 with diabetic retinopathy, and 15 with glaucoma. The high resolution of these images enables the precise identification of critical features such as blood vessels, optic discs, and lesions, which are essential for detecting conditions like diabetic retinopathy, glaucoma, and age-related macular degeneration. By utilizing the HRF dataset, we were able to implement the performance of the novel method compared with the state-of-the-art methods, enhancing our understanding of retinal diseases and advancing the development of automated diagnostic tools. The summary of all the datasets considered for retinal image data analysis is provided in Table 1.

Table 1. Characteristics of retinal image datasets.

Dataset	Type	Size
EyeQ [7,90,91]	Retinal-Training	800 × 800 pixels
	Retinal-Testing	256 × 256 pixels
Kaggle	Cataract	2464 × 1632 pixels
HRF	Diabetic Retinopathy	3304 × 1632 pixels

4. Results

In this section, we aim to present experimental simulations of the new method compared with state-of-the-art methods such as HEM [41], TLLR [64], LLIE [42], and HIEA [44]. Initially, we conducted retinal image analysis using both the testing and training datasets. Subsequently, we attempted to simulate enhancement based on cataract retinal images. Finally, we conducted experimental simulations based on HRF diabetic retinopathy images.

4.1. Degraded Retinal Image Data Analysis

First, we conduct simulations on degraded retinal images taken from the training dataset as in [7,90,91]. In this experiment, 10 degraded retinal images with size 800 × 800 pixels are considered. As a visualization, some of the improved retinal images based on the above methods are given in Figure 2, in which our novel method, shown in Figure 2e, better enhances the degraded retinal images as compared to the state-of-the-art methods [41,42,44,64]. The values of the PSNR and SSIM based on the individual images are illustrated in Figure 3, from which we note that the new approach is relatively better to improve the low-quality individual original retinal images. To validate the performance of the proposed method, we employ PSNR and SSIM. HIEA [44] has better performance than HEM, shown by [41], as HIEA [44] requires a median filter, which was employed for image denoising, which boosts the effectiveness of the new approach as compared with LLIE developed by [42]. This result resembles the results given in Table 2, and further confirms that the new method is more resilient to the degradation factors, which means that the new approach outperformed the four competitors in all evaluation metrics.

Table 2. Comparison of methods by PSNR and SSIM (training dataset).

Methods	PSNRs	SSIMs
HEM [41]	19.06	0.47
TLLR [64]	7.26	0.22
LLIE [42]	18.10	0.47
HIEA [44]	22.50	0.60
Ours	24.33	0.76

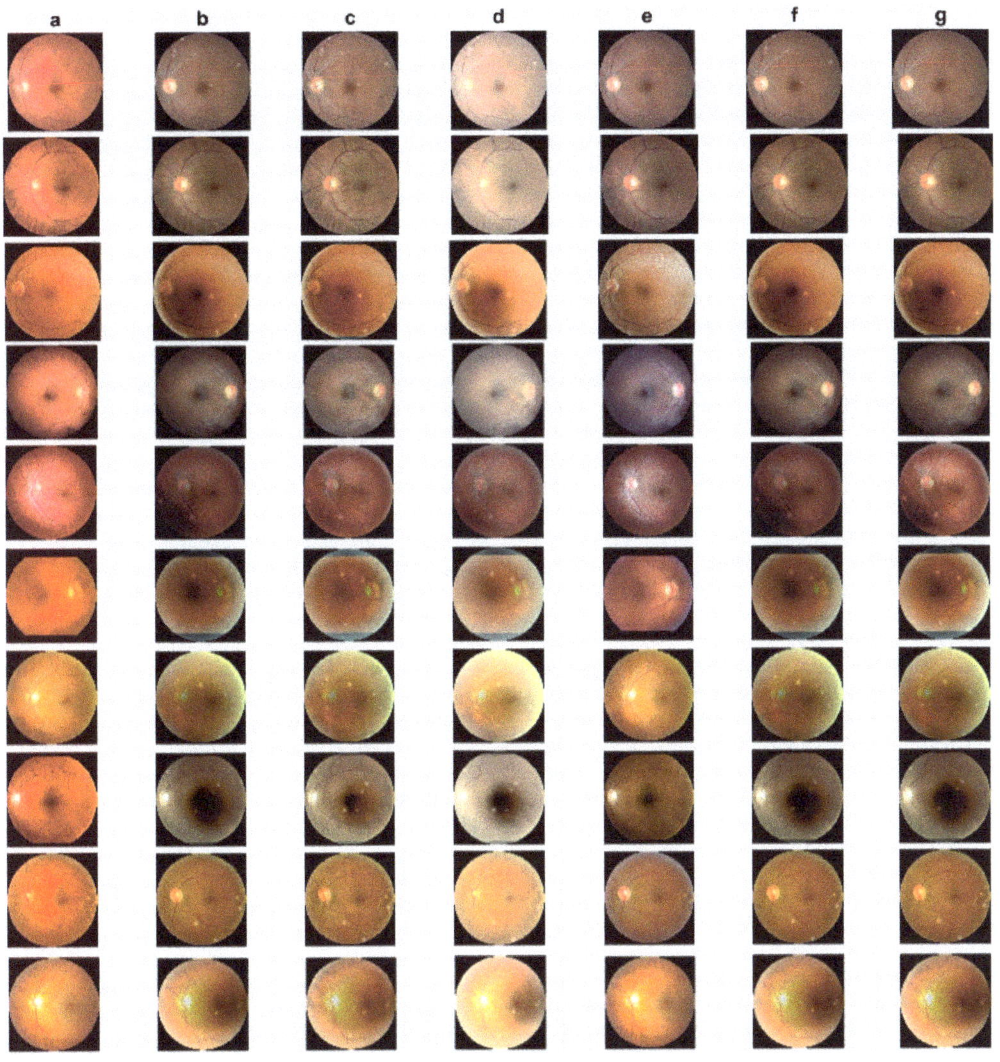

Figure 2. Degraded retinal image enhancement (training dataset): (**a**) HEM [41]; (**b**) TLLR [64]; (**c**) LLIE [42]; (**d**) HIEA [44]; (**e**) ours; (**f**) degraded image, and (**g**) ground truth.

Furthermore, we incorporate 10 degraded images from the EyeQ test dataset, which has been commonly referenced in prior studies [7,90,91], to simulate additional light interference, image blurring, and artifacts.

We check the performance of the proposed method to enhance the degraded images through visualization and numerical measures using the PSNR and SSIM between the degraded low-quality images and their high-quality counterparts. As illustrated, Figure 4e shows some visual images enhanced by the proposed method compared with existing methods. These images demonstrate that the proposed method significantly enhances degraded retinal images, bringing them closer to the ground truth. The enhanced images exhibit clearer visual quality by effectively removing light disturbance, image blurring, and retinal artifacts. To further verify the performance of the new approach based on individual retinal images, we compare it using the PSNR and SSIM with existing methods as shown in Figure 5. LLIE [42] outperforms HEA [41] by considering multi-resolution branches for a better understanding of different levels of local and global context, thus mitigating the influence of outliers and noise. HIEA [44] surpasses all three existing methods, as it is a hybrid algorithm incorporating a median filter for image denoising. Figure 4e demonstrates that the new method achieves the best performance [41,42,44,64]. The summary values of the PSNR and SSIM for ten retinal images achieved by the proposed method along with existing methods are shown in Table 3. From this table, we can see that HIEA by [44] produces the second-best performance. This aligns with the results presented in Figure 4 and further justifies that the proposed method better enhances the degraded retinal images as compared with existing methods. The summary values of the PSNR and SSIM for ten retinal images achieved by the proposed method along with existing methods is given in Table 3, from which we can observe that HIEA by [44] produces the second-best performance. The performance of the proposed method is evaluated, taking more retinal images, and it is confirmed that the performance is better than the existing methods. This is because HIEA combines median filtering to reduce variation and combines it with deep learning to minimize the $L_{2,1}$ norm of the sparse error. We can also observe from Table 3 that the new approach still outperforms all existing methods. It achieves this by including affine transformations and utilizing the $L_{2,1}$ and $L_{w,*}$ norms to render more robust degraded retinal image recovery. The visual enhancement of degraded retinal images shown in Figures 4e and 5 is more consistent with the numerical evaluation analysis provided in Table 3. This is because the new method incorporates a set of affine transformations and employs the $L_{2,1}$ and $L_{w,*}$ norms to simultaneously align and enhance the retinal images. This enables the new method to reduce the influences of outliers, heavy sparse noise, occlusions, light transmission issues, image blurring, retinal artifacts, and illuminations.

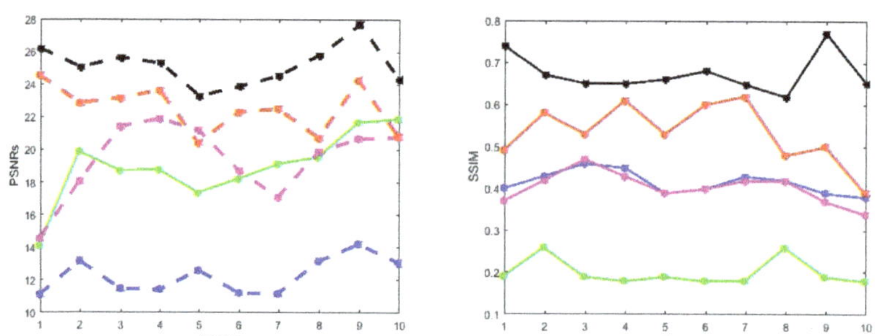

Figure 3. PSNRs and SSIMs obtained from the ground truth and enhanced images (training retinal degraded retinal images) computed by HEM [41] (blue color); TLLR [64] (green color); LLIE [42] (magenta color); HIEA [44] (red color) and ours (black color).

Table 3. Comparison of methods by the PSNR and SSIM (ground truth and enhanced images).

Methods	Based on Figure 4		Based on Figure 8	
	PSNR	SSIM	PSNR	SSIM
HEM [41]	15.78	0.43	12.28	0.52
TLLR [64]	11.40	0.24	18.94	0.50
LLIE [42]	16.73	0.39	19.44	0.55
HIEA [44]	21.57	0.57	22.52	0.65
Ours	27.87	0.72	25.15	0.78

Next, we also conduct simulations on a more challenging set of 40 large samples of low-quality degraded retinal images with size 256×256, sourced from [7,90,91], to verify the performance of the proposed method in enhancing these retinal images as shown in Figure 6e, demonstrating superior performance compared to existing methods. We also verify the effectiveness of the proposed method by taking 40 retinal images and confirm that the new method is better compared to the baselines, both through visualization and numerical analysis. Subsequently, the comparison of PSNRs and SSIMs is summarized in Table 4, revealing that LLIE [42] outperforms HEM [41] and TLLR [64]. Meanwhile, our proposed method outperforms the existing methods due to its incorporation of affine transformations, $L_{w,*}$ and $L_{2,1}$ norms. Statistically, the proposed method demonstrates superiority over the other four competitors [41,42,44,64] in terms of PSNR and SSIM as shown in Table 4. The PSNRs and SSIMs for individual images are given in Figure 7, indicating that the performance of the proposed method surpasses that of the existing methods. This improvement is attributed to the incorporation of affine transformation and $L_{w,*}$ and $L_{2,1}$ norms to further denoise the degraded retinal images.

Table 4. Comparison of methods by the PSNR and SSIM (testing dataset).

Methods	PSNR	SSIM
HEM by [41]	16.95	0.42
TLLR by [64]	10.45	0.37
LLIE by [42]	16.99	0.39
HIEA by [44]	20.82	0.56
Ours	27.17	0.68

4.2. Cataract Retinal Image Data Analysis

In this section, we present the results of the proposed method compared with existing methods [41,42,44,64] based on more challenging and high-dimensional cataract retinal images from the Kaggle dataset. In this simulation, 10 retinal images with cataract with size of 2464×1632 are considered. The visualization results show that the proposed method performs better in enhancing ten retinal images infected with cataracts, closely aligning with the ground truth as depicted in Figure 8e. To evaluate the performance of the proposed method compared with existing methods for individual retinal images, we compute PSNRs and SSIMs, from which we observe that the proposed method outperforms in enhancing cataract retinal images as depicted in the third row of Figure 9. The result of Table 3 is consistent with the results shown in image visualization. This improvement is attributed to the proposed method considering τ_i, and the $L_{w,*}$ and $L_{2,1}$ norms. This is consistent with the results in Table 3, and further confirms that the proposed method is more resilient to outliers and heavy sparse noise.

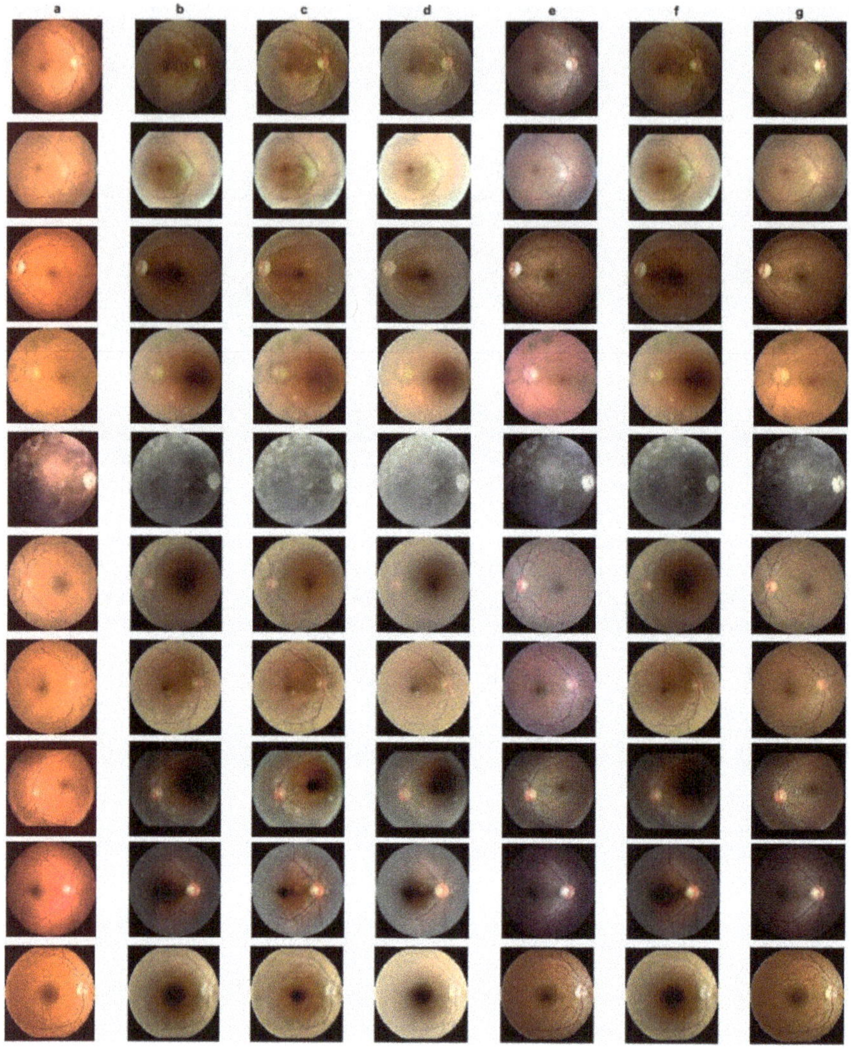

Figure 4. Degraded retinal image enhancement (testing dataset): (**a**) HEM by [41]; (**b**) TLLR by [64]; (**c**) LLIE by [42]; (**d**) HIEA by [44]; (**e**) ours; (**f**) degraded image, and (**g**) ground truth.

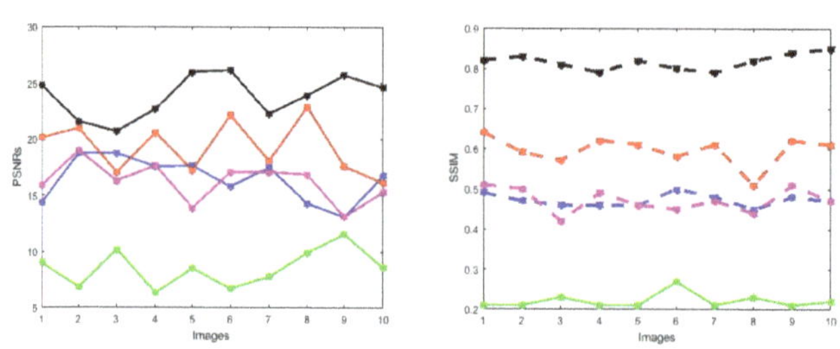

Figure 5. PSNRs and SSIMs obtained from the ground truth and enhanced images (testing retinal degraded retinal images) computed by HEM [41] (blue color); TLLR [64] (green color); LLIE [42] (magenta color); HIEA [44] (red color) and ours (black color).

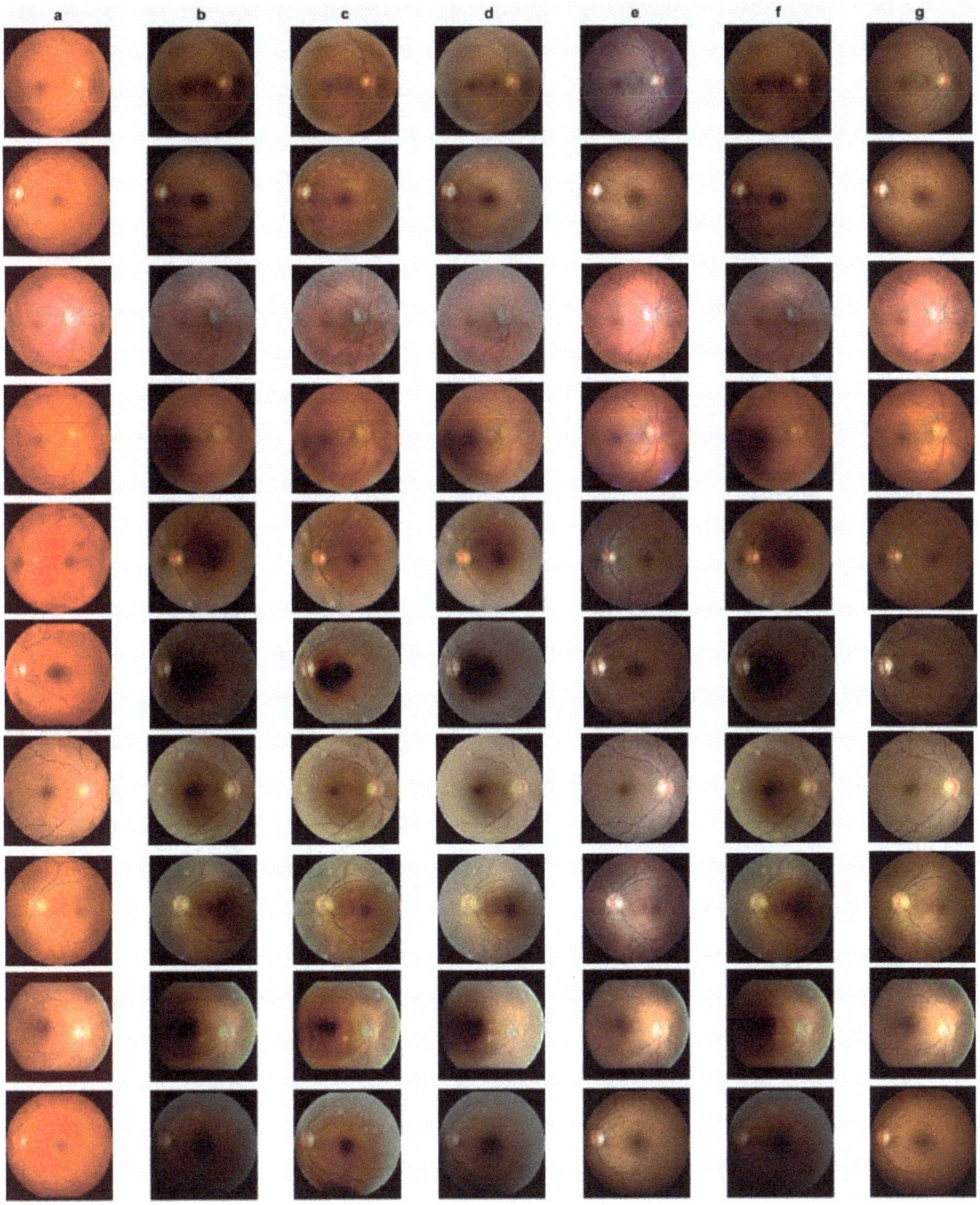
a
b
c
d
e
f
g

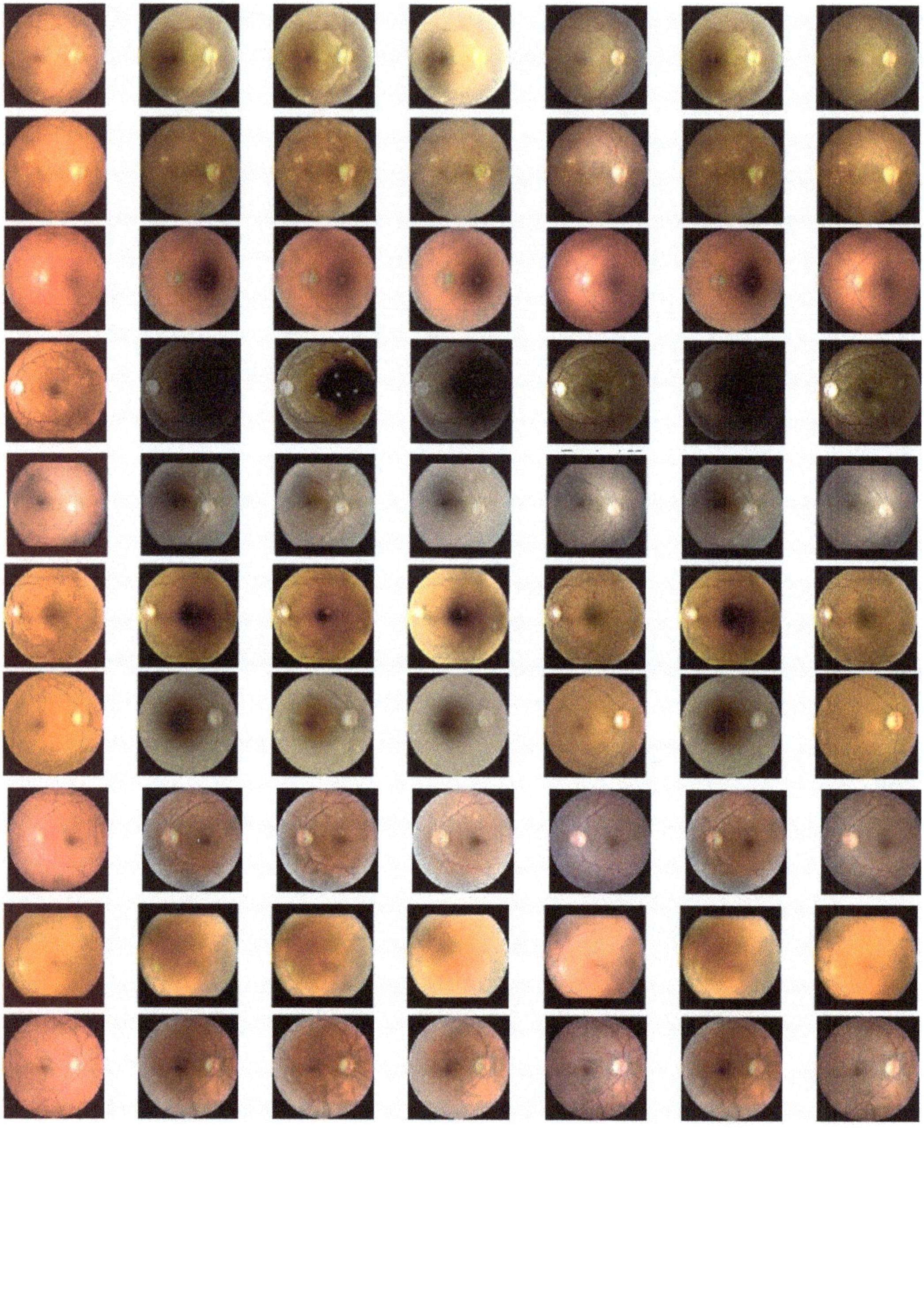

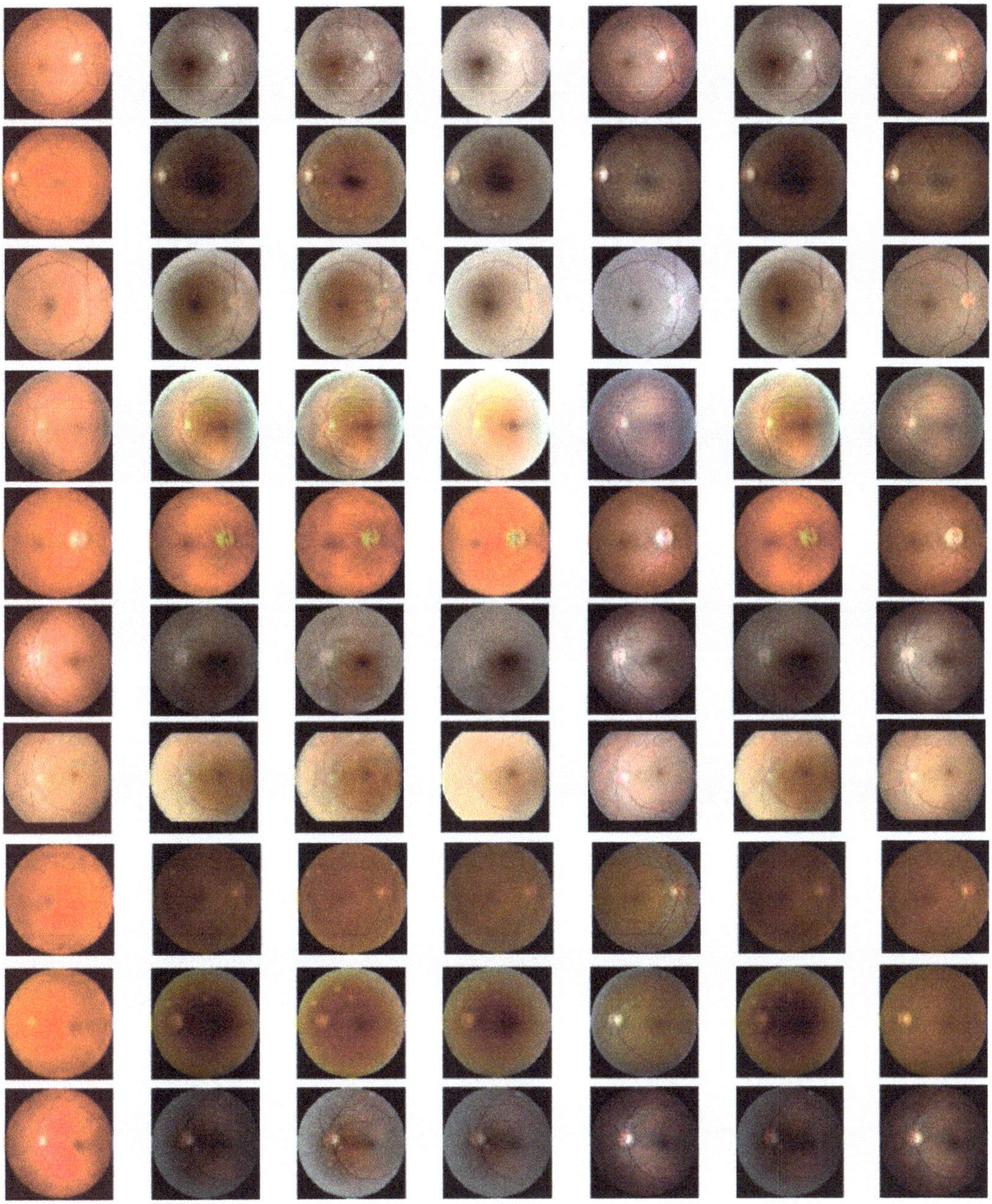

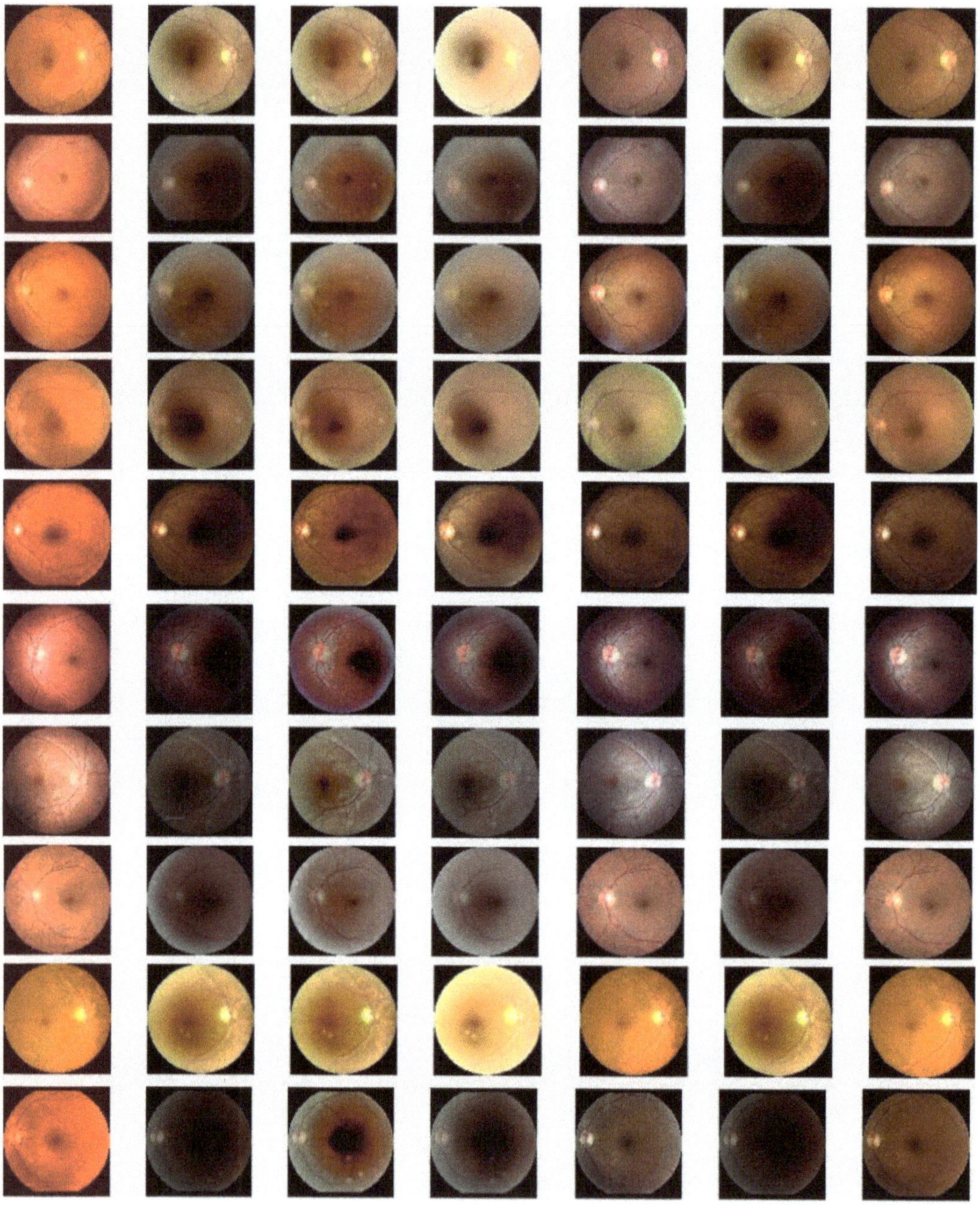

Figure 6. Degraded retinal image enhancement (training dataset): (**a**) HEM [41]; (**b**) TLLR [64]; (**c**) LLIE [42]; (**d**) HIEA [44]; (**e**) ours; (**f**) degraded image and (**g**) ground truth.

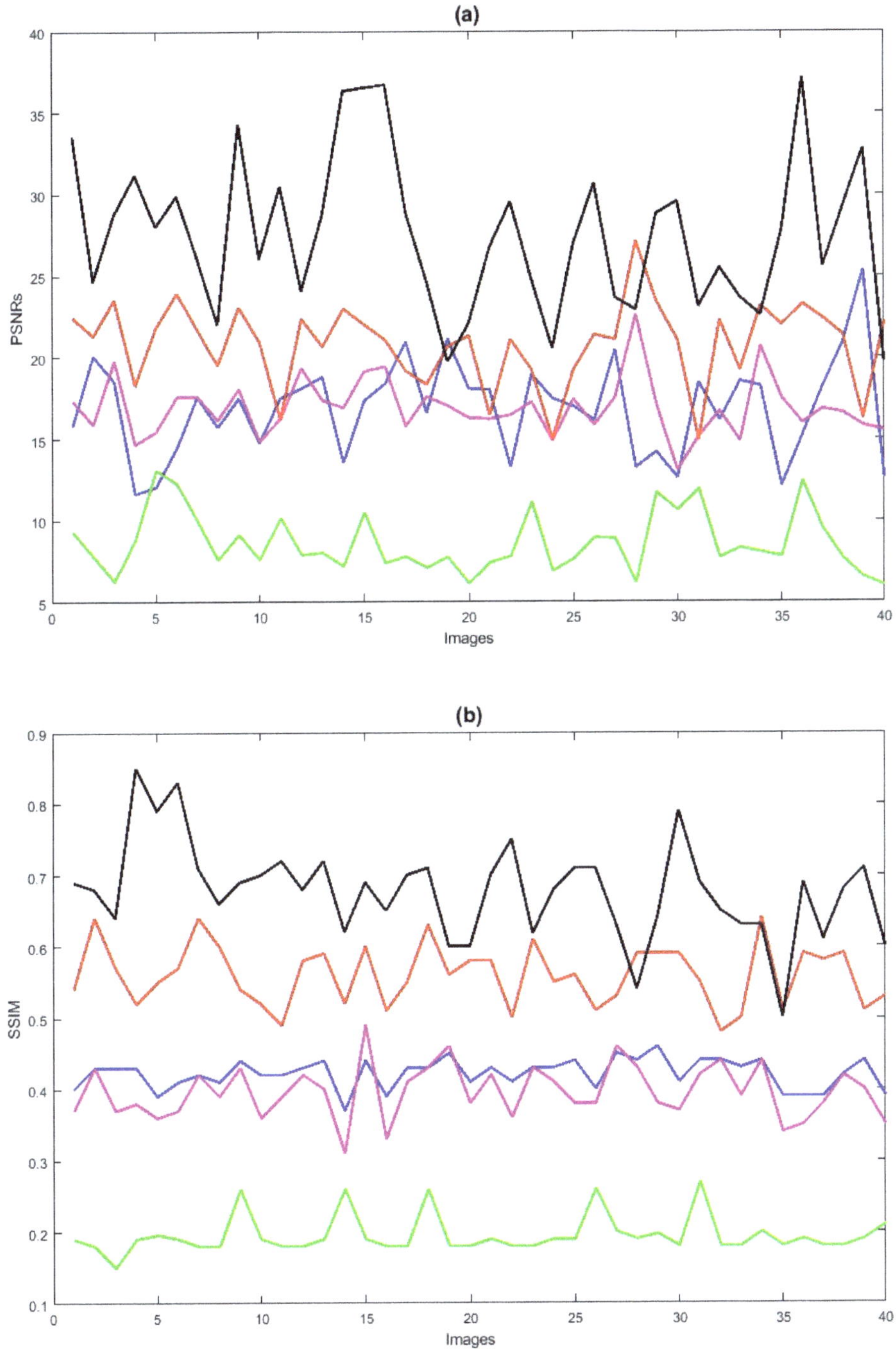

Figure 7. PSNRs (a) and SSIMs (b) vs. degraded 40 retinal images. HEM [41] (blue color); TLLR [64] (green color); LLIE [42] (magenta color); HIEA [44] (red color); ours (black color).

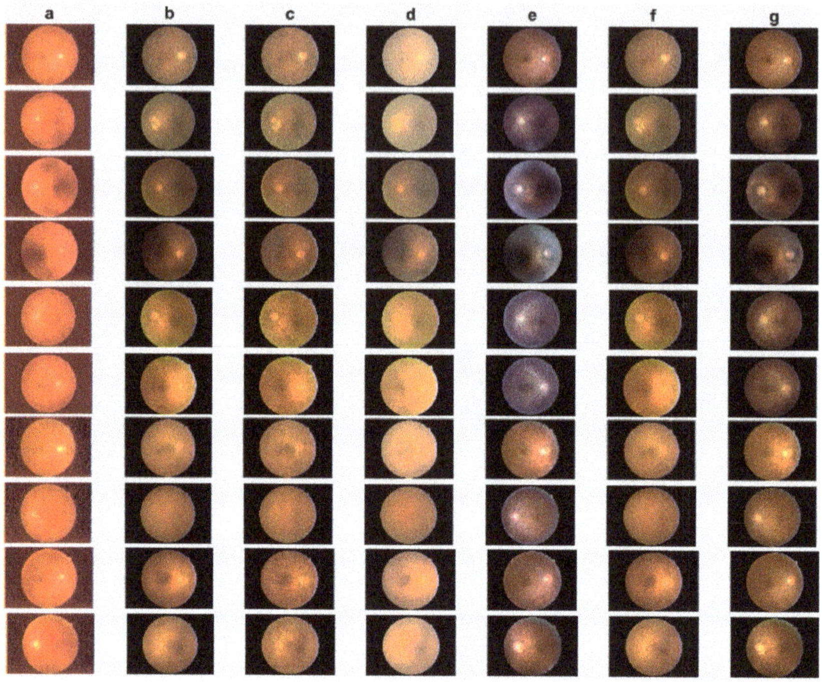

Figure 8. Cataract infected low-quality retinal image enhancement: (**a**) HEM [41]; (**b**) TLLR [64]; (**c**) LLIE [42]; (**d**) HIEA [44]; (**e**) ours; (**f**) degraded image, and (**g**) ground truth.

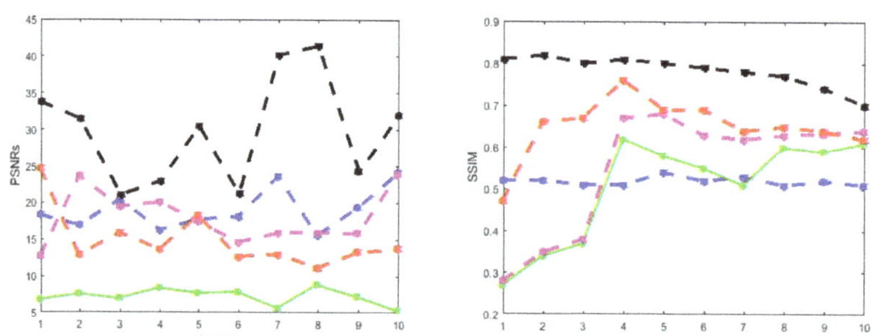

Figure 9. PSNRs and SSIMs obtained from the ground truth and enhanced images (cataract retinal images) computed by HEM [41] (blue color); TLLR [64] (green color); LLIE [42] (magenta color); HIEA [44] (red color) and ours (black color).

4.3. HRF Image Database

To further verify the performance of the new method, we also considered a complicated and high-dimensional dataset infected with various diabetic retinal images, with each having a size of 3304 × 1632 pixels, for which the visualization results are displayed in Figure 10. The result achieved by the novel method shown in Figure 10e is better in enhancing the quality of the diabetic retinal images compared with the state-of-the-art methods. This result is more consistent with the numerical simulations given in Table 5. The individual results using PSNRs, SSIMs, PCCs and VIFs are illustrated in Figure 11, in which the performance achieved by the proposed method is better compared to the existing methods.

Table 5. Comparison of methods by the PSNR, SSIM, PCC and VIF based on HRF dataset.

Methods	PSNR	SSIM	PCC	VIF
HEM [41]	24.78	0.50	0.93	0.86
TLLR [64]	6.97	0.21	0.95	0.85
LLIE [42]	20.51	0.43	0.95	0.47
HIEA [44]	14.92	0.65	0.94	0.69
Ours	28.006	0.79	0.987	0.949

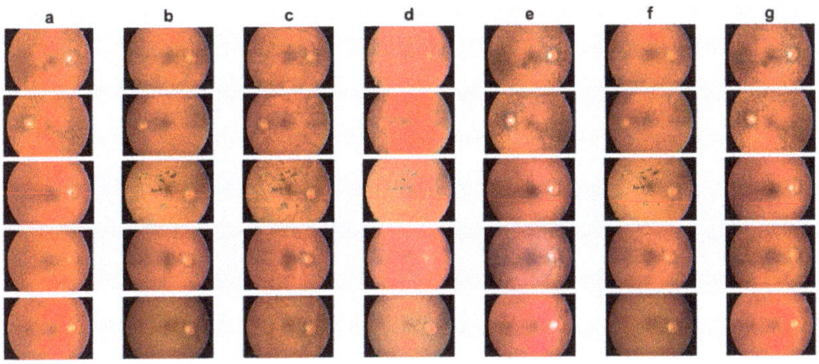

Figure 10. Diabetic retinopathy retinal image: (**a**) HEM by [41]; (**b**) TLLR by [64]; (**c**) LLIE by [42]; (**d**) HIEA by [44]; (**e**) ours; (**f**) degraded image and (**g**) ground truth.

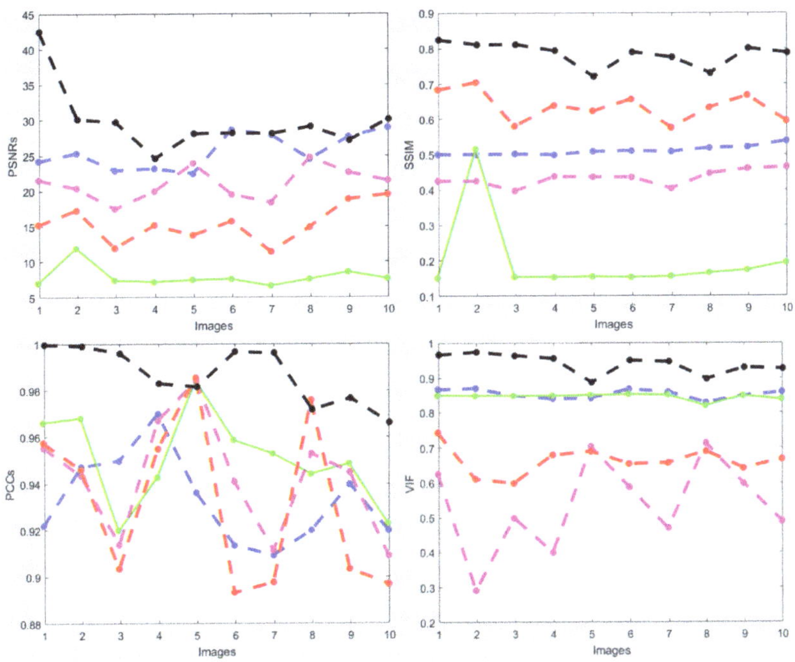

Figure 11. PSNRs and SSIMs obtained from the ground truth and enhanced images (HRF diabetic retinopathy) (first row). The values of PCCs and VIFs obtained from the ground truth and enhanced images (HRF diabetic retinopathy) (second row) computed by HEM [41] (blue color); TLLR [64] (green color); LLIE [42] (magenta color); HIEA [44] (red color) and ours (black color).

5. Discussion and Conclusions

This work is dedicated to the enhancement of retinal images by the RPCA method [53,54] through $L_{w,*}$, $L_{2,1}$ and affine transformation, aimed at improving their robustness. The development of these approaches represents a significant contribution to the field of statistics in imaging, particularly in the realm of high-dimensional medical images. Existing methods, as referenced in [53,54], often lack robustness in the presence of gross errors and outliers within high-dimensional retinal images. In response to this challenge, this article presents a novel method developed to address these issues head on.

In this paper, we propose a novel method for enhancing low-quality retinal images, which is crucial for detecting cataracts and diabetic diseases in human eyes. To ensure robustness against anomalies such as light disturbances, image blurring, and retinal artifacts, the proposed new method combines τ_i, and $L_{w,*}$ and $L_{2,1}$ norms into RPCA, thereby advancing the existing optimization techniques. The proposed method for enhancing low-quality retinal images is a multifaceted approach that integrates several mathematical frameworks to address the common challenges encountered in retinal imaging. The incorporation of τ_i, $L_{w,*}$ and $L_{2,1}$ norms into Robust Principal Component Analysis (RPCA) represents a sophisticated novel strategy to enhance the robustness of the image processing algorithm.

The utilization of τ_i for image alignment through Taylor series expansion and Jacobian transformation provides a novel application of geometric principles in the context of retinal image processing. By iteratively updating the warp parameters, this technique aims to mitigate misalignments induced by factors such as eye movement or imaging artifacts. Moreover, the introduction of the $L_{2,1}$ norm serves to address sparse adverse effects within the retinal images, leveraging sparse regularization to effectively handle noise and outliers. This norm contributes to the alignment of retinal images by minimizing the impact of correlated samples and promoting consistency across the dataset. The incorporation of the weighted nuclear norm, denoted by $L_{w,*}$, introduces a nuanced approach to singular value regularization, where weights are assigned to individual singular values based on their significance in the image processing context. By assigning appropriate weights, this technique aims to preserve important features while suppressing noise and irrelevant variations, thereby enhancing the fidelity of the reconstructed images.

The validation of this proposed new method on two widely used public datasets demonstrates its efficacy in real-world scenarios and provides empirical evidence of its superiority over existing methods. The comparative evaluation highlights the potential of this new method to significantly enhance the quality of retinal images, which is crucial for accurate disease detection and diagnosis.

As future research, this work can be extended to incorporate truncated weighted nuclear norm regularization for image denoising and its integration into Tensor RPCA. By extending the proposed new method to handle more complex data structures and incorporating additional regularization parameters, this work has the potential to enhance the quality of the retinal image processing over the baselines.

Author Contributions: Conceptualization, H.T.L. and D.-G.C.; methodology, H.T.L., D.G.C., K.C., Y.W. and W.Z; software, H.T.L.; validation, H.T.L., D.G.C., K.C., Y.W. and W.Z.; formal analysis, H.T.L.; writing—original draft preparation, H.T.L.; editing, D.-G.C., K.C., Y.W. and W.Z.; visualization, H.T.L., D.G.C., K.C., Y.W. and W.Z.; supervision, D.-G.C., K.C. and Y.W.; funding acquisition, D.-G.C. All authors have read and agreed to the published version of the manuscript.

Funding: This work is partially supported by South Africa National Research Foundation (NRF) and South Africa Medical Research Council (SAMRC) (South Africa DST-NRF-SAMRC SARChI Research Chair in Biostatistics, Grant number 114613).

Institutional Review Board Statement: Not applicable.

Informed Consent Statement: Not applicable.

Data Availability Statement: Applicable.

Acknowledgments: The authors express their gratitude to the editor and anonymous reviewers for their careful review of the manuscript and thoughtful comments. Their suggestions, including the recommendation to add more simulations, have significantly enhanced the readability and quality of this work.

Conflicts of Interest: The authors declare no conflicts of interest.

Abbreviations

The following abbreviations are used in this manuscript:

MDPI	Multidisciplinary Digital Publishing Institute
ADMM	Alternating Direction Method of Multipliers
CLAHE	Contrast Limited Adaptive Histogram Equalization
LLIEM	Low Light Image Enhancement Method
RPCA	Robust Principal Component Analysis
PSNRs	Peak Signal-to-Noise Ratio
SSIMs	Structural Similarity Index
EyeQ	Eye Quality
HIEA	Hybrid Image Enhancement Algorithm
WNNM	Weighted Nuclear Norm Minimization
TLLR	Tensor Low-Rank Representation
VIF	Visual Information Fidelity
PCC	Pearson correlation coefficient

References

1. Pratap, T.; Kokil, P. Deep neural network based robust computer-aided cataract diagnosis system using fundus retinal images. *Biomed. Signal Process. Control* **2021**, *70*, 102–985. [CrossRef]
2. Devi, Y.A.S.; Kamsali, M.C. Retinal Image Contrast Enhancement through Pixel Collaboration in Spatial Domain. *Int. J. Intell. Eng. Syst*, **2022**, *15*, 500â.
3. Gonzalez-Amador, E.; Arines, J.; Charlón, P.; Garcia-Porta, N.; Abraldes, M. Improvement of retinal images affected by cataracts. *Photonics* **2022**, *9*, 251. [CrossRef]
4. Ren, K.C.; Longdan, W.; Minjie, G.G.; Chen, Q. An improved U-net based retinal vessel image segmentation method. *Heliyon* **2022**, *8*, 10. [CrossRef] [PubMed]
5. Singh, V.; Verma, N.K.; Cui, Y. Type-2 fuzzy PCA approach in extracting salient features for molecular cancer diagnostics and prognostics. *IEEE Trans. Nanobioscience* **2019**, *18*, 482–489. [CrossRef] [PubMed]
6. Qureshi, I.; Ma, J.; Shaheed, K. A hybrid proposed fundus image enhancement framework for diabetic retinopathy. *Algorithms* **2019**, *12*, 14. [CrossRef]
7. Zhu, W.; Qiu, P.; Farazi, M.; Nandakumar, K.; Dumitrascu, O.M.; Wang, Y. Optimal transport guided unsupervised learning for enhancing low-quality retinal images. In Proceedings of the IEEE 20th International Symposium on Biomedical Imaging (ISBI), Cartagena, Colombia, 17–21 April 2023; pp. 1–5.
8. Fu, Y.; Wang, W.; Wang, C. Image change detection method based on RPCA and low-rank decomposition. In Proceedings of the 35th Chinese Control Conference (CCC), Chengdu, China, 27–29 July 2016; pp. 9412–9417.
9. Baghaie, A.; D'souza, R. M.; Yu, Z. Sparse and low rank decomposition based batch image alignment for speckle reduction of retinal OCT images. In Proceedings of the 12th International Symposium on Biomedical Imaging (ISBI), Prague, Czech Republic, 13–16 April 2016; pp. 226–230.
10. Gu, Z.; Cheng, J.; Fu, H.; Zhou, K.; Hao, H.; Zhao, Y.; Zhang, T.; Gao, S.; Liu, J. Ce-net: Context encoder network for 2nd medical image segmentation. *IEEE Trans. Med. Imaging* **2019**, *38*, 2281–2292. [CrossRef] [PubMed]
11. Fu, H.; Cheng, J.; Xu, Y.; Wong, D. W. K.; Liu, J.; Cao, X. Joint optic disc and cup segmentation based on multi-label deep network and polar transformation. *IEEE Trans. Med. Imaging* **2018**, 37, 1597–1605. [CrossRef]
12. Staal, J.; Abràmoff, M. D.; Niemeijer, M. ; Viergever, M.A.; Van Ginnek, B. Ridge-based vessel segmentation in color images of the retina. *IEEE Trans. Med. Imaging* **2004**, *23*, 501–509. [CrossRef]
13. Yang, J.; Yu, K.; Gong, Y.; Huang, T. Linear spatial pyramid matching using sparse coding for image classification. In Proceedings of the IEEE Conference on Computer Vision and Pattern Recognition, Miami, FL, USA, 20–25 June 2009; pp. 1794–1801.
14. Agarwal, S.; Roth, D. Learning a sparse representation for object detection. In Proceedings of the Computer Vision—ECCV 2002: 7th European Conference on Computer Vision, Copenhagen, Denmark, 28–31 May 2002; Volume 23, pp. 113–127.
15. Li, J.; Gsaxner, C.; Pepe, A.; Schmalstieg, D.; Kleesiek, J.; Egger, J. Sparse convolutional neural networks for medical image analysis. *TechRxiv.* **2023**. [CrossRef]
16. Bindhya, P.; Jegan, C.; Raj, V. A Review on Methods of Enhancement And Denoising in Retinal Fundus Images. *Int. J. Comput. Sci. Eng.* **2020**, *8*, 1–9.

17. Ashanand; Kaur, M. Retinal Image Enhancement for Detection of Medical Complications. *Smart Energy Adv. Power Technol. Sel. Proc. ICSEAPT*, **2022**, *2*, 667–694.
18. Zhang, R.; Nie, F.; Li, X.; Wei, X. Feature selection with multi-view data: A survey. *Inf. Fusion*, **2019**, *50*, 158–167. [CrossRef]
19. Lakshminarayanan, V.; Kheradfallah, H.; Sarkar, A.; Jothi Balaji, J. Automated detection and diagnosis of diabetic retinopathy: A comprehensive survey. *J. Imaging* **2021**, *7*, 165. [CrossRef] [PubMed]
20. Dumitrascu, O. M.; Zhu, W.; Qiu, P.; Nandakumar, K.Y. Automated Retinal Imaging Analysis for Alzheimers Disease Screening. In Proceedings of the IEEE International Symposium on Biomedical Imaging: From Nano to Macro, Chicago, IL, USA, 30 March–2 April 2011.
21. Lenka, S.; Lazarus, M.Z.; Behera, S.R. Glaucoma Detection Based on Specularity Removal Low Rank Model from Retinal Fundus Images. *ECTI Trans. Comput. Inf. Technol.* **2023**, *17*, 330–342.
22. Singh, L.K.; Pooja, G.; Hitendra, K.M.; Bhadoria, R.S. An enhanced deep image model for glaucoma diagnosis using feature-based detection in retinal fundus. *Med. Biol. Eng. Comput.* **2021**, *59*, 333–353. [CrossRef] [PubMed]
23. Fu, H.; Zhao, Y.; Yap, P.; Schönlieb, C.; Frangi, A.F. Guest Editorial Special Issue on Geometric Deep Learning in Medical Imaging. *IEEE Trans. Med. Imaging* **2023**, *42*, 332–335. [CrossRef]
24. Alom, M.Z.; Yakopcic, C.; Hasan, M.; Taha, T.M.; Asari, V.K. Recurrent residual U-Net for medical image segmentation. *J. Med. Imaging* **2019**, *6*, 014006. [CrossRef]
25. Ren, W.; Liu, S.; Ma, L.; Xu, Q.; Xu, X.; Cao, X.; Du, J.; Yang, M. Low-light image enhancement via a deep hybrid network. *IEEE Trans. Image Process.* **2019**, *28*, 4364–4375. [CrossRef]
26. Abbood, S.H.; Hamed, H.N.A.; ; Rahim, M.S.M.; Rehman, A.S.T.; Bahaj, S.A. Hybrid retinal image enhancement algorithm for diabetic retinopathy diagnostic using deep learning model. *IEEE Access* **2022**, *10*, 73079–73086. [CrossRef]
27. Zhu, W.; Qiu, P.; Dumitrascu, O.M.; Sobczak, J.M.; Farazi, M.Y.Z.; Nandakumar, K.; Wang, Y. OTRE: Where optimal transport guided unpaired image-to-image translation meets regularization by enhancing. In Proceedings of the International Conference on Information Processing in Medical Imaging, San Carlos de Bariloche, Argentina, 18–23 June 2023, pp. 415–427.
28. Liu, H.; Li, H. ; Fu, H.; Xiao, R.; Gao, Y.; Hu, Y.; Liu, J. Degradation-invariant enhancement of fundus images via pyramid constraint network. In Proceedings of the International Conference on Medical Image Computing and Computer-Assisted Intervention, Singapore, 18–22 September 2022; pp. 507–516.
29. Liu, F.; Huang, W. ESDiff: A joint model for low-quality retinal image enhancement and vessel segmentation using a diffusion model. *Biomed. Opt. Express* **2023**, *14*, 6563–6578. [CrossRef] [PubMed]
30. Kushol, R.; Kabir, M.H.; Abdullah-Al-Wadud, M.; Islam, M.S. Retinal blood vessel segmentation from fundus image using an efficient multiscale directional representation technique Bendlets. *Math. Biosci. Eng.* **2020**, *17*, 7751–7771. [CrossRef] [PubMed]
31. Oh, E.J.; Hwang, Y.; Han, Y.; Choi, T.; Lee, G.; Kim, W.H. TaeRESToring Clarity: Unpaired Retina Image Enhancement Using Scattering Transform. In Proceedings of the International Conference on Medical Image Computing and Computer-Assisted Intervention, Vancouver, BC, Canada, 8–12 October 2023; Volume 14, pp. 470–480.
32. Wang, G.; Ye, J.C.; Mueller, K.; Fessler, J.A. Image reconstruction is a new frontier of machine learning. *IEEE Trans. Med. Imaging* **2018**, *37*, 1289–1296. [CrossRef] [PubMed]
33. Albattah, W.; Nawaz, M.; Javed, A.; Masood, M.; Albahli, S. A novel deep learning method for detection and classification of plant diseases. *Complex Intell. Syst.* **2022**, *37*, 1–18. [CrossRef]
34. Aurangzeb, K.; Alharthi, R.S.; Haider, S.I.; Alhussein, M. An efficient and light weight deep learning model for accurate retinal vessels segmentation. *IEEE Access* **2022**, *11*, 23107–23118. [CrossRef]
35. Harine, M.; Kumar, A.; Gopinath, P.; Sasikala, S. Fundus Image Enhancement Using Hybrid Deep Learning Approaches. In Proceedings of the The 1st International Conference on Innovations in High Speed Communication and Signal Processing, Bhopal, India, 4–5 March 2023; pp. 347–352.
36. Goutam, B.; Hashmi, M.F.; Geem, Z.W.; Bokde, N.D. A comprehensive review of deep learning strategies in retinal disease diagnosis using fundus images. *IEEE Trans. Med. Imaging* **2022**, *10*, 57796–57823. [CrossRef]
37. Valente, J.; António, J.; Mora, C.; Jardim, S. Developments in image processing using deep learning and reinforcement learning. *J. Imaging* **2023**, *9*, 207. [CrossRef] [PubMed]
38. Fan, F.; Xiong, J.; Li, M. Wang, G. On interpretability of artificial neural networks: A survey. *IEEE Trans. Radiat. Plasma Med. Sci.* **2021**, *5*, 741–760. [CrossRef]
39. Soomro, T.A.; Afifi, A.J.; Shah, A.A.; Soomro, S.; Baloch, G.A.; Zheng, L.; Yin, M.; Gao, J. Impact of image enhancement technique on CNN model for retinal blood vessels segmentation. *IEEE Access* **2019**, *7*, 158183–158197. [CrossRef]
40. He, Y.; Zheng, Y.; Zhao, Y.; Ren, Y.; Lian, J.; Gee, J. Retinal image denoising via bilateral filter with a spatial kernel of optimally oriented line spread function. *Comput. Math. Methods Med.* **2017**, *2017*, 1769834. [CrossRef]
41. Singh, N.; Kaur, L.; Singh, K. Histogram equalization techniques for enhancement of low radiance retinal images for early detection of diabetic retinopathy. *Eng. Sci. Technol. Int. J.* **2019**, *22*, 736–745. [CrossRef]
42. Singh, A.; Chougule, A.; Narang, P.; Chamola, V.; Yu, F.R. Low-light image enhancement for UAVs with multi-feature fusion deep neural networks. *IEEE Geosci. Remote Sens. Lett.* **2022**, *19*, 1–5. [CrossRef]
43. Barros, D.M.; Moura, J.C.; Freire, C.R.; Taleb, A.C.; Valentim, R.A.; Morais, P.S. Machine learning applied to retinal image processing for glaucoma detection: Review and perspective. *Biomed. Eng. Online* **2020**, *19*, 1–21. [CrossRef] [PubMed]

44. Xian, Y.; Zhao, G.; Wang, C.; Chen, X.; Dai, Y. A Novel Hybrid Retinal Blood Vessel Segmentation Algorithm for Enlarging the Measuring Range of Dual-Wavelength Retinal Oximetry. *Photonics* **2023**, *10*, 722. [CrossRef]
45. Kandpal, A.; Jain, N. Retinal image enhancement using edge-based texture histogram equalization. In Proceedings of the International Conference on Signal Processing and Integrated Networks (SPIN), Noida, India, 27–28 February 2020; pp. 477–482.
46. Rosas-Romero, R.; Martínez-Carballido, J.; Hernández-Capistrán, J.; Uribe-Valencia, L.J. A method to assist in the diagnosis of early diabetic retinopathy: Image processing applied to detection of microaneurysms in fundus images. *Comput. Med. Imaging Graph.* **2015**, *44*, 41–53. [CrossRef] [PubMed]
47. Foracchia, M.; Grisan, E.; Ruggeri, A. Luminosity and contrast normalization in retinal images. *Med. Image Anal.* **2005**, *9*, 179–190. [CrossRef] [PubMed]
48. Gao, S.; Zhang, M.; Zhao, Q.; Zhang, X.; Li, Y. Underwater image enhancement using adaptive retinal mechanisms. *Med. Image Anal.* **2019**, 28, 5580–5595. [CrossRef] [PubMed]
49. Setiawan, A.W.; Mengko, T.R.; Santoso, O.S.; Suksmono, A.B. Color retinal image enhancement using CLAHE. In Proceedings of the International Conference on ICT for Smart Society, Bandung, Indonesia, 13–14 June 2013; pp. 1–3.
50. Joshi, G.D.; Sivaswamy, J. Colour retinal image enhancement based on domain knowledge. In Proceedings of the IEEE Sixth Indian Conference on Computer Vision, Graphics & Image Processing, Bhubaneswar, India, 16–19 December 2008; pp. 591–598.
51. Wang, S.; Jin, K.; Lu, H.; Cheng, C.; Ye, J.; Qian, D. Human visual system-based fundus image quality assessment of portable fundus camera photographs. *IEEE Trans. Med. Imaging* **2015**, *35*, 1046–1055. [CrossRef]
52. Jiang, Y.; Wang, Y.; Li, S.; Zhang, Y.; Zhao, M.; Gao, Y. Event-based low-illumination image enhancement. *IEEE Trans. Multimed.* **2023**, *26*, 1920–1931. [CrossRef]
53. Candès, E. J.; Li, X.; Ma, Y.; Wright, J. Robust principal component analysis? *J. ACM* **2011**, *58*, 1–37. [CrossRef]
54. Li, R.; Pan, J.; Si, Y.; Yan, B.; Hu, Y.; Qin, H. Specular reflections removal for endoscopic image sequences with adaptive-RPCA decomposition. *IEEE Trans. Med. Imaging* **2019**, *39*, 328–340. [CrossRef] [PubMed]
55. Zhou, X.; Yang, C.; Zhao, H.; Yu, W. Low-rank modeling and its applications in image analysis. *ACM Comput. Surv.* **2014**, *47*, 1–33. [CrossRef]
56. Hamlomo, S.; Atemkeng, M.; Brima, Y.; Nunhokee, C.; Baxter, J. Advancing Low-Rank and Local Low-Rank Matrix Approximation in Medical Imaging: A Systematic Literature Review and Future Directions. *arXiv* **2024**, arXiv:2402.14045.
57. Wright, J.; Ganesh, A.; Rao, S.; Peng, Y.; Ma, Y. Robust principal component analysis: Exact recovery of corrupted low-rank matrices via convex optimization. In *Advances in Neural Information Processing Systems 22 (NIPS 2009)*; Curran Associates, Inc.: Red Hook, NY, USA, 2009; Volume 22. Available online: https://papers.nips.cc/paper_files/paper/2009/hash/c45147dee729311ef5b5c3003946c48f-Abstract.html (accessed on 16 June 2024).
58. Kopriva, I.; Shi, F.; Chen, X. Enhanced low-rank+ sparsity decomposition for speckle reduction in optical coherence tomography. *J. Biomed. Opt.* **2016**, *21*, 076008. [CrossRef]
59. Likassa, H. T.; Fang, W. H.; Chuang, Y. Modified robust image alignment by sparse and low rank decomposition for highly linearly correlated data. In Proceedings of the 3rd International Conference on Intelligent Green Building and Smart Grid, Yilan, Taiwan, 22–25 April 2018; pp. 1–4.
60. Likassa, H.T.; Xian, W.; Tang, X. New robust regularized shrinkage regression for high-dimensional image recovery and alignment via affine transformation and tikhonov regularizationa. *Int. J. Math. Math. Sci.*, **2020**, *2020*, 1286909. [CrossRef]
61. Likassa, H.T.; Xia, Y.; Gotu, B. An Efficient New Robust PCA Method for Joint Image Alignment and Reconstruction via the Norms and Affine Transformation. *Sci. Program.* **2022**, *2022*, 5682492. [CrossRef]
62. Likassa, H.T.; Fang, W.; Leu, J. Robust image recovery via affine transformation and $L_{2,1}$ norm. *IEEE Access* **2019**, *7*, 125011–125021. [CrossRef]
63. Wahlberg, B.; Boyd, S.; Annergren, M.; Wang, Y. An ADMM algorithm for a class of total variation regularized estimation problems. *IFAC Proc. Vol.* **2012**, *45*, 83–88. [CrossRef]
64. Zhou, P.; Lu, C.; Feng, J.; Lin, Z.; Yan, S. Tensor low-rank representation for data recovery and clustering. *IEEE Trans. Pattern Anal. Mach. Intell.* **2019**, *43*, 1718–1732. [CrossRef]
65. Cheng, J. Sparse range-constrained learning and its application for medical image grading. *IEEE Trans. Med. Imaging* **2018**, *37*, 2729–27382. [CrossRef]
66. Gao, Q.; Zhang, P.; Xia, W.; Xie, D.; Gao, X.; Tao, D. Enhanced tensor RPCA and its application. *IEEE Trans. Pattern Anal. Mach. Intell.* **2020**, *43*, 2133–2140. [CrossRef] [PubMed]
67. Gong, W.; Xu, W.; Wu, L.; Xie, X.; Cheng, Z. Intrinsic image sequence decomposition using low-rank sparse model. *IEEE Access* **2018**, *7*, 4024–4030. [CrossRef]
68. Wang, L.; Wang, Y.; Wang, S.; Liu, Y.; Hu, Y.; Chen, L.; Chen, H. Global Weighted Tensor Nuclear Norm for Tensor Robust Principal Component Analysis. *arXiv*, **2022**, arXiv:2209.14084.
69. Likassa, H. T.; Fang, W. H. Robust regression for image alignment via subspace recovery techniques. In Proceedings of the 2018 VII International Conference on Network, Communication and Computing, Maui, HI, USA, 5–8 March 2018; pp. 288–293.
70. Lin, Z.; Chen, M.; Ma, Y. The augmented lagrange multiplier method for exact recovery of corrupted low-rank matrices. *arXiv* **2010**, arXiv:1009.5055.
71. Liu, G.; Lin, Z.; Yu, Y. Robust subspace segmentation by low-rank representation. In Proceedings of the 27th International Conference on Machine Learning, Haifa, Israel, 21–24 June 2010; pp. 663–670.

72. Peng, Y.; Ganesh, A.; Wright, J.; Xu, W.; Ma, Y. RASL: Robust alignment by sparse and low-rank decomposition for linearly correlated images. *IEEE Trans. Pattern Anal. Mach. Intell.* **2012**, *34*, 2233–2246. [CrossRef] [PubMed]
73. Wei, X.; Shen, H.; Li, Y.; Tang, X.; Wang, F.; Kleinsteuber, M.; Murphey, Y. L. Reconstructible nonlinear dimensionality reduction via joint dictionary learning. *IEEE Trans. Neural Netw. Learn. Syst.* **2018**, *30*, 175–189. [CrossRef] [PubMed]
74. Wei, X.; Li, Y.; Shen, H.; Xiang, W.; Murphey, Y.L. Joint learning sparsifying linear transformation for low-resolution image synthesis and recognition. *Pattern Recognit.* **2017**, *68*, 412–424. [CrossRef]
75. Bouwmans, T.; Sobral, A.; Javed, S.; Jung, S.K.; Zahzah, E.H. Decomposition into low-rank plus additive matrices for background/foreground separation: A review for a comparative evaluation with a large-scale dataset. *Comput. Sci. Rev.* **2017**, *23*, 1–71. [CrossRef]
76. Gu, S.; Zhang, L.; Zuo, W.; Feng, X. Weighted nuclear norm minimization with application to image denoising. In Proceedings of the IEEE Conference on Computer Vision and Pattern Recognition, Columbus, OH, USA, 23–28 June 2014; pp. 2862–2869.
77. Zhou, Y.; Dou, Y. Double weighted RPCA denoising algorithm for color images. In Proceedings of the IEEE 4th International Conference on Computer and Communications, Honolulu, HI, USA, 16–19 April 2018; pp. 1670–1674.
78. Xu, Z.; Lu, Y.; Wu, J.; He, R.; Wu, S.; Xie, S. Adaptive weighted robust principal component analysis. In Proceedings of the 15th IEEE Conference on Industrial Electronics and Applications, Kristiansand, Norway, 9–13 November 2020; pp. 19–24.
79. Subbarao, R.; Meer, P. Beyond RANSAC: User independent robust regression. In Proceedings of the Conference on Computer Vision and Pattern Recognition Workshop, Seattle, WA, USA, 14–19 June 2020; p. 101.
80. Sawatzky, A.; Xu, Q.; Schirra, C.O.; Anastasio, M.A. Proximal ADMM for multi-channel image reconstruction in spectral X-ray CT. *IEEE Trans. Med. Imaging* **2014**, *33*, 1657–1668. [CrossRef]
81. Boyd, S.; Parikh, N.; Chu, E.; Peleato, B.; Eckstein, J. Distributed optimization and statistical learning via the alternating direction method of multipliers. *Found. Trends Mach. Learn.* **2011**, *3*, 1–122. [CrossRef]
82. Gu, S.; Xie, Q.; Meng, D.; Zuo, W.; Feng, X.; Zhang, L. Weighted nuclear norm minimization and its applications to low level vision. *Int. J. Comput. Vis.* **2017**, *121*, 183–208. [CrossRef]
83. Yang, J.; Yin, W.; Zhang, Y.; Wang, Y. A fast algorithm for edge-preserving variational multichannel image restoration. *SIAM J. Imaging Sci.* **2009**, *2*, 569–592. [CrossRef]
84. Courrieu, P. Fast computation of Moore-Penrose inverse matrices. *arXiv* **2008**, arXiv:0804.4809.
85. Edelstein, W.; Glover, G.; Hardy, C.; Redington, R. The intrinsic signal-to-noise ratio in NMR imaging. *Magn. Reson. Med.* **1986**, *3*, 604–618. [CrossRef] [PubMed]
86. Pérez, J.; Espinosa, J.; Vázquez, C.; Mas, D. Retinal image quality assessment through a visual similarity index. *J. Mod. Opt.* **2013**, *60*, 544–550. [CrossRef]
87. Wang, Z.; Bovik, A.C.; Sheikh, H.R.; Simoncelli, E.P. Image quality assessment: From error visibility to structural similarity. *IEEE Trans. Image Process.* **2004**, *13*, 600–612. [CrossRef] [PubMed]
88. Pal, M.N.; Banerjee, M. Evaluation of effectiveness of image enhancement techniques with application to retinal fundus images. In Proceedings of the 2020 4th International Conference on Computational Intelligence and Networks (CINE), Kolkata, India, 27–29 February 2020; Volume 58, pp. 1–37.
89. Kuo, T.Y.; Su, P.C.; Tsai, C.M. Improved visual information fidelity based on sensitivity characteristics of digital images. *J. Vis. Commun. Image Represent.***2016**, *40*, 76–84. [CrossRef]
90. Shen, Z.; Fu, H.; Shen, J.; Shao, L. Modeling and enhancing low-quality retinal fundus images. *IEEE Trans. Med. Imaging* **2020**, *40*, 996–1006. [CrossRef]
91. Fu, H.; Wang, B.; Shen, J.; Cui, S.; Xu, Y.; Liu, J.; Shao, L. Evaluation of retinal image quality assessment networks in different color-spaces. In Proceedings of the Medical Image Computing and Computer Assisted Intervention–MICCAI 2019: 22nd International Conference, Shenzhen, China, 13–17 October 2019; pp. 48–56.

Journal of
Imaging

MDPI

Article

Subjective Straylight Index: A Visual Test for Retinal Contrast Assessment as a Function of Veiling Glare

Francisco J. Ávila *, Pilar Casado, Mª Concepción Marcellán, Laura Remón, Jorge Ares , Mª Victoria Collados and Sofía Otín

Departamento de Física Aplicada, Facultad de Ciencias, Universidad de Zaragoza, 50009 Zaragoza, Spain; lauremar@unizar.es (L.R.); fatxutxa@unizar.es (J.A.)

* Correspondence: avila@unizar.es

Abstract: Spatial aspects of visual performance are usually evaluated through visual acuity charts and contrast sensitivity (CS) tests. CS tests are generated by vanishing the contrast level of the visual charts. However, the quality of retinal images can be affected by both ocular aberrations and scattering effects and none of those factors are incorporated as parameters in visual tests in clinical practice. We propose a new computational methodology to generate visual acuity charts affected by ocular scattering effects. The generation of glare effects on the visual tests is reached by combining an ocular straylight meter methodology with the Commission Internationale de l'Eclairage's (CIE) general disability glare formula. A new function for retinal contrast assessment is proposed, the subjective straylight function (SSF), which provides the maximum tolerance to the perception of straylight in an observed visual acuity test. Once the SSF is obtained, the subjective straylight index (SSI) is defined as the area under the SSF curve. Results report the normal values of the SSI in a population of 30 young healthy subjects (19 ± 1 years old), a peak centered at SSI = 0.46 of a normal distribution was found. SSI was also evaluated as a function of both spatial and temporal aspects of vision. Ocular wavefront measures revealed a statistical correlation of the SSI with defocus and trefoil terms. In addition, the time recovery (TR) after induced total disability glare and the SSI were related; in particular, the higher the RT, the greater the SSI value for high- and mid-contrast levels of the visual test. No relationships were found for low contrast visual targets. To conclude, a new computational method for retinal contrast assessment as a function of ocular straylight was proposed as a complementary subjective test for visual function performance.

Keywords: ocular straylight; image processing; glare spread function; straylight index; visual acuity; recovery time; ocular aberrations

1. Introduction

Visual perception is a cognitive process that starts with photoreception at the sensitive retina that converts the electromagnetic radiation into electro-chemical impulses. Those electric signals are received by the visual primary cortex for cognitive processing. Then, the beginning of visual perception lies on the neural encoding of the retinal image by the photoreceptor cells. Perception in visual function is driven by a contrast-based pathway [1].

The light entering the natural pupil of the eye is focused on the retina after refraction through the ocular media (i.e., the cornea, crystalline lens, aqueous and vitreous humors). Then, visual performance depends primarily on the retinal image contrast and resolution of the spatial frequency of the observed world [2].

The importance of retinal image contrast and the resolution of a wide range of spatial frequencies is critical in eye growth and cognitive development in children [3,4].

The operation principle of CS tests is based on contrast modulation of grating patterns for different spatial frequencies [5]; then, the contrast sensitivity function (CSF) assesses the ability to visualize a spatial pattern as a function of its contrast and size [6] and provides a powerful tool for discriminating visual impairments due to Parkinson disease [7], Alzheimer [8] or optic neuritis [9]. In addition, the CSF shape can characterize patients with glaucoma, cataract and age-related macular degeneration [10]. Estimating the CSF is a time-consuming subjective clinical procedure that has been speed up over the last years by implementing new CS test in tablet devices [6], computerized charts [11] and computer-adaptive tools [12]. Whereas CS tests are based on the threshold measure of contrast-modulated spatial gratings, contrast modulation procedures do not involve those main physical factors degrading the retinal image quality.

Retinal image quality can be affected by diffraction, ocular aberrations and scattering effects [13,14]. Whereas optical aberrations are mainly related to central vision, ocular scattering can affect the whole visual field. In particular, forward ocular scattering gives rise to the phenomenon of disability glare [15]. This concept lies in the dispersion that the light entering the eye suffers towards the retina due to particles and/or inhomogeneities of the ocular medium that gives rise to a veil of light that covers the whole retinal field. Superimposed veiling luminance causes a loss of retinal contrast that compromises visual performance and is known as glare [16]. Glare can be understood as the reduction in CS and the loss of visual acuity as a result of the scattering or straylight throughout the structures of the eye. The main straylight sources that affect vision are the ocular media themselves (i.e., tear film, iris pigmentation, cornea, sclera, crystalline lens and the retina) [17], aging [18], pathological conditions or external conditions such as the presence of bright light sources that interfere with the field of vision [19].

Depending on the generated straylight (which in turn depends on the brightness of the inrush light source and the ocular media itself), glare vision can be classified in two main groups: discomfort and disability glare [20]. According to the International Commission on Illumination (CIE), discomfort glare is defined as "glare that causes discomfort without necessarily impairing the vision of objects" [21]. On the contrary, disability glare impairs normal vision [22] without needing to cause discomfort glare.

In the last decade, the double-pass (DP) technology [23] has been applied to wide-field optical instruments and has allowed the reconstruction of the point spread function (PSF) of the human eye at larger angles than traditional DP systems that provide the PSF of the visual field (<1°) mainly affected by ocular aberrations [24].

Visual disability glare assessment should consider not only the objective measure of the glare covering the retinal field but also the neural processing of the retinal image affected by veiling glare (i.e., final visual perception). Over the past 50 years, many glare tests and straylight meters have been developed under a psychophysical methodology such as the Miller-Nadler glare instrument [25], Berkeley Glare Tester [26], brightness acuity tester [27] or Vistech MTC-8000 [28].

Although there seems to be no well-stablished standard instrument in clinical practice yet, the "C-Quant (Oculus, GmbH, Wetzlar, Germany)" device based on the compensation comparison method [29] is the most studied in the literature and is commercially available.

The aim of this work was to develop a complementary computational methodology to evaluate the subjective perception of straylight without the need for any optical instrument. The glare spread function (GSF) is employed to generate a bidimensional kernel for image convolution processing with visual acuity charts and then to generate visual tests affected by veiling glare. A new metric is introduced for the subjective visual evaluation of the spatial resolution affected by ocular straylight.

The subjective straylight function (SSF) evaluates the maximum tolerated straylight affecting visual testing based on the subject's visual acuity. From the *SSF*, a new parameter defined as the subjective straylight index (SSI) is tested, establishing the average value in a sample of 30 young healthy volunteers. SSI is also correlated with ocular wavefront measures and temporal aspects of vision. The results show a relationship between *SSI* and high-order aberrations and with recovery time (RT) after inducing retinal photostress.

2. Materials and Methods

2.1. Computational Straylight Generator: The Glare Spread Function

Straylight can be measured by the equivalence luminance generated by an external source that causes the same visual veiling luminance perception as the bright disabling source at a given angular distance, θ [29]. Physically, straylight can be defined as the outer angular region (>1° outside the center) of the optical aberrations domain of the point spread function (PSF) [30] and can be quantified by the straylight parameter, s [31]: $s = \theta^2 \cdot \text{PSF}$.

In the reported literature, s is specified at the glare angle θ at which the straylight is measured and given on a logarithmic scale with normal values around log (s) = 0.9 [32,33]. The *CIE* adopted the standard glare observer given by the total glare equation proposed by Vos and van den Berg [31] as:

$$\text{GSF}(\theta) = \left[1 - 0.08 * \left(\frac{\text{Age}}{70}\right)^4\right] * \left[\frac{9.2*10^6}{\left[1+\left(\frac{\theta}{0.046}\right)^2\right]^{1.5}} + \frac{1.5*10^5}{\left[1+\left(\frac{\theta}{0.045}\right)^2\right]^{1.5}}\right] + \left[1 + 1.6 * \left(\frac{\text{Age}}{70}\right)^4\right] * \left\{\left[\frac{400}{1+\left(\frac{\theta}{0.1}\right)^2} + 3 * 10^{-8} * \theta^2\right] + \text{PF}\left[\frac{1300}{\left[1+\left(\frac{\theta}{0.1}\right)^2\right]^{1.5}} + \frac{0.8}{\left[1+\left(\frac{\theta}{0.1}\right)^2\right]^{0.5}}\right]\right\} + 2.5 * 10^{-3} * \text{PF}[Sr^{-1}] \quad (1)$$

where Age is the age of the observer and PF the eye pigment factor. The glare spread function (GSF) can be easily computed as a bidimensional kernel from Equation (1) as reported in [34]. The top row of Figure 1 shows computer-generated GSFs for different θ values. Then, any digital image (i) with known spatial characteristics can be characterized by its straylight (i_s) through a mathematical convolutional processing given by:

$$i_s = i \otimes GSF(\theta) \quad (2)$$

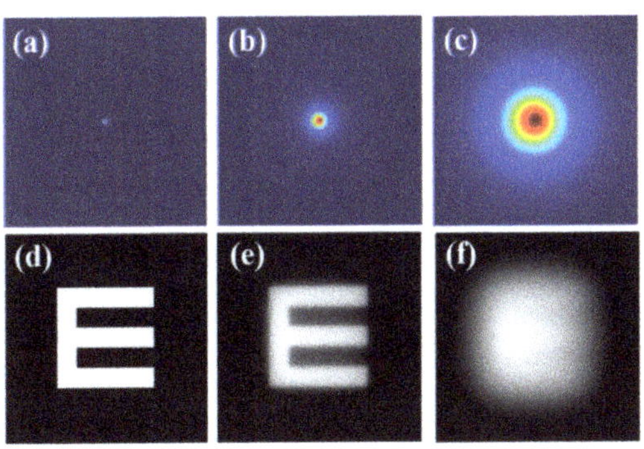

Figure 1. Top row: (**a**) diffraction-limited PSF (i.e., no straylight contribution); (**b**) *GSF* for θ = 4°; (**c**) *GSF* for θ = 8°. Bottom row: retinal images simulated for the PSF and GSFs shown at the upper row (**d**–**f**).

As an example, the bottom row of Figure 1 shows an E-Snellen chart free of straylight (Figure 1d), and the same test affected by two different θ values of the GSF operator. GSF generation (Figure 1 top row) and image simulation (Figure 1 bottom row) were processed using a custom script written in Matlab 2019b (the MathWorks Inc., Natick, MA, USA) (a detailed description of the bidimensional GSF generation can be found elsewhere [34]).

2.2. Computational Straylight Detector: Straylight Index

A previous work published by our group reported a direct measurement of the wide-angle ocular straylight from retinal images [35]. The principle of operation is summarized as follows: The retina is illuminated by an extended bright annular ring that subtends 22° of the retinal field (b), the central part of the ring (a = 1.6° angular field) blocks the light acting as the measurement area (See Figure 2a). Figure 2b shows a retinal image from an artificial eye without any straylight effect. If a straylight source (corneal edema, cataracts or an external bright source, for instance) characterizes the PSF of the eye with intraocular diffusion, as shown in Figure 1, the retinal image of Figure 2b is covered by a uniform veil of light called veiling glare (Figure 1c–e), as a consequence of the increased scattered light throughout the ocular medium.

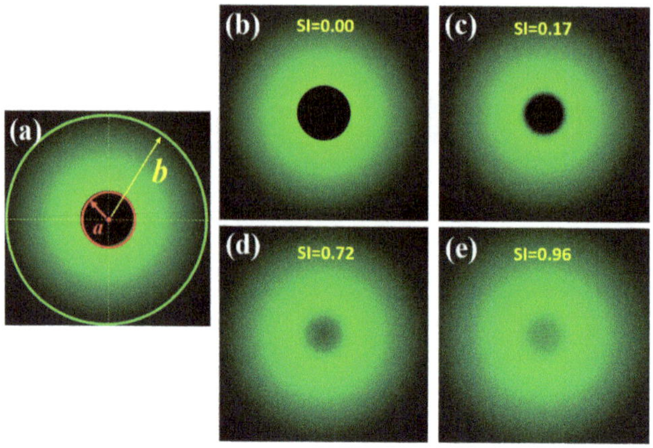

Figure 2. Scheme of the SI principle (**a**); simulated retinal images affected by different *SI* values (**b–e**). Retinal images were simulated in a straylight eye model (reported in [35]) using Zemax 13 optical design software (Zemax OpticStudio, LCC, Arlington Capital Partners, Washington, DC, USA).

Then, the straylight index (SI) parameter can be computed for a given visual angle α, as follows [35]:

$$\mathrm{SI} = \frac{\int_0^a I(\alpha)\cdot d\alpha}{\int_0^b I(\alpha)\cdot d\alpha} \tag{3}$$

2.3. Straylight Index as a Measure of the Glare Spread Function

If the *GSF* (Equation (1)) as a function of the glare angle (θ) and the working principle of the wide-field straylight meter [36] (Figure 2a) are combined into a convolution procedure shown in Figure 3, any retinal image affected by the GSF (θ) can be characterized by a numeric SI value. Figure 4 shows the computed straylight index (SI) values in simulated retinal images affected by GSF as a function of the glare angle (θ). The numerical results shown in Figure 4 can be fitted by the following expression:

$$\theta = \mathrm{A{\cdot}SI^3 + B{\cdot}SI^2 + C{\cdot}SI + D} \tag{4}$$

where A = 130.25, B = −85.92, C = 40.46 and D = −0.58.

Given the relationship between the SI parameter and the glare angle (θ) of the GSF, any visual chart can be processed to be affected by straylight as shown in Figure 5. First, an SI value is chosen and its corresponding θ is found using the Equation (3). The test image size is then read out to generate a bidimensional kernel GSF with the same spatial resolution. Finally, a convolutional image processing returns the test image affected by straylight.

As an example of operation, Figure 6 shows computational results obtained from the convolutional procedure described in Figure 3. Figure 6a shows a Snellen E-letter optotype and the same image affected by different *SI* values (Figure 6b–d). Veiling glare effects are easily visible as the SI increases.

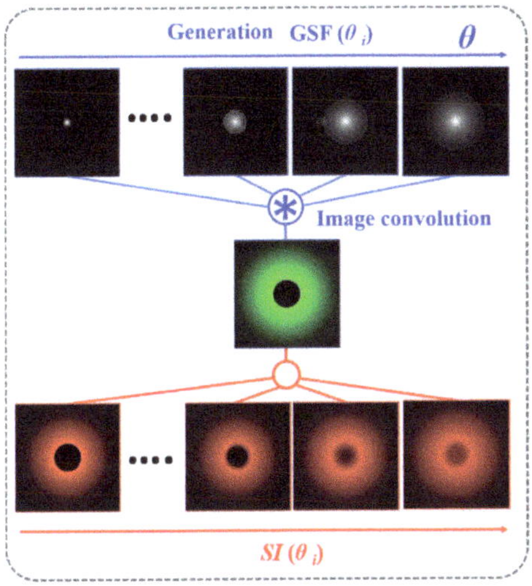

Figure 3. Convolutional image processing to obtain the SI parameter as a function of the glare angle (θ) of the glare spread function.

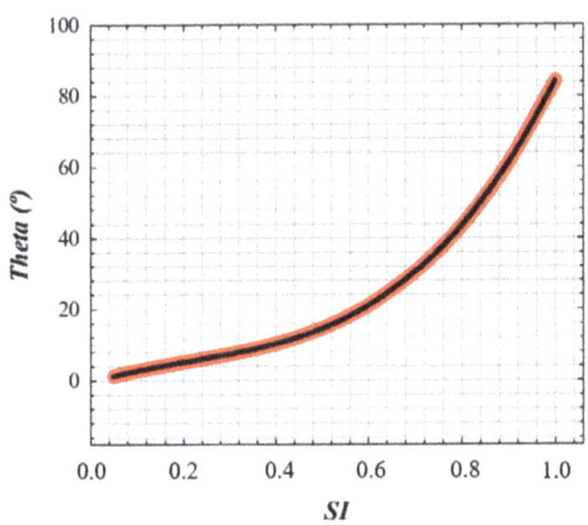

Figure 4. Numerical results of SI values obtained from computer generated retinal images affected by straylight (i.e., characterized by the GSF (θ)).

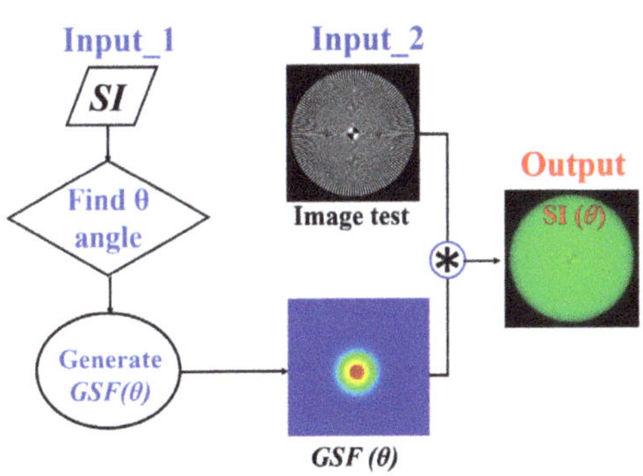

Figure 5. Image processing procedure to generate veiling glare in any digital image quantified by SI (θ).

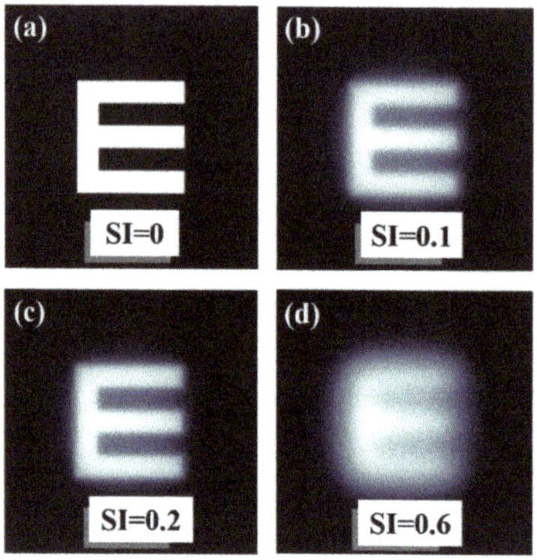

Figure 6. Examples of an E-letter optotype affected by different values of SI. Straylight-free (**a**) and optotypes affected by SI = 0.1 (**b**), SI = 0.2 (**c**) and SI = 0.6 (**d**) values.

2.4. Subjective Straylight Function

A Sloan visual acuity chart optotype script [36] was modified to generate visual acuity (VA) charts ranging from 0.3 to 1.2 (decimal notation). For each VA chart generated, image convolution processing was performed to generate VA optotypes affected by straylight. Then, a total of 10 convolutional processes were computed for each VA value (i.e., 110 VA optotypes as a function of the SI value, all files are available in Supplementary Materials). Figure 7 shows examples of generated optotypes for different values of VA and SI.

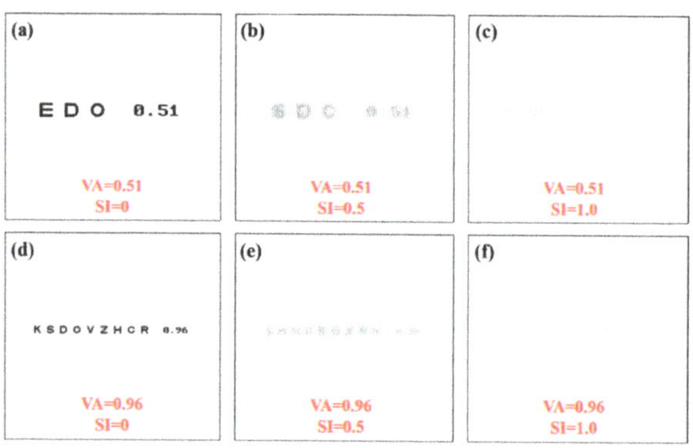

Figure 7. Visual acuity charts for different values of visual acuity and straylight index. Straylight-free VA charts (**a**,**d**) and for SI = 0.5 (**b**,**e**) and SI = 1.0 (**c**,**f**) values.

The subjective straylight function (SSF) is given by the maximum tolerated SI value for each VA Sloan chart. The subjective straylight index (SSI) is then defined as the area under the SSF. The larger the area under the SSF, the better the tolerance of the visual system to preserve contrast sensitivity when ocular straylight veils the retinal field. SSI can be computed as:

$$\mathrm{SSI} = \int_{VA_{min}}^{VA_{max}} \mathrm{SSF}\cdot \mathrm{d}(\mathrm{VA}) \tag{5}$$

Here, $\mathrm{VA}_{\mathrm{min}} = 0.3$ and $\mathrm{VA}_{\mathrm{max}} = 1.2$.

2.5. Subjects

Thirty young adult subjects (19 $\pm$ 1 years old) participated in this study. Participants with ocular pathologies, visual impairment or uncorrected refractive errors were excluded from the sample. All of them were asked to use their optometric correction (i.e., wear contact lenses or spectacles) during the visual test.

2.6. Wavefront Aberrometry

Ocular aberrometry was measured using a laser ray tracing system (iTrace, Tracey Technologies, Houston, TX, USA). A total of 24 Zernike terms were calculated for a pupil size of 6 mm (diameter). Root-mean-square (RMS) values for total low- and high-order aberrations (LOA and HOA, respectively) as well as defocus, spherical aberration, coma and trefoil terms were obtained monocularly (right eye) for each subject.

2.7. Time Recovery from Total Disability Glare

An optical disability glare instrument [37] was employed to measure the recovery time after retinal photostress for different levels of Michelson contrast of the visual target: 100%, 50%, 25%, 10% and 5%. Briefly, the experimental procedure consisted in perceiving (monocularly) the visual test with the given contrast sensitivity, and then total disability glare (i.e., retinal photobleaching) was induced by triggering a bright glare source during a short exposure time of 240 milliseconds (a detailed description of the instrument and measurements' protocol can be found elsewhere [37]). As mentioned in the Introduction, disability glare can range between two main stages from normal vision: discomfort glare and total disability glare. Figure 8a depicts the time course of disability glare after retinal photostress induced by a dazzling light. At the moment of flashing, the straylight causes retinal veiling glare which makes it impossible to discriminate any spatial frequency pattern (i.e., temporal blindness). Before recovering normal vision (baseline), the visual system goes through the intermediate regime between total disability glare and normal vision: discomfort glare. The time required to return to baseline (recovery time, RT) is the clinical parameter used by the disability glare instrument. For a sense of completeness, Figure 8b shows retinal image simulations for a visual target display at different levels of Michelson contrast and how contrast sensitivity degrades as straylight increases. Retinal images were generated by optical simulation on a previously reported straylight eye model [35].

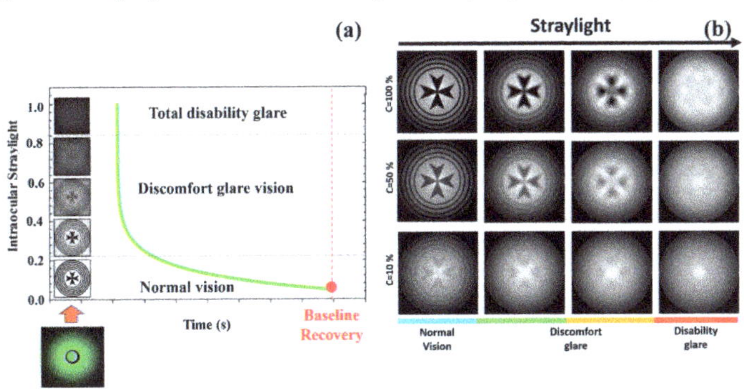

Figure 8. Temporal phases of disability glare vision after flash-lighting until baseline vision is recovered (**a**); simulated retinal images as a function of the contrast of the visual target and ocular straylight (**b**). Blue bar corresponds to normal vision; green and orange to medium and high straylight in discomfort glare regime and finally red bar corresponds to a severe degree of straylight in disability glare vision.

2.8. Image Processing, Data Analysis and Statistics

GSF kernel generation image convolution and *VA* optotypes' algorithms for the *SSF* and *SSI* evaluation (see available files in Supplementary Material) were custom-programmed using Matlab2019b (the MathWorks Inc., Natick, MA, USA). The statistical analysis consisted of the Pearson correlation coefficient to establish significant correlation between the *SSI* parameter and the data of the aberrometric and visual temporal aspects. Graphical representation and statistics were performed in Origin Lab software (OriginPro 2024, Origin Lab Corp., Northampton, MA, USA). Optical simulations were implemented in Zemax optical design software (Zemax OpticStudio, LCC, Arlington Capital Partners, Washington, DC, USA).

3. Results

3.1. Subjective Straylight Index in Young Healthy Subjects

SSI values calculated from the 30 subjects ranged from 0.365 to 0.680. Figure 9 shows the histogram representation of the SSI (for four intervals), which is normally distributed with an asymmetric gaussian shape ($R^2 = 0.82$). The gaussian peak was found at a value of SSI = 0.46.

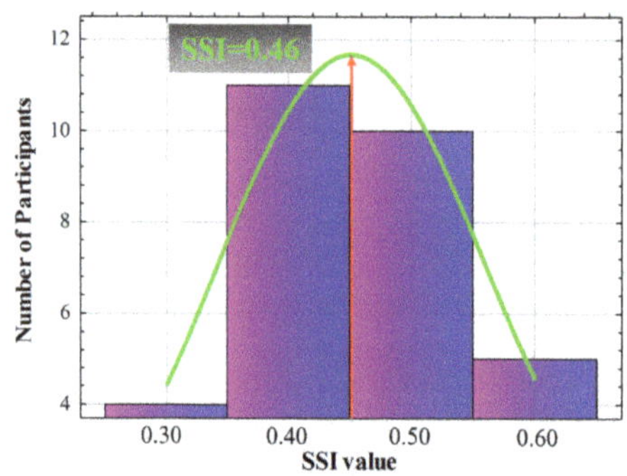

Figure 9. Histogram representation of the *SSI* value for all participants. A gaussian fit ($R^2 = 0.82$) was found with a peak centered at SSI = 0.46.

Figure 10 compares the SSF for two different subjects with extreme SSI values of the Gaussian distribution shown in Figure 10 above. The SSF curves showed similar behavior for the range between VA values of 0.3 and 0.6. However, starting from a VA higher than 0.6, the SSI values decreased drastically, while the subject with SSI = 0.63 remained with a relatively constant tolerance to induced straylight in the optotypes shown. According to the data, the higher the SSI value, the greater the area under the SSF curve, which may translate into the visual system's ability to preserve contrast sensitivity in the presence of external straylight sources.

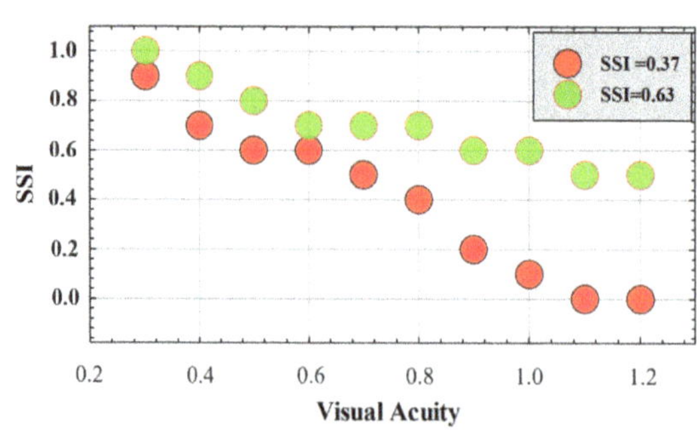

Figure 10. *SSFs* for two subjects with different *SSI* values.

3.2. Effect of SSI on the Spatial Resolution of the Eye: Ocular Wavefront

The influence of straylight on spatial vision is well reported, and as a relevant aspect, its capability as a predictor of visual performance in night-driving conditions [38] or as an indicator of cataract surgery [39]. This section studies how the SSI can be affected by ocular aberrations and then the influence of the eye optics on the final visual perception of veiling glare. Table 1 shows the averaged root-mean-square (RMS) values for low- and high-order aberrations and for the specific low-order defocus and high-order coma, spherical aberration and trefoil terms.

Table 1. RMS values for low (LOA) and high-order aberrations (HOA) and for defocus, coma, spherical aberration (spherical ab.) and trefoil. The asterisk marks those terms for which statistical correlations with *SSI* were found.

RMS LOA	RMS HOA	Defocus *	Coma	Spherical Ab.	Trefoil *
1.82 ± 1.50	0.44 ± 0.59	−1.29 ± 1.73	0.25 ± 0.40	0.01 ± 0.33	0.18 ± 0.14

Figure 11 shows the SSI values as a function of low (Figure 11a) and high-order terms (Figure 11b). A statistical analysis revealed a significant dependence of the SSI parameter on defocus ($R^2 = 0.39$, $p = 0.03$) and trefoil ($R^2 = 0.46$, $p = 0.012$).

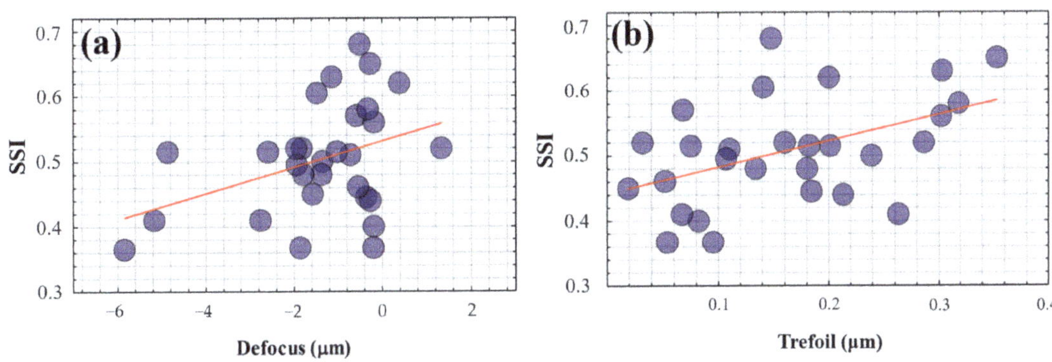

Figure 11. *SSI* as a function of defocus (**a**) and trefoil (**b**) terms. The red lines correspond to the linear statistical correlations of *SSI* with both defocus ($R^2 = 0.39$, $p = 0.03$) and trefoil ($R^2 = 0.46$, $p = 0.012$) terms.

On the one hand, these results corroborate the evident impact of optical blur on visual spatial resolution but under the influence of glare effects. In addition, the modulation of ocular trefoil appears to play an important role in visual tolerance to straylight.

3.3. Effect of SSI on Visual Contrast Sensitivity Recovery Time

A photostress measure was carried out for each volunteer as a function of the Michelson contrast of the visual target. No significant relationships were found between SSI and recovery time (RT) for low contrast levels (i.e., for 25%, 10% and 5% Michelson contrast values). However, for medium and high contrast levels, the *RT* from disability glare depended on the subject's SSF. Figure 12 shows the correlations found between SSI and RT after total induced disability glare. The average RTs for 100% and 50% contrast levels were 4.41 and 4.38 s, respectively; these small differences were not statistically significant ($p = 0.982$).

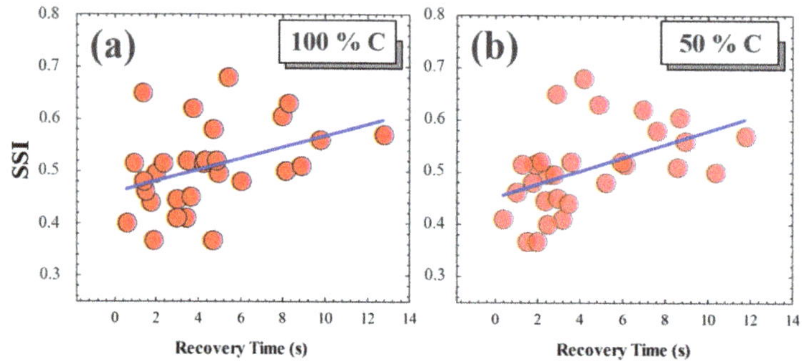

Figure 12. SSI as a function of RT for 100% (**a**) and for 50% visual test contrast levels (**b**). Blue lines correspond to linear statistical correlations.

4. Discussion and Conclusions

We developed a new methodology for computational modelling of ocular straylight in visual acuity tests using convolutional image processing. The CIE disability glare formula [31] and direct straylight meter methodology [35] were combined to generate and measure the effects of veiling glare on visual perception. The generation of straylight in visual tests was performed by convolving the images (veiling glare) with the glare spread function [34] based on the Straylight Index [35].

Visual acuity was assessed in 30 healthy young subjects as a function of the induced SI. The subjective measure of the maximum resolved SI value for each value of VA provides the Subjective Straylight Function (SSF) that addresses the neural and optical attenuation thresholds of VA due to ocular straylight effects. The area under the SSF curve is defined as the Subjective straylight index (SSI).

The SSI parameter was tested in a young population and showed a normal distribution with a peak centered at SSI = 0.46. Once the normal value was established, the influence of the spatial and temporal aspects of vision on SSI was analyzed.

Wavefront aberrometry and RT after induced total disability glare were measured and correlated with SSI. LOA played a role in SSF and in particular the SSI value was found to be negatively correlated with the defocus term, similar to the degradation found in edge contrast sensitivity (CS) due to optical blur [40].

Fernández-Sanchez et al. [41] investigated the effect of high-order aberrations on human vision; they found relationships between visual CS with coma and third-order trefoil terms. In particular, they found a significant reduction in CS with induced coma and trefoil values of around (or more than) 1.03 and 0.96 μm, respectively. In agreement with their results, our findings revealed a negative statistical correlation between SSI and trefoil values (mean values 0.18 ± 0.14 μm); however, no relationship was found between SSI and coma, probably due to the relatively low values and mean of our sample (0.25 ± 0.40 μm).

Finally, it has been reported that RT after induced total disability glare is significantly increased in myopic eyes [42]. Our finding revealed a significant correlation between RT and SSI values for high (100%) and medium (50%) Michelson contrast levels of visual targets. However, SSI was independent of RT for low levels of visual target contrast.

To conclude, a new computational visual test was proposed for the evaluation of retinal contrast based on ocular straylight. VA charts characterized by the straylight index allowed us to define the SSF and the SSI parameter as a neuro-optical measure of visual perception under glare vision conditions. The SSI depends not only on optical blur (defocus) but also on high-order aberrations (trefoil). Furthermore, higher SSI values are related to longer recovery times after retinal photostress for high and medium contrast levels.

Future work will include a clinical study in patients affected by ocular straylight sources and the incorporation into the visual computational method of a functionality for the modulation of ocular aberrations in combination with straylight effects.

Supplementary Materials: The visual acuity charts based on SI to obtain the SSF can be found in this repository: https://github.com/favilago/SSI (accessed on 9 April 2024).

Author Contributions: Conceptualization, F.J.Á.; methodology, F.J.Á., LR., J.A., M.V.C. and P.C.; software, F.J.Á., L.R. and M.C.M.; validation, F.J.Á.; formal analysis, F.J.Á.; investigation, F.J.Á., P.C., L.R., M.C.M., J.A., S.O. and M.V.C.; resources, F.J.Á.; data curation, F.J.Á.; writing—original draft preparation, F.J.Á.; writing—review and editing, F.J.Á., P.C., L.R., M.C.M. and J.A.; visualization, F.J.Á., P.C., L.R., M.C.M., J.A., S.O. and M.V.C.; supervision, F.J.Á.; project administration, F.J.Á.; funding acquisition, F.J.Á. and J.A. All authors have read and agreed to the published version of the manuscript.

Funding: Departamento de Ciencia, Universidad y Sociedad del Conocimiento del Gobierno de Aragon (research group E44−23R).

Institutional Review Board Statement: The study was conducted in accordance with the Declaration of Helsinki, and approved by the Ethics Committee of the Health Sciences Institute of Aragon, Spain. Protocol code: C.P.-C.I. PI20/377, date of approval: 14 July 2020.

Informed Consent Statement: Informed consent was obtained from all subjects involved in the study.

Data Availability Statement: Visual charts are available in the Github repository https://github.com/favilago/SSI (accessed on 9 April 2024).

Acknowledgments: The authors thank the students of the subject Visual Optics (II) of the degree in Optometry at the Faculty Sciences of the University of Zaragoza for their participation in the study. The authors also thank Paul Taylor and Patrick Bannon for reviewing and editing the English grammar.

Conflicts of Interest: The authors declare no conflicts of interest.

References

1. Xiao, N.; Xu, S.; Li, Z.-K.; Tang, M.; Mao, R.; Yang, T.; Ma, S.-X.; Wang, P.-H.; Li, M.-T.; Sunilkumar, A.; et al. A single photoreceptor splits perception and entrainment by cotransmission. *Nature* **2023**, *623*, 562–570. [CrossRef] [PubMed]
2. Liu, L.; Wang, Y.; Liu, J.; Liu, W. Retinal-image quality and contrast sensitivity function in eyes with epiretinal membrane: A cross-sectional observational clinical study. *BMC Ophthalmol.* **2018**, *18*, 290. [CrossRef] [PubMed]
3. Maurer, D.; Mondloch, C.J.; Lewis, T. L Effects of early visual deprivation on perceptual and cognitive development. *Prog. Brain Res.* **2007**, *164*, 87–104. [PubMed]
4. Tran, N.; Chiu, S.; Tian, Y.; Wildsoet, C.F. The significance of retinal image contrast and spatial frequency composition for eye growth modulation in young chicks. *Vis. Res.* **2008**, *48*, 1655–1662. [CrossRef] [PubMed]
5. Jindra, L.F.; Zemon, V. Contrast sensitivity testing: A more complete assessment of vision. *J. Cataract. Refract. Surg.* **1989**, *15*, 141–148. [CrossRef] [PubMed]
6. Dorr, M.; Lesmes, L.A.; Elze, T.; Wang, H.; Lu, Z.L.; Bex, P.J. Evaluation of the precision of contrast sensitivity function assessment on a tablet device. *Sci. Rep.* **2017**, *7*, 46706. [CrossRef] [PubMed]
7. Ming, W.; Palidis, D.J.; Spering, M.; McKeown, M.J. Visual Contrast Sensitivity in Early-Stage Parkinson's Disease. *Investig. Ophthalmol. Vis. Sci.* **2016**, *57*, 5696–5704. [CrossRef] [PubMed]
8. Cormack, F.K.; Tovee, M.; Ballard, C. Contrast sensitivity and visual acuity in patients with Alzheimer's disease. *Int. J. Geriatr. Psychiatry* **2000**, *15*, 614–620. [CrossRef] [PubMed]
9. Mahayana, I.T.; Sakti, D.H.; Gani, T.T. Automated grating contrast-sensitivity: The easy test for hidden visual loss in recovered optic neuritis patient. *Taiwan. J. Ophthalmol.* **2021**, *11*, 301–304. [CrossRef] [PubMed]
10. Kara, S.; Gencer, B.; Ersan, I.; Arikan, S.; Kocabiyik, O.; Tufan, H.A.; Comez, A. Repeatability of contrast sensitivity testing in patients with age-related macular degeneration, glaucoma, and cataract. *Arq. Bras. Oftalmol.* **2016**, *79*, 323–327. [CrossRef] [PubMed]
11. Savini, G.; Calossi, A.; Schiano-Lomoriello, D.; Barboni, P. Precision and Normative Values of a New Computerized Chart for Contrast Sensitivity Testing. *Sci. Rep.* **2019**, *9*, 16537. [CrossRef] [PubMed]
12. Rosenkranz, S.C.; Kaulen, B.; Zimmermann, H.G.; Bittner, A.K.; Dorr, M.; Stellmann, J.P. Validation of Computer-Adaptive Contrast Sensitivity as a Tool to Assess Visual Impairment in Multiple Sclerosis Patients. *Front. Neurosci.* **2021**, *23*, 591302. [CrossRef]
13. Martínez-Roda, J.A.; Vilaseca, M.; Ondategui, J.C.; Giner, A.; Burgos, F.J.; Cardona, G.; Pujol, J. Optical quality and intraocular scattering in a healthy young population. *Clin. Exp. Optom.* **2011**, *94*, 223–229. [CrossRef] [PubMed]
14. Liao, X.; Lin, J.; Tian, J.; Wen, B.; Tan, Q.; Lan, C. Evaluation of optical quality: Ocular scattering and aberrations in eyes implanted with diffractive multifocal or Monofocal intraocular lenses. *Curr. Eye Res.* **2018**, *43*, 696–701. [CrossRef] [PubMed]
15. Donald, R.B. Veiling glare reduction methods compared for ophthalmic applications. *Appl. Opt.* **1981**, *20*, 3787–3791.

16. Van den Berg, T.J.T.P.; Franssen, L.; Kruijt, B.; Coppens, J.E. History of ocular straylight measurement: A review. *Med. Phys.* **2013**, *23*, 6–20. [CrossRef] [PubMed]
17. Van den Berg, T.J.T.P.; IJspeert, J.K.; De Waard, P.W. Dependence of intraocular straylight on pigmentation and light transmission through the ocular wall. *Vis. Res.* **1991**, *31*, 1361–1367. [CrossRef] [PubMed]
18. Mueller-Schotte, S.; Van der Schouw, Y.T.; Schuurmans, M.J. Ocular Straylight: A Determinant of Quality of Life in the Elderly? *Gerontol. Geriatr. Med.* **2015**, *9*, 2333721415610193. [CrossRef] [PubMed]
19. Ayama, M.; Yamazaki, R.; Nakanoya, S.I.; Tashiro, T.; Ishikawa, T.; Ohnuma, K.; Shinoda, H.; Araki, K. Estimation of straylight in the eye and its relation to visual function. *Opt. Rev.* **2015**, *22*, 185–196. [CrossRef]
20. Mainster, M.A.; Turner, P.L. Glare's causes, consequences, and clinical challenges after a century of ophthalmic study. *Am. J. Ophthalmol.* **2012**, *153*, 587–593. [CrossRef] [PubMed]
21. Commission Internationale de l'Éclairage. *CIE e-ILV Term 17-333 Discomfort Glare*; CIE: Vienna, Austria, 2019; Available online: http://eilv.cie.co.at/term/333 (accessed on 16 October 2019).
22. Commission Internationale de l'Éclairage. *CIE e-ILV Term 17-492 Glare*; CIE: Vienna, Austria, 2019; Available online: http://eilv.cie.co.at/term/492 (accessed on 16 October 2019).
23. Galliot, F.; Patel, S.R.; Cochener, B. Objective Scatter Index: Working Toward a New Quantification of Cataract? *J. Refract. Surg.* **2016**, *32*, 96–102. [CrossRef] [PubMed]
24. Ginis, H.; Pérez, G.M.; Bueno, J.M.; Artal, P. The wide-angle point spread function of the human eye reconstructed by a new optical method. *J. Vis.* **2012**, *12*, 20. [CrossRef] [PubMed]
25. Miller, D.; Jernigan, M.E.; Molnar, S.; Wolf, E.; Newman, J. Laboratory evaluation of a clinical glare tester. *Arch. Opthalmol* **1972**, *87*, 324–332. [CrossRef]
26. Bailey, I.L.; Bullimore, M.A. A new test for the evaluation of disability glare. *Optom. Vis. Sci.* **1991**, *68*, 911–917. [CrossRef]
27. Holladay, J.T.; Prager, T.C.; Trujillo, J.; Ruiz, R.S. Brightness acuity test and outdoor visual acuity in cataract patients. *J. Cataract. Refract. Surg.* **1987**, *13*, 67–69. [CrossRef] [PubMed]
28. Elliott, D.B.; Bullimore, M.A. Assessing the reliability, discriminative ability, and validity of disability glare tests. *Investig. Ophthalmol. Vis. Sci.* **1993**, *34*, 108–119.
29. Franssen, L.; Coppens, J.E.; Van den Berg, T.J. Compensation comparison method for assessment of retinal straylight. *Investig. Ophthalmol. Vis. Sci.* **2006**, *47*, 768–776. [CrossRef] [PubMed]
30. Vos, J.J.; Van den Berg, T.J.T.P. On the course of the disability glare function and its attribution to components of ocular scatter. *CIE Collect.* **1997**, *124*, 11–29.
31. Vos, J.J.; Van den Berg, T.J.T.P. Report on disability glare. *CIE Collect.* **1999**, *135*, 1–9.
32. Van den Berg, T.J.T.P. Analysis of intraocular straylight, especially in relation to age. *Optom. Vis. Sci.* **1995**, *72*, 52–59. [CrossRef] [PubMed]
33. Van Den Berg, T.J.; Van Rijn, L.R.; Michael, R.; Heine, C.; Coeckelbergh, T.; Nischler, C.; Wilhelm, H.; Grabner, G.; Emesz, M.; Barraquer, R.I.; et al. Straylight effects with aging and lens extraction. *Am. J. Ophthalmol.* **2007**, *144*, 358–363. [CrossRef] [PubMed]
34. Ávila, F.J.; Ares, J.; Marcellán, M.C.; Collados, M.V.; Remón, L. Iterative-Trained Semi-Blind Deconvolution Algorithm to Compensate Straylight in Retinal Images. *J. Imaging* **2021**, *16*, 73. [CrossRef] [PubMed]
35. Ávila, F.; Collados, M.V.; Ares, J.; Remón, L. Wide-field direct ocular straylight meter. *Opt. Express* **2020**, *28*, 11237–11242. [CrossRef] [PubMed]
36. SergioBonaqueGonzalez/Sloan-Visual-Acuity-Eye-Chart-Optotype. Available online: https://github.com/SergioBonaqueGonzalez/Sloan-Visual-Acuity-Eye-Chart-Optotype (accessed on 1 December 2022).
37. Ávila, F.J.; Casado, P. Optical instrument for the study of time recovery from total disability glare vision. *Appl. Opt.* **2022**, *61*, 2438–2443. [CrossRef] [PubMed]
38. Ungewiss, J.; Schiefer, U.; Eichinger, P.; Wörner, M.; Crabb, D.P.; Jones, P.R. Does intraocular straylight predict night driving visual performance? Correlations between straylight levels and contrast sensitivity, halo size, and hazard recognition distance with and without glare. *Front. Hum. Neurosci.* **2022**, *16*, 910620. [CrossRef] [PubMed]
39. Van Bree, M.C.J.; Pierrache, L.; Zijlmans, B.L.M.; Reus, N.J.; Van den Born, L.I.; Van den Berg, T.J.T.P. Straylight as an Indicator for Cataract Extraction in Patients with Retinal Dystrophy. *Ophthalmol. Retin.* **2017**, *1*, 531–544. [CrossRef] [PubMed]
40. Jansonius, N.M.; Kooijman, A.C. The effect of defocus on edge contrast sensitivity. *Ophthalmic Physiol. Opt.* **1997**, *17*, 128–132. [CrossRef] [PubMed]
41. Fernández-Sánchez, V.; Ponce, M.E.; Lara, F.; Montés-Micó, R.; Castejón-Mochón, J.F.; López-Gil, N. Effect of 3rd-order aberrations on human vision. *J. Cataract. Refract. Surg.* **2008**, *34*, 1339–1344. [CrossRef] [PubMed]
42. Ávila, F.J.; Casado, P.; Ares, J. Photostress Recovery Time after Flash-Lighting Is Increased in Myopic Eyes. *Photonics* **2023**, *10*, 86. [CrossRef]

Journal of *Imaging*

MDPI

Article

When Sex Matters: Differences in the Central Nervous System as Imaged by OCT through the Retina

Ana Nunes [1], Pedro Serranho [1,2], Pedro Guimarães [1], João Ferreira [3], Miguel Castelo-Branco [1,4] and Rui Bernardes [1,4,*]

1 Coimbra Institute for Biomedical Imaging and Translational Research (CIBIT), Institute of Nuclear Sciences Applied to Health (ICNAS), University of Coimbra, 3000-548 Coimbra, Portugal; anunes@fmed.uc.pt (A.N.)
2 Department of Sciences and Technology, Universidade Aberta, 1269-001 Lisboa, Portugal
3 Faculty of Sciences and Technology, University of Coimbra, 3030-201 Coimbra, Portugal
4 Clinical Academic Center of Coimbra (CACC), Faculty of Medicine, University of Coimbra, 3000-548 Coimbra, Portugal
* Correspondence: rmbernardes@fmed.uc.pt

Abstract: Background: Retinal texture has gained momentum as a source of biomarkers of neurodegeneration, as it is sensitive to subtle differences in the central nervous system from texture analysis of the neuroretina. Sex differences in the retina structure, as detected by layer thickness measurements from optical coherence tomography (OCT) data, have been discussed in the literature. However, the effect of sex on retinal interocular differences in healthy adults has been overlooked and remains largely unreported. Methods: We computed mean value fundus images for the neuroretina layers as imaged by OCT of healthy individuals. Texture metrics were obtained from these images to assess whether women and men have the same retina texture characteristics in both eyes. Texture features were tested for group mean differences between the right and left eye. Results: Corrected texture differences exist only in the female group. Conclusions: This work illustrates that the differences between the right and left eyes manifest differently in females and males. This further supports the need for tight control and minute analysis in studies where interocular asymmetry may be used as a disease biomarker, and the potential of texture analysis applied to OCT imaging to spot differences in the retina.

Keywords: texture analysis; neuroretina; central nervous system; sex differences; interocular asymmetry; optical coherence tomography

1. Introduction

In light of the thoroughly researched relationship between the eye and the brain [1–4], sex differences in brain size have been hypothesised [5] to be associated with reports of sexual dimorphism in visual perception. Interestingly, lateral hemisphere activation during visual perception tasks was found to be independent of sex, handedness, and ocular dominance [6], the latter being associated with interocular retinal thickness asymmetries [7].

The difference between the right and left eyes of the same individual is a phenomenon of interest where retinal biomarkers are concerned. The difference between the two eyes is a key factor to consider when interpreting retinal thickness values in ocular disorders such as glaucoma [8,9], particularly in its early stages, when both eyes still yield measurements within normative values.

When it comes to the healthy population, a certain degree of retinal thickness symmetry is often assumed. While some studies confirm this symmetry assumption [8,10], others report statistically significant differences between healthy individuals' right and left eyes [9,11]. Notably, Jacobsen et al. [12] found a slight age and sex effect on retinal thickness interocular differences but advised caution in considering the magnitude of such an effect. Furthermore, a few studies in pediatric subjects have reported no sex effects on interocular thickness symmetry [13–15]. Overall, sex remains largely overlooked when comparing right and left eyes in healthy adults.

The literature on sex differences found in retinal structure is focused mostly around single-layer or full retina thickness, as measured by optical coherence tomography (OCT) [10,16–19]. OCT is a non-invasive imaging technique used for the visualization of the microstructure of various biological tissues in vivo and in situ. In clinical settings, OCT imaging plays an important role in the diagnosis of ocular disorders. Furthermore, this medical imaging technique is increasingly employed in neuroimaging research, where the potential of using the retina as a window to the brain [1–4] for the diagnosis of neurodegenerative disorders such as Alzheimer's disease [20,21], Parkinson's disease [22,23], and multiple sclerosis [24,25], is being explored.

Several OCT-based studies have reported differences in retinal thickness between women and men. Men are often reported to have thicker retinas than women [10,16,17], although some conflicting results have been found [8,17]. In their study, Ooto and Yoshimura [18] hypothesised that the increased retinal thickness in men might be associated with a relatively larger eye size. On the other hand, Delori et al. [19] discusses how sex differences in the shape and size of the fovea may affect thickness measurements. However, non-anatomical factors may play an even larger role in the sexual dimorphism of the human retina. The relationship between various retinopathies and gonadal hormones has been addressed, as in [26,27], where the authors highlighted the role of sex hormones on the pathophysiology of ocular disorders, such as age-related macular degeneration and diabetic retinopathy. Despite the possible connection between gonadal hormones and the health/disease status of the retina, the intricate interplay between sex and vision both in healthy individuals and across different disease processes [5] suggests that observed sex differences in the retina cannot be attributed to a single mechanism [28].

Ultimately, studying the differences between the right and left eyes in the healthy population helps determine the normal demographic variations, allowing pathological deviations to be correctly identified. Indeed, for some disorders, interocular asymmetries have been used as a biomarker of disease [29,30]. Although thickness measurements are usually the metric of choice, interocular differences have also been identified in non-thickness measurements, such as vessel diameter [11]. Another example is the topography of the foveal centre mosaic, where a slight sex asymmetry was found [31].

In recent years, the use of yet another type of information–image texture–from OCT data has been introduced [32,33]. Texture plays an important role in identifying objects and understanding complex scenes [34,35]. The concept of image or visual texture is challenging to define, and different definitions have been proposed in the literature. A generally agreed-upon description for image texture defines it as the spatial arrangement of an image's pixels' grey-level/intensity/brightness values or colour [36–38].

Texture analysis is thus an umbrella term encompassing different approaches aimed at quantifying or describing the spatial distribution of grey levels or colour within an image. The field of texture analysis has been gaining momentum as a source of valuable biomarkers for neurodegeneration. The use of texture-based methods as a tool to assess the central nervous system (CNS) status from medical neuroimaging data is based on the premise that medical images contain a quantifiable texture "signature" [37] that is specific to a particular biological process or state, be it healthy (normal) or pathological. Where retinal OCT imaging is concerned, texture analysis methods have been employed to explore structural alterations resulting from specific ophthalmological (e.g., glaucoma [32]) and neurodegenerative disorders (e.g., Alzheimer's disease and Parkinson's disease [33], and multiple sclerosis [24,25]), and to characterise the healthy (normal) ageing process [39].

Out of the wide variety of texture-based methods available, the grey-level co-occurrence matrix (GLCM) method stands out for its popularity and broad applicability. This method was first proposed by Haralick et al. [36] in 1973. Despite its early development, it remains, to this day, a benchmark method in comparative studies that employ texture ensemble approaches.

The GLCM texture analysis method starts by computing one or more co-occurrence matrices from the images under study. In a co-occurrence matrix, pairs of grey-level values of image pixels with a specific spatial relationship (e.g., two adjacent pixels in the horizontal direction) are tabulated. Typically, four different matrices are computed for the same inter-pixel distance, corresponding to four different pixel pair orientations: $0°$, $45°$, $90°$, and $135°$. Texture features are then computed from these matrices to characterise the local grey-level variations within the image.

We propose using texture metrics derived from the GLCM to assess the sex effect on interocular differences in the healthy population, as texture-based methods can detect discrepancies in the retina that are not conveyed by thickness measurements [24,25,32,33].

In this study, we investigate the differences in texture of the neuroretina between healthy individuals' right and left eyes. More specifically, we demonstrate how interocular differences manifest differently in men and women, as quantified by retinal GLCM-based texture features. Furthermore, focusing specifically on the six layers of the neuroretina, these variations highlight sex differences in the CNS.

2. Materials and Methods

All image processing and statistical analysis were performed using Matlab R2022a (The MathWorks Inc., Natick, MA, USA) running on a Ubuntu operating system desktop computer.

2.1. Data Collection

The data were gathered from the authors' institutional database. The studies from which data were collected were approved by the Faculty of Medicine of the University of Coimbra Ethics Committee, performed according to the tenets laid out in the Declaration of Helsinki [40], and informed consents were obtained from all participants. All data herein were anonymized.

Both eyes of 98 age-matched and sex-balanced healthy controls with no known retinal pathology were imaged by Cirrus HD-OCT 5000 (Carl Zeiss Meditec, Dublin, CA, USA) scanner, using the 512×128 macular cube protocol centred on the macula. The same operator performed all acquisitions. All participants had normal visual acuity ($\geq$0.8 in each eye), a sphere between $\pm$5 diopters, and a cylinder between $\pm$3 diopters. None of the participants had a medical history of ophthalmological, neurological, or other systemic diseases. For more details on group demographics, please refer to Table 1.

Table 1. Demographic data.

	Female	Male
N	49	49
Age (years): mean(std)	42.5(16.3)	42.0(16.0)
Age (years): min(max)	19(74)	20(74)
Right (left) eyes	49(49)	49(49)
Total acquisitions	98	98

2.2. Image Processing

The OCT Explorer software (Retinal Image Analysis Lab, Iowa Institute for Biomedical Imaging, Iowa City, IA, USA) [41–43] was used to segment the six layers of the neuroretina: the retinal nerve fibre layer (RNFL), the ganglion cell layer (GCL), the inner plexiform layer (IPL), the inner nuclear layer (INL), the outer plexiform layer (OPL), and the outer nuclear layer (ONL).

A 2D mean value fundus (MVF) reference image for each of the six neuroretina layers was computed from the 3D OCT data of each eye scan, following the approach in [44]. In this image, each pixel is the depth-wise average of the A-scan values located between the two retinal layer interfaces that delineate the layer of interest (see Figures 1 and 2).

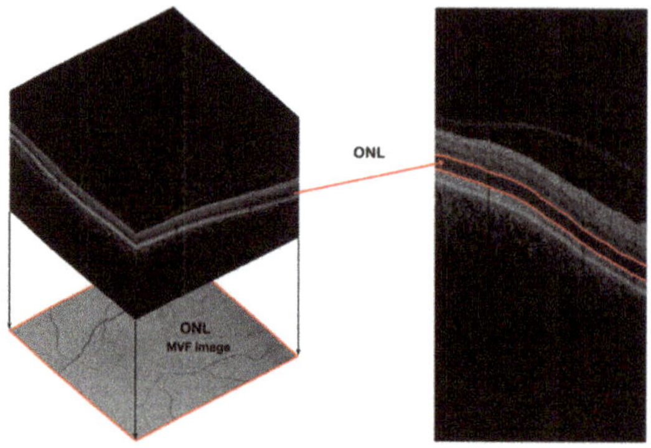

Figure 1. Representation of a mean value fundus image (MVF) computed for the outer nuclear layer (ONL) of the right eye of a 54-year-old female participant regarding the optical coherence tomography volumetric data acquired. These images are shown for reference purposes only; the MVF image projection was intensity-corrected for ease of visualisation.

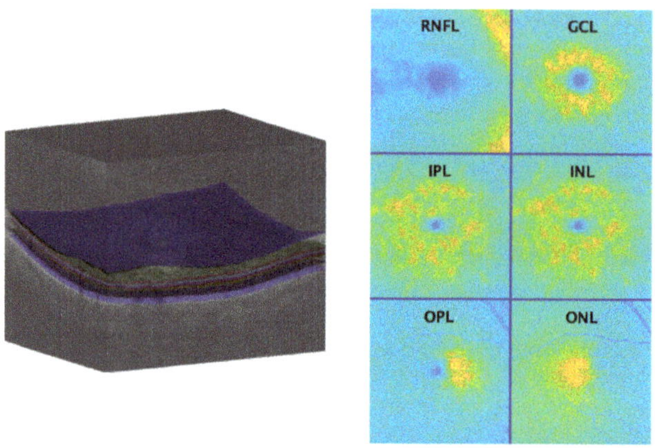

Figure 2. An optical coherence tomography volume (**left**) is shown with interfaces displayed in colour, e.g., the interface vitreous-retina in blue. Pseudo-colour fundus images are (**right**) shown for retinal layers of the right eye of a 54-year-old female participant: the retinal nerve fibre layer (RNFL), the ganglion cell layer (GCL), the inner plexiform layer (IPL), the inner nuclear layer (INL), the outer plexiform layer (OPL), and the outer nuclear layer (ONL). Pseudo-colour is used for the ease of visualisation only.

MVF images of left eyes were flipped horizontally so that the temporal and nasal regions were consistent across all eye scans. This ensures that comparative analysis of the computed texture features considers metrics originating from the same relative position across eyes (see Figure 3).

2.3. Texture Analysis

In this work, GLCM-based texture features were computed for each of six neuroretina MVF images, following the previously used approach of [33,39]. The computation of texture metrics from fundus reference images is a unique approach to OCT data analysis–one that is significantly distinct from the traditional approach of measuring the thickness of the retina (or that of specific retinal layers). A prior study [33] showed that texture-based metrics computed from retinal MVF images are not simply a surrogate for OCT thickness measurements. It was also discussed how these texture-based features could arguably be more discriminative than retinal thickness measurements, as they are able to capture multivariate information; thickness-based measurements, on the other hand, only allow for univariate comparisons.

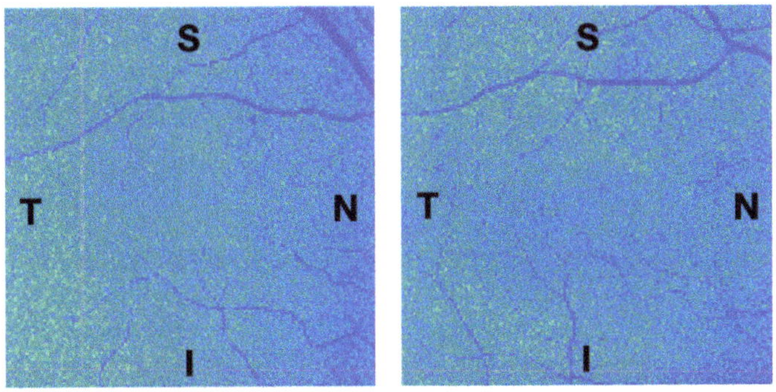

(a) (b)

Figure 3. Example of the mean value fundus (MVF) images computed for the outer nuclear layer of a 54-year-old female participant: (**a**) the right eye; and (**b**) the left eye, flipped to match the regions of the right eye. S, T, N, and I stand for the eye's superior, temporal, nasal, and inferior regions, respectively. These images are shown for reference purposes only; they were intensity-corrected and pseudo-colour-coded for ease of visualisation.

The MVF images, originally of 512 × 128 pixels, were first down-sampled to 128 × 128 to obtain isotropic sampling in both the horizontal and vertical directions. These images were also converted to 16 grey levels to reduce the size of the GLCM matrices.

Similar to our prior works [33,39], the down-sampled images were split into 7 × 7 non-overlapping blocks (block size: 18 pixels in both directions), which were independently analyzed. In each image, the central (4th) row and column of the 49-block grid square were discarded from the analysis to exclude the image region corresponding to the fovea, where some of the six retinal layers examined are reduced or inexistent (see Figure 4).

For each of the remaining 36 blocks on each image, four GLCMs were computed for the pixel-pair orientations of 0°, 45°, 90°, and 135°, with angles 180° apart considered the same, e.g., the pixel-pairs (i, j)/(i, j + d) and (i, j)/(i, j − d), corresponding to the angles 0° and 180°, respectively, are considered to contribute to the same angle (0°). All matrices were computed for a pixel distance of one (d = 1).

For each block and pixel-pair orientation, 21 features were computed: (1) Autocorrelation, (2) Cluster Prominence, (3) Cluster Shade, (4) Contrast, (5) Correlation, (6) Difference Entropy, (7) Difference Variance, (8) Dissimilarity, (9) Energy, (10) Entropy, (11) Homogeneity, (12) Information Measure of Correlation 1 (IMC1), (13) Information Measure of Correlation 2 (IMC2), (14) Inverse Difference, (15) Inverse Difference Moment Normalised (IDMN), (16) Inverse Difference Normalized (IDN), (17) Maximum Probability, (18) Sum Average, (19) Sum Entropy, (20) Sum of Squares–Variance (SSV), and (21) Sum Variance. The definition of the texture features (1) to (3), (8), (10), and (17) can be found in [45], feature (14) in [46], and the remaining features in [36]—except features (15) and (16), which are simply normalized versions of features (11) and (14), respectively.

The maximum value across the four orientations for each feature in a given block was selected as the sole feature value for the block. Blocks were then aggregated into four groups of 3 × 3, corresponding to the temporal-superior (Q1), nasal-superior (Q2), temporal-inferior (Q3), and nasal-inferior (Q4) quadrants. Finally, the average value of each feature per retinal quadrant was used as the feature value for the quadrant. This resulted in a total of 504 texture features (21 GLCM features × 6 layers × 4 quadrants) being computed for each eye.

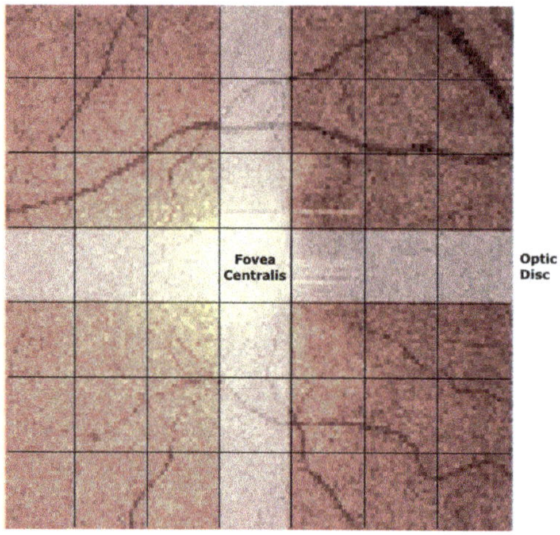

Figure 4. Computed mean value fundus image of the outer nuclear layer of the right eye of a 54-year old female participant. Each of the 7 × 7 blocks show the individually analysed areas which results were later aggregated into larger regions (shaded areas). Image axes are: x-axis (horizontal)—temporal (**left**) to nasal (**right**) and y-axis (vertical)—superior (**top**) to inferior (**bottom**).

2.4. Statistical Analysis

The computed texture features were tested for group mean differences between female and male participants' right (OD) and left (OS) eyes. A pairwise correlation analysis was first performed to exclude similar features.

Each feature was correlated (Pearson correlation) with all other features of the same eye and group (female OD and female OS; male OD and male OS). First, all feature pairs with a correlation coefficient $|r| \geq 0.5$ in both the right (r_{OD}) and the left (r_{OS}) eyes were identified separately for the female and male groups. Given the list of correlated features, they were then ordered by the decreasing number (n) of correlations with other features for which $|r| \geq 0.5$. In the case of a draw with respect to n, the largest sum of explained variance, given by $\sum_{i=1}^{n} r_{OD,i}^2 + r_{OS,\ i}^2$, was considered. Features were selected following this sorted list, while all their correlated features ($|r| \geq 0.5$) were discarded. This process was repeated until two subsets of texture features (one for the female and the other for the male group) were obtained, where all pairwise correlation coefficients satisfied $-0.5 < r < 0.5$.

Each non-correlated feature was first tested for normality within the same eye and group. The Shapiro–Wilk test was used at a 10% significance level for a more conservative normality decision. If the feature presented a normal distribution for both the right- and the left-eye groups, a paired-sample t-test was used to assess the group mean differences; otherwise, a Wilcoxon signed rank test was applied.

Due to the large number of independent tests performed, we considered three corrections for multiple comparisons: Bonferroni, Benjamini–Hochberg, and the False Discovery Rate (FDR), as proposed by Storey [47].

3. Results

3.1. Pairwise Correlations

Out of 504 texture features, 146 (29.0%) and 148 (29.4%) were found to be non-correlated for the female and male groups, respectively. These features are listed in an additional file (see Supplementary Table S1).

3.2. Normality Testing

For female participants, a non-parametric test was applied to 73 out of 146 non-correlated texture features (50.0%), while for male participants, it was applied to 74 out of 148 features (50.0%).

3.3. Hypothesis Testing

Of the 146 non-correlated texture features in the female group, 21 (14.4%) present statistical differences ($p < 0.05$ for each test) between the right and left eyes (left side of Figure 5). Seven (33.3%) of these 21 features are distributed in Q3, and 9 (42.9%) in Q4. Layer-wise, the GCL and IPL gather the highest number of differences: 6 (28.6%) and 4 (19.1%) features, respectively. Of the 21 features, two stand out: Energy (IPL/Q4) and Entropy (IPL/Q4), as they present results with a p-value < 0.001. In addition, six distinct features show a spread-out effect across different layers and quadrants of the retina: Dissimilarity (RNFL/Q4 and OPL/Q4), SSV(GCL/Q2 and IPL/Q1), Entropy (GCL/Q2, IPL/Q4, and OPL/ Q2), IMC2 (GCL/Q3 and INL/Q4), Difference Variance (GCL/Q3 and ONL/Q3), and IMC1 (INL/Q4 and OPL/Q3).

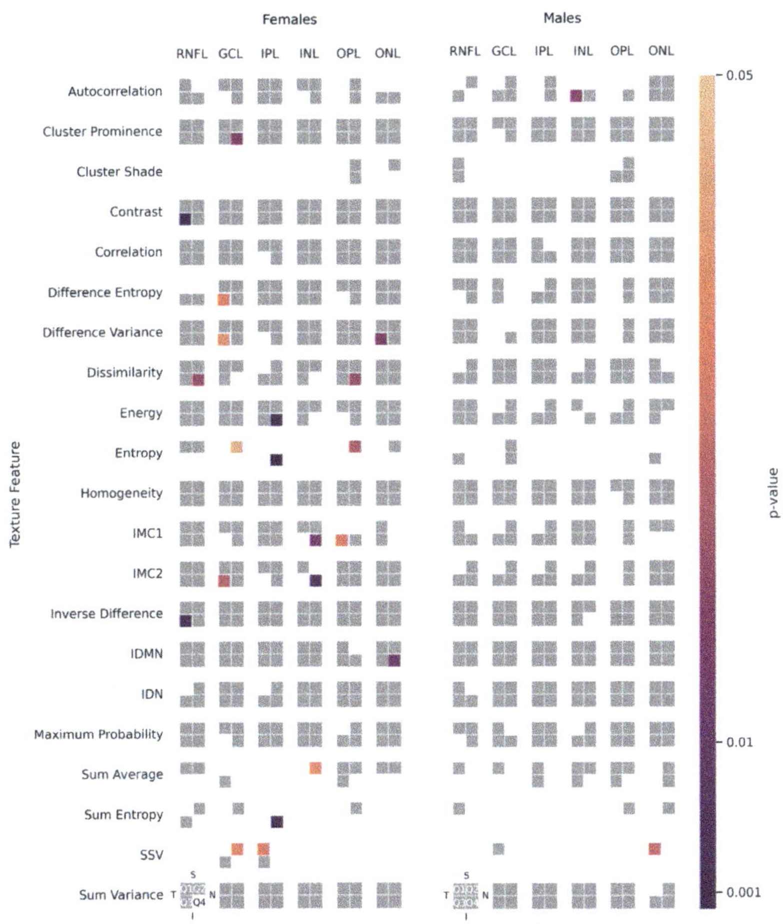

Figure 5. Texture features under study, for the female (**left**) and male (**right**) groups, for the retinal nerve fibre layer (RNFL), the ganglion cell layer (GCL), the inner plexiform layer (IPL), the inner nuclear

layer (INL), the outer plexiform layer (OPL), and the outer nuclear layer (ONL). Q1 stands for the superior-temporal quadrant, Q2 stands for the superior-nasal quadrant, Q3 stands for the inferior-temporal quadrant, and Q4 stands for the inferior-nasal quadrant. In the features IMC1 and IMC2, IMC stands for Information Measure of Correlation; IDMN stands for Inverse Difference Moment Normalised; IDN stands for Inverse Difference Normalised, and SSV stands for Sum of Squares–Variance. The grey-coloured cells indicate features that can be disregarded (for they are correlated to other features). The colour-filled cells indicate non-correlated texture features with a p-value < 0.05 (significant difference). In contrast, the white/empty cells represent the non-correlated features with a p-value ≥ 0.05 (non-significant difference).

Only two (1.4%) out of the 148 non-correlated texture features in the male group show statistical differences (right side of Figure 5): Autocorrelation (INL/Q3) and SSV (ONL/Q1).

3.4. Multiple Comparison Corrections

When correcting the pairwise tests using the Bonferroni and the Benjamini–Hochberg correction, two statistically significant results are found in the female group: the Energy (IPL/Q4) and Entropy (IPL/Q4). No significant differences were found for the male group.

Using Storey's [47] method, the FDR was estimated to be $\hat{FDR}_F = 13.8\%$ (female group) and $\hat{FDR}_M = 100\%$ (male group). By calculating the q-values [47] and considering a 5% cut-off, six statistically significant differences (see Figure 6 for sample distributions) can be found in the female group (out of the 21 non-correlated features, i.e., 28.6%): (1) Contrast (RNFL/Q3), (2) Inverse Difference (RNFL/Q3), (3) Energy (IPL/Q4), (4) Entropy (IPL/Q4), (5) Sum Entropy (IPL/Q4), and (6) IMC2 (INL/Q4). No significant differences were found for the male group.

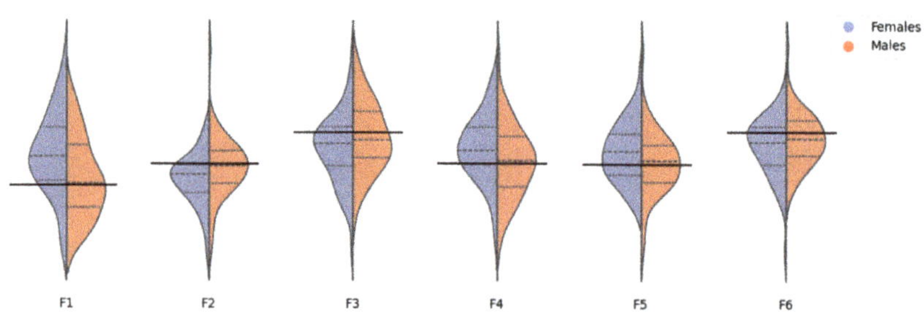

Figure 6. Kernel density estimation of the inter-eye differences for six features presenting differences between the left and right eye for women, but not for men. These features yield the most significant texture differences in the female group, even after Storey's multiple comparisons correction. Features are: F1—Entropy (IPL/Q4); F2—Energy (IPL/Q4); F3—Contrast (RNFL/Q3); F4—Sum Entropy (IPL/Q4); F5—Inverse Difference (RNFL/Q3); and F6—IMC2 (INL/Q4). Features are ordered by increasing p-value, as per the female group analysis. The thick black line marks the zero.

The number of statistically significant texture features in the female and male groups, before and after the three multiple comparisons correction methods were applied, are summarised in Table 2.

Table 2. Number of statistically significant texture features before and after correcting for multiple comparisons.

	Females	Males
Non-corrected results	21	2
Bonferroni	2	0
Benjamini-Hochberg	2	0
False Discovery Rate	6	0

Initial uncorrected results showed that interocular retinal texture asymmetries were mainly present in women and showed a larger effect size than in men. After correcting for multiple comparisons, all three correction methods revealed that interocular texture differences were, in fact, only present in the female group, while the eyes of men were similar.

4. Discussion

The present work focuses on sex differences between the right and left eye in healthy adults, as identified by neuroretina tissue texture analysis. The decision to specifically address sex differences was informed by our previous work [39], in which texture features revealed statistically significant differences between the retinas of the two sexes. Furthermore, texture features computed from the retina have been shown to be able to inform about the CNS status in different neurodegenerative disorders [24,25,33].

The rationale behind the analysis of these texture features is that each conveys quantitative information on the arrangement of the retina. Although a single texture feature, on its own, does not translate into a meaningful interpretation per participant/eye scan, we can quantify statistical differences between groups by combining several of these features.

The hypothesis tests performed illustrate a clear pattern of significant differences between the two eyes in the female group, with texture differences spreading across all six layers of the neuroretina. In contrast, only a few significant differences were found in the male group.

However, one must correct the pairwise tests due to the high number of comparisons. Considering the Benjamini–Hochberg correction, one finds the higher k, such that $\widetilde{p}_k < \frac{k\alpha}{m}$, where $\widetilde{p}_k$ are the ascending sorted p-values p_k. Both this correction and the highly conservative Bonferroni method resulted in no statistically significant differences in the male group and two in the female group.

We also considered the correction by FDR introduced by Storey [47], which is regarded as the most robust method of the three herein. In this approach, one starts by estimating the FDR in each group, resulting in the estimates $\widehat{FDR}_F = 13.8\%$ and $\widehat{FDR}_M = 100\%$. Although both estimates are above the desired 5%, it is clear that in the female group, the $\widehat{FDR}_F$ is closer to the desired 5% than the $\widehat{FDR}_M$, which is much larger. This indicates that some of the 21 significant differences ($p < 0.05$) found in the female group are indeed statistically significant.

In contrast, all the significant differences found in the male group are false positives. This conjecture is confirmed by calculating the q-values in both cases and considering a 5% cut-off, which results in six significant differences in the female group (from the 21 pairwise ones) and none in the male group. This way, the corrected results confirm that texture differences are found only in the female group.

To the authors' knowledge, the literature regarding the differences between the right and left eyes is mainly based on thickness assessments, and sex differences are seldom investigated in the healthy adult population—apart from [12], where a slight sex- and age-effect on retinal differences between the right and left eyes was reported.

The strong point of this study is the reporting of a previously undocumented difference between women and men, which offers a unique contribution to determining normative sex differences. Texture metrics appear to be sensitive to these differences while raising the question of why such differences occur, how they might vary over the lifespan, and any potential evolutive advantage. A limitation of the present study is the absence of a stratified analysis per age group. However, such an analysis could not be performed because a reduced sample size in the study would negatively impact the statistical robustness of the reported results.

This work highlights the potential of texture analysis to spot differences in the retina. Our findings suggest that special care should be taken when designing research protocols that assess retina differences, particularly when texture-based assessments are involved. On the one hand, the participant groups being compared should be sex-balanced—a good practice that was shown to be relevant in vision research [28]. On the other hand, beyond accounting for sex differences, studies should also control the interocular discrepancies that may be present in healthy individuals and patients diagnosed with ophthalmological or neurodegenerative disorders.

Supplementary Materials: The following supporting information can be downloaded at: https://www.mdpi.com/article/10.3390/jimaging10010006/s1, Table S1: Non-correlated texture features in the female and male groups, sorted by increasing *p*-value.

Author Contributions: Conceptualization, A.N., P.S., P.G., R.B. and M.C.-B.; data curation. A.N.; formal analysis, A.N. and P.S.; funding acquisition, A.N. and M.C.-B.; investigation, A.N. and J.F.; supervision, M.C.-B. and R.B.; visualization, A.N. and P.G.; writing—original draft, A.N.; writing—review and editing, A.N., P.S., P.G., M.C.-B. and R.B. All authors have read and agreed to the published version of the manuscript.

Funding: This work was supported by the Portuguese Foundation for Science and Technology (FCT) under UIDP&B/04950/2020 (M.C.-B.) and Ph.D. grant 2020.05578.BD (A.N.).

Institutional Review Board Statement: Data were gathered from the authors' institutional database. The multiple research projects from which data were collected were approved by the Ethics Committee of the Faculty of Medicine of the University of Coimbra. Ethics Committee of the Faculty of Medicine of the University of Coimbra Approval Code: CE-119/2018 Approval Date: 29 April 2018.

Informed Consent Statement: Informed consent was obtained from all subjects involved in the study. Written informed consent for publication has not been obtained but is not applicable, as all published data has been anonymized, and participants cannot be identified.

Data Availability Statement: The data presented in this study are available on request from the corresponding author. The data are not publicly available because they were originally collected in the scope of different research projects.

Acknowledgments: The authors would like to thank Hugo Quental for the OCT image acquisition.

Conflicts of Interest: The authors declare no conflict of interest. The funders had no role in the study's design, in the collection, analyses, or interpretation of data, in the writing of the manuscript, or in the decision to publish the results.

References

1. London, A.; Benhar, I.; Schwartz, M. The retina as a window to the brain—From eye research to CNS disorders. *Nat. Rev. Neurol.* **2013**, *9*, 44–53. [CrossRef]
2. Svetozarskiy, S.; Kopishinskaya, S. Retinal optical coherence tomography in neurodegenerative diseases (review). *Sovrem. Tehnol. Med.* **2015**, *7*, 116–123. [CrossRef]
3. Alves, C.; Jorge, L.; Canário, N.; Santiago, B.; Santana, I.; Castelhano, J.; Ambrósio, A.F.; Bernardes, R.; Castelo-Branco, M. Interplay between macular retinal changes and white matter integrity in early Alzheimer's disease. *J. Alzheimers Dis.* **2019**, *70*, 723–732. [CrossRef] [PubMed]
4. Czakó, C.; Kovács, T.; Ungvari, Z.; Csiszar, A.; Yabluchanskiy, A.; Conley, S.; Csipo, T.; Lipecz, A.; Horváth, H.; Sándor, G.L.; et al. Retinal biomarkers for Alzheimer's disease and vascular cognitive impairment and dementia (VCID): Implication for early diagnosis and prognosis. *GeroScience* **2020**, *42*, 1499–1525. [CrossRef] [PubMed]
5. Vanston, J.E.; Strother, L. Sex differences in the human visual system. *J. Neurosci. Res.* **2017**, *95*, 617–625. [CrossRef] [PubMed]
6. Hougaard, A.; Jensen, B.H.; Amin, F.M.; Rostrup, E.; Hoffmann, M.B.; Ashina, M. Cerebral asymmetry of fMRI-BOLD responses to visual stimulation. *PLoS ONE* **2015**, *10*, e0126477. [CrossRef] [PubMed]
7. Liu, S.; Zhao, B.; Shi, C.; Ma, X.; Sabel, B.A.; Chen, X.; Tao, L. Ocular dominance and functional asymmetry in visual attention networks. *Investig. Ophthalmol. Vis. Sci.* **2021**, *62*, 9. [CrossRef]
8. El-Ashry, M.; Hegde, V.; James, P.; Pagliarini, S. Analysis of macular thickness in British population using optical coherence tomography (OCT): An emphasis on interocular symmetry. *Curr. Eye Res.* **2008**, *33*, 693–699. [CrossRef]
9. Altan, C.; Arman, B.H.; Arici, M.; Urdem, U.; Solmaz, B.; Pasaoglu, I.; Basarir, B.; Onmez, F.; Taskapili, M. Normative posterior pole asymmetry analysis data in healthy Caucasian population. *Eur. J. Ophthalmol.* **2019**, *29*, 386–393. [CrossRef]
10. Çubuk, M.; Kasım, B.; Koçluk, Y.; Sukgen, E.A. Effects of age and gender on macular thickness in healthy subjects using spectral optical coherence tomography/scanning laser ophthalmoscopy. *Int. Ophthalmol.* **2018**, *38*, 127–131. [CrossRef]
11. Ly, A.; Banh, J.; Luu, P.; Huang, J.; Yapp, M.; Zangerl, B. Interocular asymmetry of the superonasal retinal nerve fibre layer thickness and blood vessel diameter in healthy subjects. *PLoS ONE* **2019**, *14*, e0226728. [CrossRef] [PubMed]

12. Jacobsen, A.G.; Bendtsen, M.D.; Vorum, H.; Bøgsted, M.; Hargitai, J. Normal value ranges for central retinal thickness asymmetry in healthy Caucasian adults measured by SPECTRALIS SD-OCT posterior pole asymmetry analysis. *Investig. Ophthalmol. Vis. Sci.* **2015**, *56*, 3875–3882. [CrossRef] [PubMed]
13. Al-Haddad, C.; Antonios, R.; Tamim, H.; Noureddin, B. Interocular symmetry in retinal and optic nerve parameters in children as measured by spectral domain optical coherence tomography. *Br. J. Ophthalmol.* **2014**, *98*, 502–506. [CrossRef] [PubMed]
14. Pawar, N.; Maheshwari, D.; Ravindran, M.; Ramakrishnan, R. Interocular symmetry of retinal nerve fiber layer and optic nerve head parameters measured by Cirrus high-definition optical coherence tomography in a normal pediatric population. *Indian J. Ophthalmol.* **2017**, *65*, 955–962. [CrossRef] [PubMed]
15. Song, M.Y.; Hwang, Y.H. Interocular symmetry of optical coherence tomography parameters in healthy children and adolescents. *Sci. Rep.* **2022**, *12*, 653. [CrossRef] [PubMed]
16. Adhi, M.; Aziz, S.; Muhammad, K.; Adhi, M.I. Macular thickness by age and gender in healthy eyes using spectral domain optical coherence tomography. *PLoS ONE* **2012**, *7*, e37638. [CrossRef] [PubMed]
17. Ooto, S.; Hangai, M.; Tomidokoro, A.; Saito, H.; Araie, M.; Otani, T.; Kishi, S.; Matsushita, K.; Maeda, N.; Shirakashi, M.; et al. Effects of age, sex, and axial length on the three-dimensional profile of normal macular layer structures. *Investig. Ophthalmol. Vis. Sci.* **2011**, *52*, 8769–8779. [CrossRef]
18. Ooto, S.; Hangai, M.; Yoshimura, N. Effects of sex and age on the normal retinal and choroidal structures on optical coherence tomography. *Curr. Eye Res.* **2015**, *40*, 213–225. [CrossRef]
19. Delori, F.C.; Goger, D.G.; Keilhauer, C.; Salvetti, P.; Staurenghi, G. Bimodal spatial distribution of macular pigment: Evidence of a gender relationship. *J. Opt. Soc. Am. A* **2006**, *23*, 521–538. [CrossRef]
20. Haan, J.; Verbraak, F.D.; Visser, P.J.; Bouwman, F.H. Retinal thickness in Alzheimer's disease: A systematic review and meta-analysis. *Alzheimers Dement. Diagn. Assess. Dis. Monit.* **2017**, *6*, 162–170. [CrossRef]
21. Hart, N.J.; Koronyo, Y.; Black, K.L.; Koronyo-Hamaoui, M. Ocular indicators of Alzheimer's: Exploring disease in the retina. *Acta Neuropathol.* **2016**, *132*, 67–787. [CrossRef] [PubMed]
22. Archibald, N.K.; Clarke, M.P.; Mosimann, U.P.; Burn, D.J. The retina in Parkinson's disease. *Brain* **2009**, *132*, 1128–1145. [CrossRef] [PubMed]
23. Tian, T.; Zhu, X.H.; Liu, Y.H. Potential role of retina as a progression of Parkinson's disease. *Int. J. Ophthalmol.* **2011**, *4*, 433–438. [CrossRef] [PubMed]
24. Varga, B.E.; Gao, W.; Laurik, K.L.; Tátrai, E.; Simó, M.; Somfai, G.M.; DeBuc, D.C. Investigating tissue optical properties and texture descriptors of the retina in patients with multiple sclerosis. *PLoS ONE* **2015**, *10*, e0143711. [CrossRef] [PubMed]
25. Tazarjani, H.D.; Amini, Z.; Kafieh, R.; Ashtari, F.; Sadeghi, E. Retinal OCT texture analysis for differentiating healthy controls from multiple sclerosis (MS) with/without optic neuritis. *BioMed Res. Int.* **2021**, *2021*, 5579018. [CrossRef] [PubMed]
26. Nuzzi, R.; Scalabrin, S.; Becco, A.; Panzica, G. Gonadal hormones and retinal disorders: A review. *Front. Endocrinol.* **2018**, *9*, 66. [CrossRef]
27. Nuzzi, R.; Caselgrandi, P. Sex hormones and their effects on ocular disorders and pathophysiology: Current aspects and our experience. *Int. J. Mol. Sci.* **2022**, *23*, 3269. [CrossRef]
28. Shaqiri, A.; Roinishvili, M.; Grzeczkowski, L.; Chkonia, E.; Pilz, K.; Mohr, C.; Brand, A.; Kunchulia, M.; Herzog, M.H. Sex-related differences in vision are heterogeneous. *Sci. Rep.* **2018**, *8*, 7521. [CrossRef]
29. Schuman, J.; Orgul, S.; Gugleta, K.; Dubler, B.; Flammer, J. Interocular difference in progression of glaucoma correlates with interocular differences in retrobulbar circulation. *Am. J. Ophthalmol.* **2000**, *129*, 728–733. [CrossRef]
30. Xu, G.; Hu, Y.; Zhu, S.; Guo, Y.; Xiong, L.; Fang, X.; Liu, J.; Zhang, Q.; Huang, N.; Zhou, J.; et al. A multicenter study of interocular symmetry of corneal biometrics in Chinese myopic patients. *Sci. Rep.* **2021**, *11*, 5536. [CrossRef]
31. Cava, J.A.; Allphin, M.T.; Mastey, R.R.; Gaffney, M.; Linderman, R.E.; Cooper, R.F.; Carroll, J. Assessing interocular symmetry of the foveal cone mosaic. *Investig. Ophthalmol. Vis. Sci.* **2020**, *61*, 23. [CrossRef] [PubMed]
32. Anantrasirichai, N.; Achim, A.; Morgan, J.E.; Erchova, I.; Nicholson, L. SVM-based texture classification in optical coherence tomography. In Proceedings of the 2013 IEEE 10th International Symposium on Biomedical Imaging, San Francisco, CA, USA, 7–11 April 2013; pp. 1332–1335. [CrossRef]
33. Nunes, A.; Silva, G.; Duque, C.; Januário, C.; Santana, I.; Ambrósio, A.F.; Castelo-Branco, M.; Bernardes, R. Retinal texture biomarkers may help to discriminate between Alzheimer's, Parkinson's, and healthy controls. *PLoS ONE* **2019**, *14*, e0218826. [CrossRef] [PubMed]
34. Julesz, B. Visual pattern discrimination. *IRE Trans. Inf. Theory* **1962**, *8*, 84–92. [CrossRef]
35. Bergen, J.R.; Landy, M.S. Computational modeling of visual texture segregation. In *Computational Models of Visual Processing*; Landy, M., Movshon, J.A., Eds.; MIT Press: Cambridge, MA, USA, 1991; pp. 253–271.
36. Haralick, R.M.; Shanmugam, K.; Dinstein, I.H. Textural features for image classification. *IEEE Trans. Syst. Man Cybern.* **1973**, *SMC-3*, 610–621. [CrossRef]
37. Tourassi, G.D. Journey toward computer-aided diagnosis: Role of image texture analysis. *Radiology* **1999**, *213*, 317–320. [CrossRef]
38. Castellano, G.L.; Bonilha, L.; Li, L.M.; Cendes, F. Texture analysis of medical images. *Clin. Radiol.* **2004**, *59*, 1061–1069. [CrossRef]
39. Nunes, A.; Serranho, P.; Quental, H.; Ambrósio, A.F.; Castelo-Branco, M.; Bernardes, R. Sexual dimorphism of the adult human retina assessed by optical coherence tomography. *Health Technol.* **2020**, *10*, 913–924. [CrossRef]

40. World Medical Association. World Medical Association Declaration of Helsinki: Ethical principles for medical research involving human subjects. *JAMA* **2013**, *310*, 2191–2194. [CrossRef]
41. Li, K.; Wu, X.; Chen, D.; Sonka, M. Optimal surface segmentation in volumetric images—A graph-theoretic approach. *IEEE Trans. Pattern Anal. Mach. Intell.* **2005**, *28*, 119–134. [CrossRef]
42. Garvin, M.K.; Abramoff, M.D.; Wu, X.; Russell, S.R.; Burns, T.L.; Sonka, M. Automated 3-D intraretinal layer segmentation of macular spectral-domain optical coherence tomography images. *IEEE Trans. Med. Imaging* **2009**, *28*, 1436–1447. [CrossRef]
43. Abràmoff, M.D.; Garvin, M.K.; Sonka, M. Retinal imaging and image analysis. *IEEE Rev. Biomed. Eng.* **2010**, *3*, 169–208. [CrossRef] [PubMed]
44. Guimarães, P.; Rodrigues, P.; Lobo, C.; Leal, S.; Figueira, J.; Serranho, P.; Bernardes, R. Ocular fundus reference images from optical coherence tomography. *Comput. Med. Imaging Graph.* **2014**, *38*, 381–389. [CrossRef] [PubMed]
45. Soh, L.-K.; Tsatsoulis, C. Texture analysis of SAR sea ice imagery using gray level co-occurrence matrices. *IEEE Trans. Geosci. Remote Sens.* **1999**, *37*, 780–795. [CrossRef]
46. Clausi, D.A. An analysis of co-occurrence texture statistics as a function of grey level quantization. *Can. J. Remote Sens.* **2002**, *28*, 45–62. [CrossRef]
47. Storey, J.D. A direct approach to false discovery rates. *J. R. Stat. Soc. Ser. B* **2002**, *64*, 479–498. [CrossRef]

Journal of
Imaging

MDPI

Article

Assessment: A Novel Method for Evaluating Retinal Vasculature Morphology and Its Diagnostic Potential in Hypertensive Retinopathy and Other Eye-Related Diseases. Arteriovenous Length Ratio: A Novel Method for Evaluating Retinal Vasculature Morphology and Its Diagnostic Potential in Eye-Related Diseases

Sufian A. Badawi [1,*], Maen Takruri [2], Mohammad Al-Hattab [3], Ghaleb Aldoboni [2], Djamel Guessoum [2,4], Isam ElBadawi [5], Mohamed Aichouni [5], Imran Ali Chaudhry [5], Nasrullah Mahar [6] and Ajay Kamath Nileshwar [7,8]

1 Department of Computer Science, Faculty of Information Technology, Applied Science Private University, P.O. Box 541350, Amman 11937, Jordan
2 Center of Information, Communication and Networking Education and Innovation (ICONET), American University of Ras Al Khaimah, Ras Al Khaimah 72603, United Arab Emirates; maen.takruri@aurak.ac.ae (M.T.); ghaleb.aldoboni@aurak.ac.ae (G.A.); djamel.guessoum.1@ens.etsmtl.ca (D.G.)
3 College of Engineering, Al Ain University, Al Ain 64141, United Arab Emirates; mohammad.alhattab@aau.ac.ae
4 Electrical Engineering Department, Ecole de Technologie Superieure, Montreal, QC H3C 1K3, Canada
5 Industrial Engineering Department, College of Engineering, University of Ha'il, Ha'il 81481, Saudi Arabia; i.elbadawi@uoh.edu.sa (I.E.); m.aichouni@uoh.edu.sa (M.A.); i.chaudhry@uoh.edu.sa (I.A.C.)
6 Computer Science Department, Bahauddin Zakariya University, Multan 60800, Pakistan; nasr@wizinsights.com
7 Department of Ophthalmology, RAK Medical and Health Sciences University, Ras Al Khaimah 11172, United Arab Emirates; ajaykamath@rakmhsu.ac.ae
8 Saqr Hospital, Ministry of Health and Prevention, P.O. Box 5450, Ras Al Khaimah 72603, United Arab Emirates
* Correspondence: s_badawi@asu.edu.jo

Abstract: Retinal imaging is a non-invasive technique used to scan the back of the eye, enabling the extraction of potential biomarkers like the artery and vein ratio (AVR). This ratio is known for its association with various diseases, such as hypertensive retinopathy (HR) or diabetic retinopathy, and is crucial in assessing retinal health. HR refers to the morphological changes in retinal vessels caused by persistent high blood pressure. Timely identification of these alterations is crucial for preventing blindness and reducing the risk of stroke-related fatalities. The main objective of this paper is to propose a new method for assessing one of the morphological changes in the fundus through morphometric analysis of retinal images. The proposed method in this paper introduces a novel approach called the arteriovenous length ratio (AVLR), which has not been utilized in previous studies. Unlike commonly used measures such as the arteriovenous width ratio or tortuosity, AVLR focuses on assessing the relative length of arteries and veins in the retinal vasculature. The initial step involves segmenting the retinal blood vessels and distinguishing between arteries and veins; AVLR is calculated based on artery and vein caliber measurements for both eyes. Nine equations are used, and the length of both arteries and veins is measured in the region of interest (ROI) covering the optic disc for each eye. Using the AV-Classification dataset, the efficiency of the iterative AVLR assessment is evalutaed. The results show that the proposed approach performs better than the existing methods. By introducing AVLR as a diagnostic feature, this paper contributes to advancing retinal imaging analysis. It provides a valuable tool for the timely diagnosis of HR and other eye-related conditions and represents a novel diagnostic-feature-based method that can be integrated to serve as a clinical decision support system.

Keywords: inflection count metric; tortuosity; morphological changes; fundus images; arteriovenous length ratio

1. Introduction

Various morphological abnormalities in the retinal vasculature, such as increased artery tortuosity, altered arteriovenous width and ratio (AVR), arteriovenous (AV) nicking, and the onset of macular edema in severe cases, characterize hypertensive retinopathy (HR) [1] and diabetic retinopathy [2]. These abnormalities can lead to reduced visual acuity and an increased risk of stroke later in life, suggesting inflammation in the retinal vasculature. To avoid issues like stroke, visual loss, and cardiovascular disease, it is essential to diagnose and treat HR promptly [1,3,4].

Current ophthalmology practices involve using fundus cameras and computerized images to analyze morphometric changes in retinal blood vessels for identifying and grading HR. However, early HR symptoms can be subtle and challenging to detect, leading to delays in diagnosis and complicating treatment [1,3,4]. Therefore, encouraging hypertensive patients to undergo regular follow-ups with ophthalmologists is crucial to halt the progression of retinopathy and prevent potential stroke or other severe outcomes. Developing an automated system capable of evaluating morphometric alterations in blood vessels for HR diagnosis is essential to aid in efficient treatment.

This paper introduces a novel approach called the arteriovenous length ratio (AVLR), which has not been utilized in previous studies, to address the morphological abnormalities caused by hypertensive retinopathy (HR) in the retinal vasculature or which could also be used to observe the morphological changes caused by diabetic retinopathy. Unlike commonly used measures such as the arteriovenous width ratio or tortuosity, AVLR focuses on assessing the relative length of arteries and veins. By harnessing deep learning techniques, this innovative method accurately quantifies the proposed AVLR, providing a unique perspective on the morphological changes associated with HR. Additionally, the proposed approach incorporates the analysis of vascular tortuosity and utilizes a combination of nine length feature measurements. The algorithm is trained on the AV-Classification dataset [5], consisting of 504 retinal images meticulously annotated at the pixel level for vascular segmentation, artery/vein classification, and optic disc localization. The dataset also includes image-level labels provided by ophthalmologists for vascular tortuosity indices and HR grading. This automated approach to measuring this feature holds great promise in efficiently diagnosing and grading HR, enabling timely treatment and improving patient outcomes.

The main contributions of this research work are listed below:

1. We propose a unique automated method that combines vascular tortuosity assessment with AVLR analysis, which could be used to detect and grade HR. Since this is a novel approach and is not present in existing medical literature, our study represents a substantial development in the area.
2. We provide a novel iterative approach to compute the AVLR inside the ROI (region of interest) around the optic disc using vessel caliber data. This technique improves the AVLR assessment's precision and dependability.
3. In the assessment of vessel AVLR, we employ a unique combination of nine features, which helps with accurately quantifying the severity of AVLR and effectively minimizing the subjectivity and variability associated with manual assessments conducted by ophthalmologists.
4. The AV-Classification dataset has been expanded with new labels for AVLR metrics, creating the AV-Classification dataset extension. This extension includes vessel segments and image-level AVLR metrics, quantified using the proposed methodology.

The rest of this paper is organized as follows: Section 2 elaborates on the fundus image processing steps related to this research. Then, Section 3 presents the used material, performance measures, and the proposed method. The results, the newly prepared dataset, and the extension are explained in Section 4.7, followed by discussions in Section 4.5.4 and conclusions in Section 5.

2. Related Work

The evaluation of retinal vascular morphology has been studied as a means to quantify biomarkers for ocular diseases. Two key quantitative measures that have been investigated are the arteriovenous width ratio (AVR) and the arteriovenous tortuosity index. Accurately measuring these metrics requires robust vessel segmentation and classification [6].

For AVR analysis, several semi-automated approaches relying on manual input have been explored. Akbar et al. calculated AVR from manually labeled vessels [7], demonstrating its potential clinical utility. Similarly, proprietary software for semi-automated arteriolar and venular caliber measurements was used by Akbar et al. to compute AVR. However, the manual component makes these methods time-consuming, subjective, and difficult to reproduce.

To enable high-throughput screening, fully automated vessel segmentation and classification techniques are needed. Narasimhan et al. proposed a method using median filtering, top hat transform, and ROI detection to calculate AVR. However, performance was only demonstrated in a small dataset of 75 images without disease grading, thereby limiting clinical validation. The authors of [8] developed an automated technique using multi-scale linear structure enhancement and thresholding for vessel segmentation. However, AVR was only calculated within a limited region of 0.5–1 optic disc diameter around the optic disc, which may bias results and not capture overall vascular morphology. Researchers in [9] classified vessels and calculated AVR using Gabor wavelets and multilayered thresholding but did not describe the AVR measurement methodology sufficiently enough for it to be evaluated and reproduced.

More recently, advances in machine learning have enabled more reliable automated vessel analysis. Badawi et al. optimized vessel segmentation and artery/vein classification by developing B-COSFIRE filters and training convolutional neural networks (CNNs). This enabled accurate pixel-wise vessel labeling to derive vascular morphology metrics. Similarly, their proposed automated pipeline using unsupervised learning for vessel classification and wavelet transforms for diameter measurement to quantify AVR demonstrated accurate AVR measurements compared to manual grading. Further research on interpretable and robust machine learning techniques is critical in order to employ automated retinal vascular analysis for clinical use.

Additionally, Badawi et al. developed a method to automatically grade the tortuosity severity of retinal images into four levels: normal, mild, moderate, and severe. They calculated 14 tortuosity metrics for each vessel segment and used these metrics to train machine learning models to classify images. The best model, distributed random forest (DRF), achieved 99.4% accuracy on the AV-Classification dataset. This demonstrates the feasibility of automatic tortuosity severity grading.

Beyond AVR, quantifying vessel tortuosity could provide complementary information about vascular health. However, fewer studies have investigated automated measurement of the arteriovenous tortuosity index. The authors of [10] segmented vessels using k-means clustering and measured tortuosity based on the arc-chord ratio and stationary points along vessel centerlines. While promising, the methodology was only validated in two public datasets, necessitating wider evaluation. Kanski et al. proposed an index combining local and global tortuosity metrics for optical coherence tomography (OCT) images. However, only eleven OCT volumes were analyzed, limiting clinical significance.

A major limitation of the existing techniques is that most of them focus exclusively on either AVR or tortuosity, whereas simultaneously quantifying both could improve disease detection. Towards this goal, the authors of [11] presented a multi-task CNN to measure AVR and the tortuosity index end-to-end from fundus images jointly. However, its performance did not exceed single-task models, indicating that further architectural refinements are needed to learn complementary feature representations effectively.

Incorporating domain knowledge to derive interpretable biomarkers is an active area of research. Poplin et al. developed a deep learning system to predict cardiovascular risk factors by extracting interpretable vascular features, mimicking how clinicians qualitatively assess retinal images. However, the model was trained on only two datasets, requiring broad validation. Developing and rigorously evaluating hybrid machine learning systems that integrate vascular domain knowledge could advance clinical translation and adoption [12].

Emerging imaging modalities also provide new opportunities for vascular analysis. OCT angiography enables depth-resolved visualization of the retinal capillary network. However, few automated algorithms have been developed to quantify morphological changes in these volumetric scans. A key challenge is accurately segmenting the capillaries, which have low contrast and discontinuity compared to major vessels in enface projections [13].

In summary, the robust quantification of retinal vascular morphology shows strong promise as a non-invasive biomarker for systemic and ocular diseases. While progress has been made, especially in AVR measurement, substantial opportunities exist to improve interpretability, incorporate multimodal data, and move these technologies toward clinical integration. Key priorities for further research include: (1) developing unified algorithms that simultaneously quantify multiple morphology measures; (2) advancing machine learning techniques that integrate anatomical domain knowledge to derive explanatory biomarkers; (3) expanding rigorous validation across diverse patient populations and eye diseases; and (4) innovating volumetric analysis methods for emerging modalities like OCT angiography. Advances in these areas could enable more accurate disease risk stratification and unlock the full potential of retinal vascular imaging as a precision medicine tool.

3. Methodologies and Materials

3.1. Used Materials

The evaluation of AVLR has undergone rigorous examination using the AV-Classification dataset [5] (See Figure 1). This comprehensive dataset comprises 504 retinal images meticulously annotated at the pixel level for vessel segmentation, artery/vein classification, and optic disc localization. The meticulous annotations were carried out by highly skilled ophthalmologists. Of the 504 images, 22 were from healthy subjects while 482 were from subjects with various forms of retinopathy, with a majority of the pathological images showing signs of hypertensive retinopathy. Other retinal vascular pathologies, like diabetic retinopathy, were also included in this dataset. All 504 images were acquired, using the same model of non-mydriatic fundus camera (Topcon), from 50 patients in the middle-aged category, ensuring consistency in image capture settings like field-of-view. There are 504 labels in total for each label type (vessel segmentation label, AV-classification label). The original retinal image from the AV-Classification dataset is fed to a deep learning optimized method to enable vessel segmentation and classification using the colored vessel segmentation labels. The type-1 and type-2 pictures and labels are 2002 by 2000 pixels in size. Each original retinal picture used in vascular segmentation and AV classification investigations has two labels: colored and monochromatic.

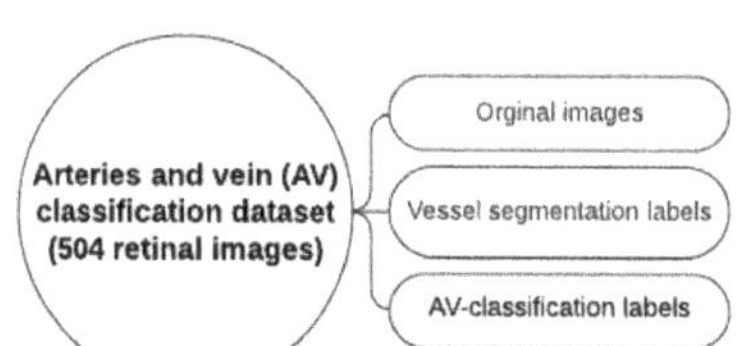

Figure 1. The dataset for AV-Classification utilized in this paper.

3.2. Methodology

The methodology for calculating the AVLR involves the application of nine metrics. Figure 2 depicts a summary of the procedure, where the workflow steps are displayed. The initial trial took place in the ROI-1 zone, specifically focusing on an area ranging from twice the size of the optic disc radius to thrice its size. The subsequent trial occurred in the R2 region, targeting a region spanning from twice the radius of the optic disc to five times its radius. The segmented and skeletonized retinal images were initially created, as depicted in Figure 3, to prepare them for further analysis. Once the segmentation was completed, the images were divided into segments. Next, every segment was subjected to the nine length metrics. These measures were created particularly to measure the length properties of the retinal arteries and veins. The AVLR may be estimated by measuring the lengths of these vessels, offering important insights into the morphological alterations connected to hypertensive retinopathy (HR) and other eye-related disorders. A thorough review procedure was carried out to verify the precision of the AVLR readings. This entailed comparing the estimated AVLR values with expert assessments and ground truth annotations. The evaluation was carried out on the AV-Classification dataset, which comprises retinal pictures that ophthalmologists have carefully tagged and categorized.

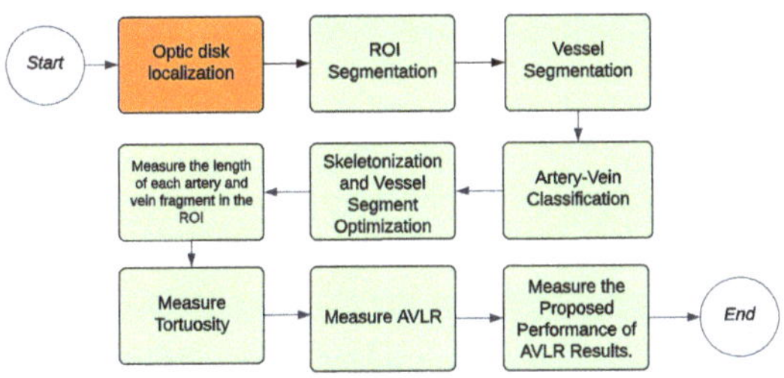

Figure 2. The process of measuring the proposed AVLR metric (green denotes fully automated, and orange denotes the semi-automated steps).

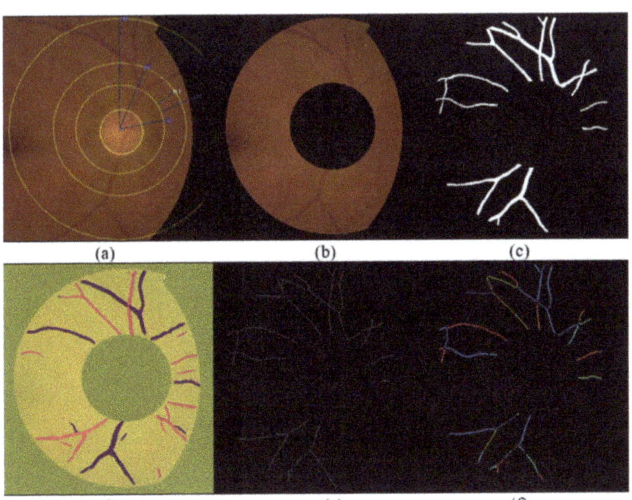

Figure 3. Illustration of the proposed AVLR calculation steps: (**a**) a representation of the fundus image where the optic disc is localized, (**b**) ROI segmentation, (**c**) extraction of blood vessels, (**d**) artery–vein classification (**e**) conversion to a skeletonized form, (**f**) segmentation of vessel segments into fragments where the computation for the nine metrics happens for each vessel segment; all those figures are the input to compute the AVLR for each retinal image.

Firstly, branch points were detected using a morphological operation to identify the locations where the vessel tree branches [14]. This step is crucial for measuring the AVLR. It enables the analysis and quantification of AVLR by segmenting the detected branch points into different vessel segments. After that, another morphological process that removes the inner pixels while keeping the pixels that indicate the vessel structure was used to acquire the vessel tree's vasculature edges [14]. Following this, vessel skeletonization was performed using an optimized vessel fragments extraction technique, outlined in [15]. This method improves the skeletonization results by progressively eliminating spur dots and smoothing the skeleton. The vascular tree generates eleven vessel branch segments after an innovative approach is used to remove bogus "L"-shaped (junction) segment sections. For the best accuracy, we used the optimal technique for extracting vessel fragments as outlined in [15]. This approach optimizes vessel segments for segment and image-wise morphometric analysis, allowing us to proceed with AVLR evaluation. The proposed methodology focuses on measuring the AVLR. It can be utilized in conjunction with other methods for comprehensive analysis, like tortuosity or arteriovenous width ratio, enabling a thorough assessment of retinal vasculature morphology. In addition, the AVLR is compared for both eyes to identify any potential differences. By examining the AVLR values from the left and right eyes, the methodology aims to detect inter-eye variations that may provide valuable insights into the asymmetry or bilateral involvement of retinal vascular changes. This comparative analysis of AVLR between the eyes adds an important dimension to the assessment, enabling a comprehensive evaluation of the retinal vasculature in relation to hypertensive retinopathy (HR) and other eye-related conditions. The table presented below provides a concise overview of the methodology employed in the paper, outlining the key steps involved in the process:

1. Optic Desk Localization: The fundus image undergoes a process to detect the location of the optic disk.
2. ROI Segmentation: The fundus images are subjected to ROI segmentation to detect and extract the ROI-2-3 ROD and 2-5 ROD from the rest of the image.
3. Vessel segmentation: The retinal images are subjected to a vessel segmentation process to separate the blood vessels from the background.
4. Artery–vein classification: The segmented vessels are analyzed to differentiate between arteries and veins in the vascular structure of the ROI.
5. Vessel skeletonization: The segmented vessels are further processed using skeletonization techniques to obtain a thin representation of the vascular structure.
6. Identification of the bifurcation/intersection points: Key points along the skeletonized vessels are identified to extract the vessel segments between them.
7. Generate the optimized vessel segments: Optimized vessel segments are generated based on the verified points, focusing on specific regions of interest.
8. Calculate each of the nine metrics long feature vectors: The nine length/tortuosity metrics are calculated for each optimized vessel segment, providing quantitative measurements for AVLR assessment.
9. Finalize the new AVLR feature set: The calculated AVLR values from each vessel segment are compiled to form a new feature set specifically for AVLR analysis.

3.3. Artery–Vein Classification

A deep learning strategy is employed using the author's method in [16] to categorize retinal blood vessels as either arteries or veins. Specifically, an encoder-decoder fully convolutional neural network (FCN) architecture is utilized for this purpose.

The encoder part of the FCN is responsible for acquiring high-level distinctive features from the input retinal images. It comprises several convolutional layers, each followed by a ReLU (Rectified Linear Activation) function to introduce nonlinearity. Additionally, max-pooling layers are inserted to gradually reduce spatial dimensions and encode the most salient features.

The decoder component of the FCN operates on the encoded feature representation and generates dense pixel-wise predictions to produce a segmentation mask output. It incorporates upsampling layers to restore the original input dimensions, ensuring the output segmentation masks match the input image size. Convolutional layers and ReLU activations are also employed in the decoder.

Unlike typical convolutional neural networks, this FCN includes only convolutional layers in both the encoder and decoder, omitting any fully connected layers. This design allows it to accept input images of varying dimensions and produce corresponding output segmentation masks.

The network is trained end-to-end, utilizing the raw retinal images as input without any preprocessing steps like vessel segmentation. A multi-loss function is developed to optimize the learning of arterial and venous labels at the pixel level. This function incorporates both pixel-wise and segment-wise loss components.

The pixel-wise loss assesses predictions and ground truth labels individually for each pixel. On the other hand, the segment-wise loss introduces contextual information by considering the predominant label within each vessel segment. It adapts pixel-level predictions based on segment-wise vascular labeling. This combined approach enhances performance compared to relying solely on pixel-wise loss.

By directly learning highly distinctive features from the retinal images, the FCN can accurately classify individual pixels as artery, vein, or background, eliminating the need for hand-crafted features or pre-segmented vessel maps. The inclusion of a multi-loss function with both pixel and segment-level components further elevates the performance of artery/vein classification.

3.4. Vasculature Segmentation and Skeletonization

The segmentation of blood vessels in retinal images is a complex task due to various challenges, such as poor contrast, uneven lighting, and background artifacts. Despite these difficulties, several supervised and unsupervised methods have been developed for retinal vessel segmentation. In this study, we optimize the parameters of a filter to effectively segment retinal vessels based on the work in [5]. Our proposed technique uses multi-objective optimization to enhance the results obtained from the B-COSFIRE algorithm used in [17] to address issues like the central light reflex. The method successfully segments the vasculature, including the vessels, with the center light reflex. Artery–vein segmentation results are also provided using a method previously proposed in [15]. Furthermore, a thinning process removes the vascular skeleton, eliminating noise and refining the skeleton's shape. The process involves iterative detection and elimination of endpoints, followed by removing extra pixels that create L-shaped angles and fragment the vessels.

3.5. Optimized Retinal Images

To generate the optimized retinal image, we utilized a blood vessel segmentation method that builds on the trainable B-COSFIRE filter. To find additional optimum parameters, the selection of the thresholding parameter was thoroughly analyzed, and background artifact removal methods were used, according to the process description in [17]. The findings from the suggested strategy outperformed those from other cutting-edge techniques used in vessel segmentation. We employed the ANOVA analysis with a p-value threshold of less than 0.05 to determine the significance of the parameters influencing the performance. This analysis allowed us to identify the most influential parameters in the process.

3.6. AVLR Metrics

The process begins by traversing the extracted skeleton segments of each image. The feature extraction procedure begins by calculating the straight-line and geodesic distances inside each vascular segment. Subsequently, all nine measures of AVLR that are given in Equations (1)–(9) are calculated. This generates a record for the segment within a dedicated feature set designed to capture the length attributes of each vessel segment. In addition, the tortuosity calculation is incorporated into our metrics to capture the degree of curvature or winding in blood vessels. The structural characteristics of arteries and veins are quantitatively assessed by including tortuosity measurements, providing insights into vessel elongation, branching, kinking, and other abnormalities associated with retinal vascular diseases and systemic conditions. This integration enhances the accuracy and comprehensiveness of our assessment of vessel length attributes, enabling the derivation of the AVLR for evaluating the asymmetry and differences between arteries and veins. To provide a comprehensive overview, statistical summaries are then computed for the segments within each image, forming a row in the feature set file that captures image-level attributes. Sample metrics showcasing distance-based measures for vessel length evaluation are depicted in Figure 4, leading to the final calculation of AVLR. Below is a comprehensive description and definition of all nine length metrics along with their corresponding attributes, which are essential for the calculation of the AVLR.

1. **Arteriovenous Chord Length Ratio:** A ratio of the average Euclidean length of all veins to all arteries in the retinal picture is represented by this measure. It measures the proportional variations in length between veins and arteries.

$$Arterivenous\_SLD\_Length\_Ratio = \frac{average(Artery\_SDLE\_Length)}{average(Vein\_Chord\_Length)}. \quad (1)$$

2. **Arteriovenous Arc Length Ratio:** The ratio of all the arteries' mean geodesic distance to all the veins' mean geodesic distance in the retinal picture is represented by this measure. It sheds light on the differences and curving routes of veins and arteries.

$$Arterivenous\_Arc\_Length\_Ratio = \frac{average(Artery\_Arc\_Length)}{average(Vein\_Arc\_Length)}. \quad (2)$$

3. **Arteriovenous Distance Metric Ratio:** The ratio of the average tortuosity length metric of all veins to all arteries in the retinal picture is represented by this measure. The overall tortuosity and variations in vein and artery curvature are quantified.

$$Arterivenous\_DM\_Length\_Ratio = \frac{average(Artery\_DM\_Length)}{.} average(Vein\_DM\_Length) \quad (3)$$

4. **Arteriovenous Inflection Count Metric Ratio:** This measure shows the proportion of the average tortuosity. The total inflections measurement of each artery to the average tortuosity total inflections measurement of every vein in the retinal picture. The quantity of inflection points is counted, which represents the degree of vascular tortuosity.

$$Arterivenous\_ICM\_Length\_Ratio = \frac{average(Artery\_ICM\_Length)}{average(Vein\_ICM\_Length)}. \quad (4)$$

5. **Arteriovenous Inflection Count Metric Binomial Ratio:** The mean tortuosity inflection count measure binomial of all arteries divided by the average tortuosity inflection_Count_Metric_Binomial of all veins in the retinal picture is represented by this measure. It evaluates the veins' and arteries' binomial distribution of inflection sites.

$$Arterivenous\_ICMb\_Length\_Ratio = \frac{average(Artery\_ICMb\_Length)}{average(Vein\_ICMb\_Length)}. \quad (5)$$

6. **Arteriovenous Sum of Angles Metric Ratio:** This metric is the average tortuosity inflection count divided by its ratio. Comparing the average inflection count metric of

all arteries to the binomial of all veins in the retinal image; it assesses the binomial distribution of artery and vein inflection points.

$$Arterivenous\_SOAM\_Length\_Ratio = \frac{average(Artery\_SOAM\_Length)}{average(Vein\_SOAM\_Length)}. \quad (6)$$

7. **Arteriovenous Norm of Curvature Ratio:** This measure represents the proportion of the average curvature of all vein segments to all artery segments in the retinal picture. It gauges the vessel segments' curvature.

$$Arterivenous\_NC\_Length\_Ratio = \frac{average(Artery\_NC\_Length)}{average(Vein\_NC\_Length)}. \quad (7)$$

8. **Arteriovenous of Average Curvature Ratio Standard Deviation:** This statistic calculates the difference between the average curvature of all arteries' mean standard deviation and the average curvature of all veins' mean standard deviation. the retinal picture. Quantification is used to the curvature variation along vessel segments.

$$AV\_SDAC\_Length\_Ratio = \frac{average(Artery\_SDAC\_Length)}{average(Vein\_SDAC\_Length)}. \quad (8)$$

9. **Arteriovenous of Centerline Length Ratio:** This metric represents The mean centerline length of all the arteries divided by the mean centerline length of all the veins in the retinal image. It determines the distances between the arteries' and veins' centerlines.

$$AV\_CL\_Length\_Ratio = \frac{average(Artery\_CL\_Length)}{average(Vein\_CL\_Length)}. \quad (9)$$

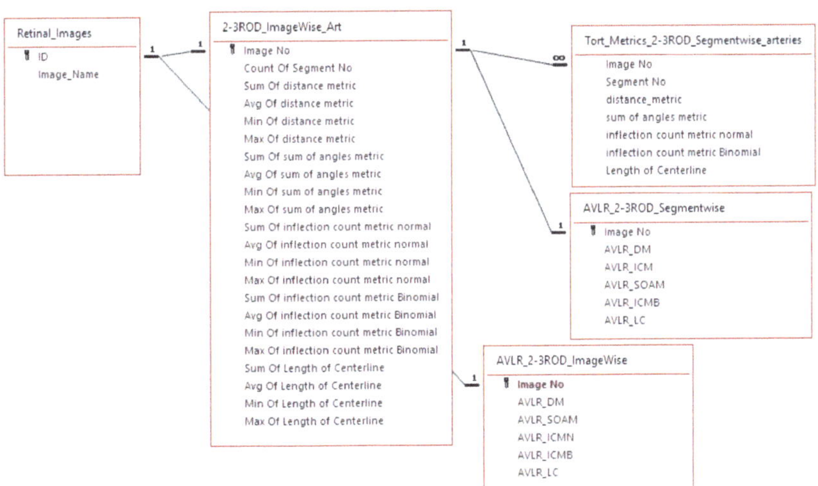

Figure 4. ERD diagram of the image-level and segment-level feature sets.

3.7. The New Feature-Set Preparation

The AVLR calculation for the entire AV-Classification dataset (504 images) is performed using the proposed method. The initial step described in Figure 2 involves dividing a binary image that contains the retinal vessels into segments using an optimized method [17]. After that, iterative thinning is applied to extract the vessel skeleton. The skeleton is then broken into fragments and optimized into vessel fragments, which connect intersections, bifurcations, or endpoints in the skeleton, as explained in [5]. Each of these vessel fragments is treated as a curve, and nine mathematical formulas (Formulas (1)–(9)) are utilized to measure the length ratios and quantify the attributes of the AVLR for each fragment. This process generates a feature set at the fragment level, consisting of the AVLR metrics. Additionally, summary statistics are computed for each image to create the image-level feature set as depicted in Figure 4. The process involves evaluating the nine-length ratio metrics and incorporating the newly labeled AVLR feature set as an extension to the AV-Classification dataset, which encompasses both image-wise and vessel segment-wise tortuosity features. Two types of AVLR features are introduced: segment-level AVLR features and image-level statistics AVLR features. The image-level statistics include the count of segments in the image and, for each length metric, statistical summaries such as average, minimum, and maximum values are calculated. Figure 4 illustrates the entity-relationship diagram (ERD), representing the feature set.

3.8. Right and Left Eyes Comparison

The comparison of AVLR values between the left and right eyes is the final step in our analysis, providing a comprehensive assessment of any asymmetry or significant differences. This comparison allows for a detailed evaluation of the retinal characteristics of each eye and serves as a valuable tool in identifying potential abnormalities or discrepancies.

To conduct this comparison, AVLR values are obtained separately for the left and right eyes. We apply the AVLR calculation methodology outlined in the previous steps to the retinal images of each eye, recording and compiling the AVLR values for further analysis.

Once we have obtained the AVLR values for both eyes, we conduct a thorough comparison using statistical methods and techniques to assess the level of asymmetry or significant differences between the AVLR values of the left and right eyes.

The results of this comparison provide valuable insights into retinal health and potential variations between the eyes. Clinicians and researchers carefully examine any notable discrepancies or asymmetry in AVLR values and interpret these findings. These results contribute to a deeper understanding of retinal conditions and may have implications for diagnosis, treatment, and overall eye health management.

Incorporating this step into the methodology enables a comprehensive evaluation of retinal health and asymmetry between the left and right eyes, enhancing the diagnostic process and assisting in developing personalized treatment plans tailored to the specific needs of each individual.

4. Results

4.1. Optic Disc Localization

Figure 5 illustrates the first step in generating a ring cut image, which identifies the optic disk's center and radius. This involves converting the image to grayscale and monochrome to isolate the circular shape. The calculated geometric center and optic disc radius serve as the basis for further image analysis and metrics.

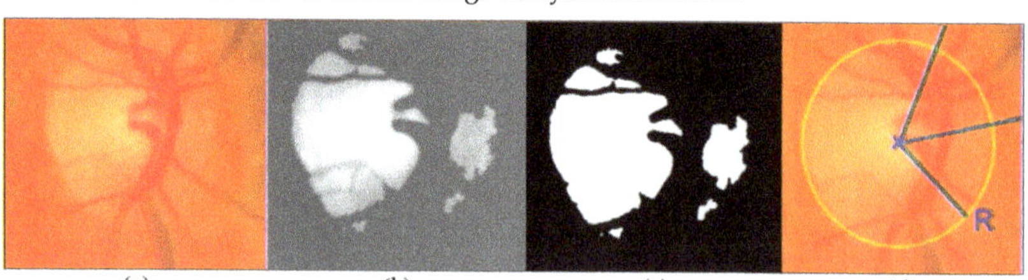

(a) (b) (c) (d)

Figure 5. Optic disc localization illustration: (**a**) OD in the fundus image (**b**) converted to grayscale (**c**) and monochrome (**d**) the image annotated with the OD center and radius.

4.2. ROI Segmentation

In the second stage, we create ring-cut images by segmenting the regions of interest (ROIs) with radii ranging from 2ROD to 3ROD and an extended range up to 5ROD (See

Figure 6). Each segment corresponds to a new image, allowing targeted analysis of specific circular regions around the optic disk.

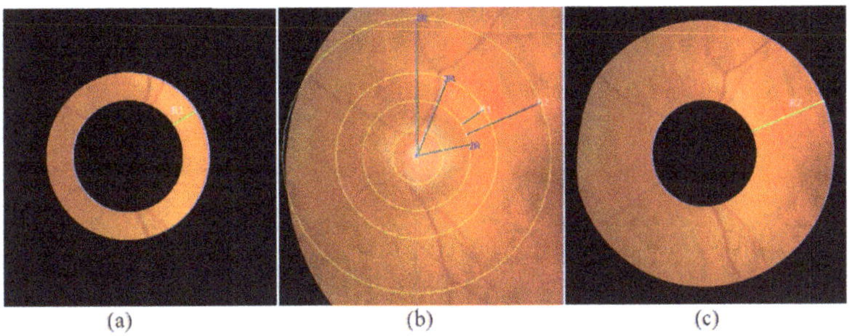

Figure 6. ROI segmentation: (**a**) Ring cut from 2ROD to 3ROD; (**b**) annotated original image (**c**); ring cut from 2ROD to 5ROD.

4.3. Vessel Segmentation

The third step utilizes the previous work of [17] to optimize blood vessel segmentation. By extending the trainable B-COSFIRE filter, more optimal parameters are identified, resulting in improved vessel segmentation accuracy; see Figure 7.

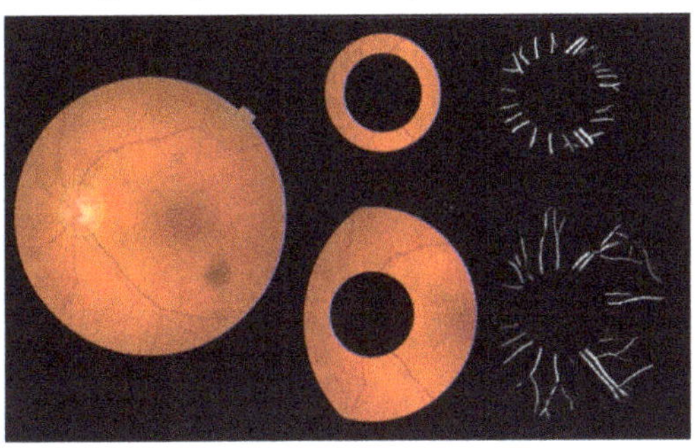

Figure 7. Fundus image and vessel segmentation for the two ROI areas with improved accuracy using the trainable B-COSFIRE filter and optimized parameters.

4.4. Artery Vein Classification

In the fourth step, deep learning is applied using the method in Section 3.3 to segment arteries and veins in each ring cut (Figures 8 and 9). The implementation of the multi-loss technique elevates pixel accuracy from 93.5% to 97% on the dataset [16]. Following training on this extensively annotated dataset, the method attains exceptional accuracy, underscoring its efficacy in robust artery/vein classification.

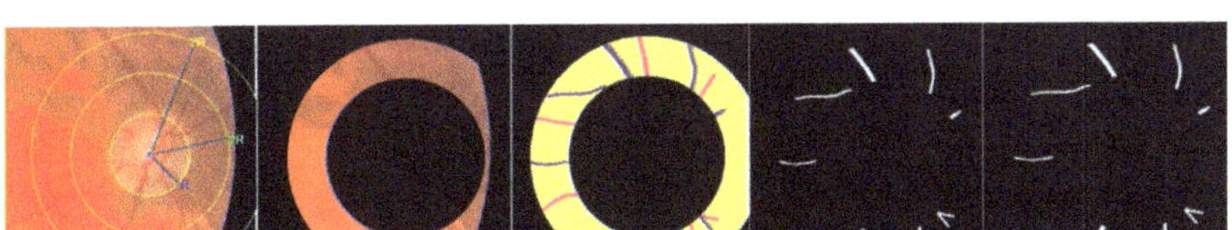

Figure 8. Artery–vein classification and segmentation of ROI of the segmented ring from 2ROD to 3ROD.

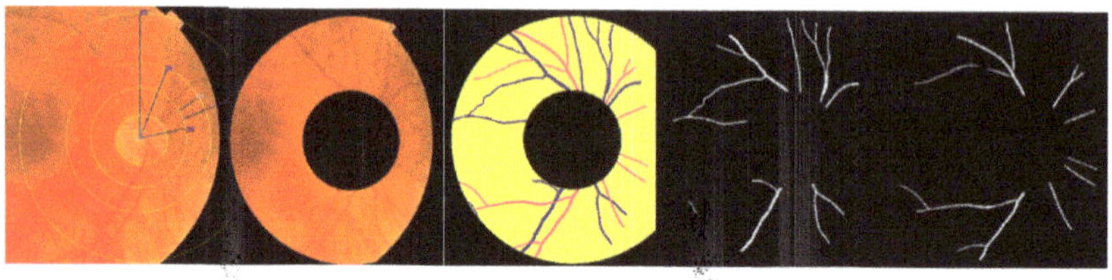

Figure 9. Artery–vein classification and segmentation of ROI of the segmented ring from 2ROD to 5ROD.

4.4.1. Skeletonization and Vessel Segment Optimization

In the next step, the segmented vasculature is skeletonized and optimized using the method from [18], preparing the vessel segments for further processing (Figures 10 and 11). These improvements result in a more accurate vessel tortuosity calculation, with significant enhancements in sigma level and confirming yield.

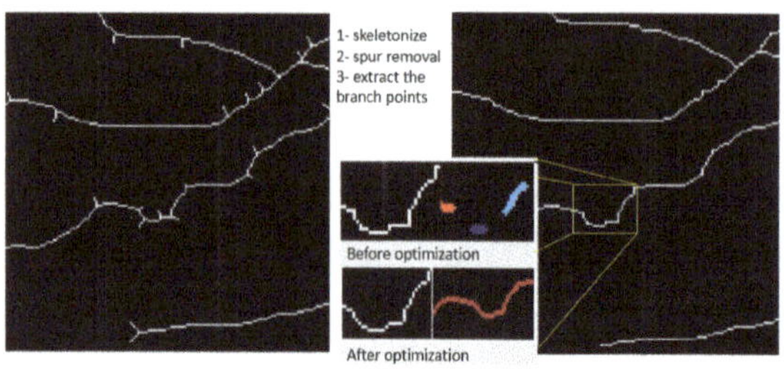

Figure 10. Illustration of improving the extraction of vessel segments by removing spurs after skeletonization and branch points removal.

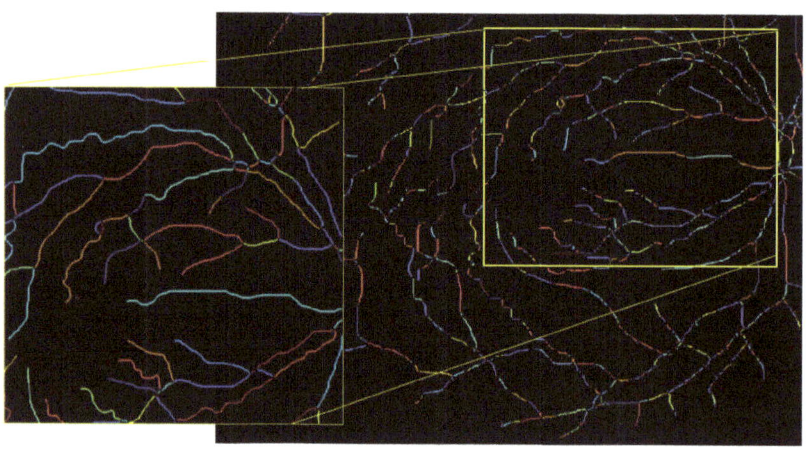

Figure 11. View of the enhanced vessel segments before and after the enhancement.

4.4.2. Measure the Length of Each Artery and Vein Fragment in the ROI

Each extracted vessel segment is passed to calculate its length to be ready as input for tortuosity metrics calculations using the method in [5] (see Figure 12).

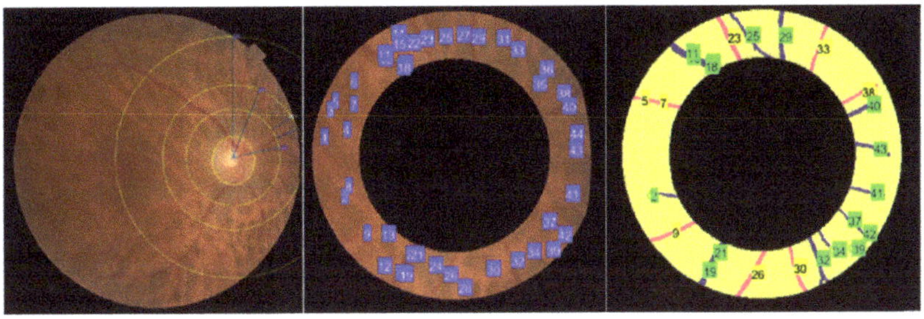

Figure 12. Illustration of the vessel segments that have been extracted, along with the calculation of their respective lengths.

4.5. Observations on Fundus Images and Their Corresponding Metric Values

The intricate analysis of fundus images, along with their corresponding metric values, not only provide valuable insights into the complex domain of retinal morphology but also unveil important characteristics of arteriovenous length and the AVLR. In this section, we delve into a detailed discussion of selected fundus images, highlighting our observations and findings.

4.5.1. Arteriovenous Length Analysis

Upon a meticulous analysis of the selected fundus image metrics results, we discerned a noteworthy pattern: the reported length in the arteries was generally consistent with that of the veins, thereby reinforcing the hypothesis that the lengths of arteries and veins in the retina exhibit a considerable degree of similarity in the normal situation [18]. Figure 13 showcases a fundus image from the dataset that has this case.

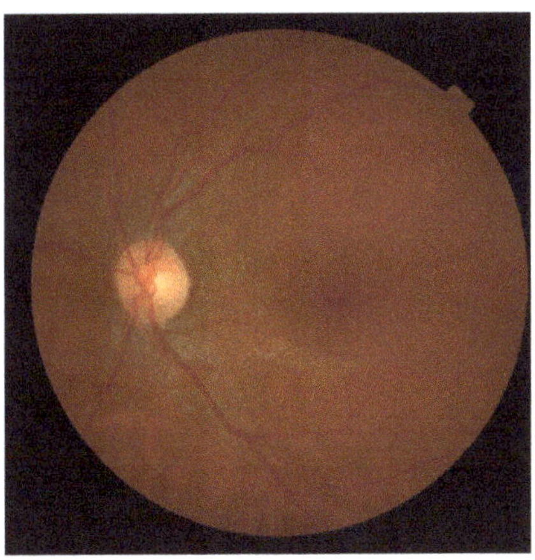

Figure 13. A healthy fundus image showcasing balanced vasculature with metric values: AVLR_DM (1.013), AVLR_SOAM (1.15), AVLR_ICMN (1.055), AVLR_ICMB (1.004), and AVLR_LC (0.977), signifying normalcy as the arteries' metric values closely match the mean of the veins.

4.5.2. An Intriguing Observation

We made an interesting observation while curating the AV-Classification dataset, which inspired further research. We noticed that the fundus images showed clear patterns of tortuosity in the arteries and veins, which could be classified into one of three categories.

- *Atypically Elongated Arteries and Normal Veins:* In a subset of retinal images within the AV-Classification dataset, we encountered instances where the arteries displayed significant elongation while the veins retained a relatively normal appearance. This divergence in vascular morphology raises intriguing questions about the potential factors driving this asymmetry and may help physicians to efficiently diagnose retinal abnormalities. Figure 14 showcases that.

- *Atypically Elongated Veins and Normal Arteries:* On the other hand, a contrasting pattern was evident in certain other fundus images. The veins showcased noticeable elongation, while the arteries preserved a normal appearance. This polarized pattern adds another layer of complexity to our understanding of retinal vascular abnormalities; such cases may be signs of specific further diagnoses.
- *Mixed Elongation:* A subset of images revealed that both the arteries and veins exhibited varying degrees of elongation or, conversely, both appeared to be normal. This variability within individual retinas underscores the intricate nature of vascular architecture and emphasizes the need for a more holistic investigation; Figure 15 showcases this.

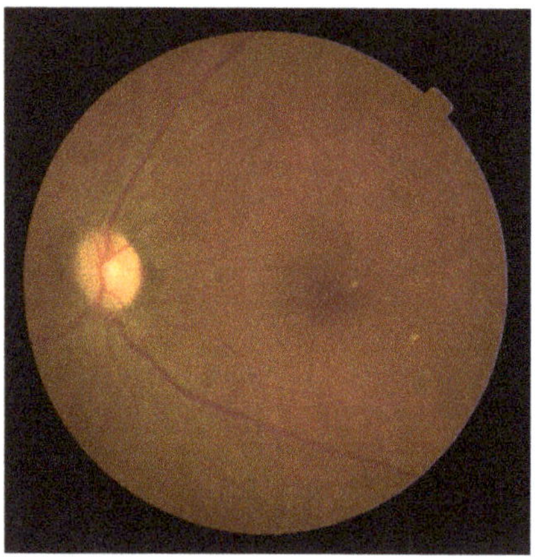

Figure 14. A fundus image displaying atypically elongated arteries and normal veins with metric values: AVLR_DM (7.194), AVLR_SOAM (0.843), AVLR_ICMN (29.811), AVLR_ICMB (13.785), and AVLR_LC (8.579).

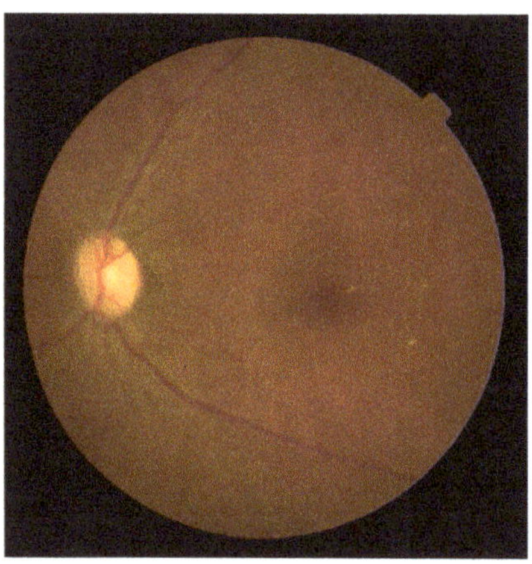

Figure 15. A fundus image displaying atypically elongated arteries and normal veins with metric values: AVLR_DM (0.424), AVLR_SOAM (0.942), AVLR_ICMN (0.27), AVLR_ICMB (0.615), and AVLR_LC (0.487).

4.5.3. Asymmetry in in Eye Pairs

Furthermore, our analysis led us to a specific pair of fundus images (Figure 16) that showcased a striking difference in the AVLR between the right and left eyes. The right eye displayed a severe degree of AVLR in contrast to the left eye. This pronounced asymmetry in AVLR invokes questions about its potential impact on retinal health and functionality, which warrants further investigation.

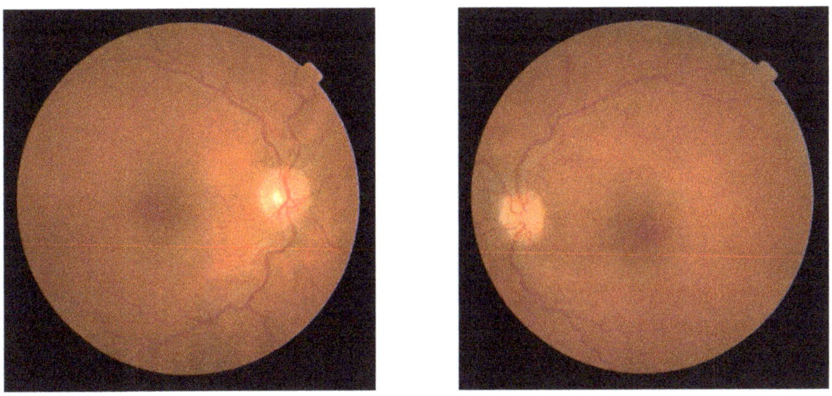

(**a**) (**b**)

Figure 16. Comparison of AVLR between the right and left eyes. (**a**) Metric values for Image 1: AVLR_DM (0.102), AVLR_SOAM (1.097), AVLR_ICMN (0.042), AVLR_ICMB (0.102), AVLR_LC (0.136). (**b**) Metric values for Image 2: AVLR_DM (2.186), AVLR_SOAM (0.789), AVLR_ICMN (3.626), AVLR_ICMB (1.331), AVLR_LC (1.699).

4.5.4. Discussion

The myriad variations in vascular AVLR elucidate the complexities of retinal vasculature and its potential implications for ocular health. The underlying reasons for these asymmetries remain enigmatic and necessitate further exploration. Potential contributing factors could range from genetic predispositions and systemic diseases to localized pathologies impacting specific regions of the retina.

The findings put forth in this study provide pivotal insights into the diversity of retinal vascular morphologies, indicating that a generic, one-size-fits-all approach to retinal analysis may not be adequate. A comprehensive understanding of these variations could be instrumental in the early detection of ocular diseases and the development of personalized treatment strategies.

Our observations underscore the critical role of the in-depth analysis of fundus images and corresponding metric values in augmenting our understanding of ophthalmology and propelling advancements in ocular health diagnosis and care. To unravel the mysterious mechanisms governing these observed patterns and their clinical implications, further research and interdisciplinary collaboration are indispensable.

4.6. Observations on the Box Plot Graphs

Box plots of the generated AVLR metrics—specifically the ratio of the mean artery to a mean vein—for each image across the AV-Classification dataset are given in Figures 17–20. The AVLR ratios are generated for the following metrics: inflection count metric normal (ICMN), inflection count metric binomial (ICMNB), distance metric (DM), sum of angles metric (SOAM), and length centerline (LC). The equations are explained in Section 3.6.

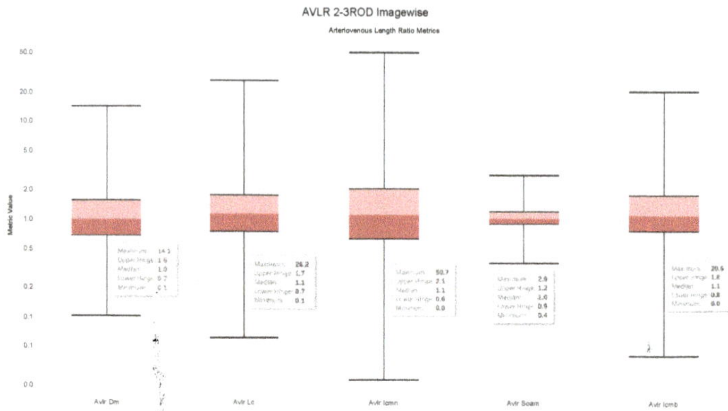

Figure 17. AVLR 2-3ROD Imagewise.

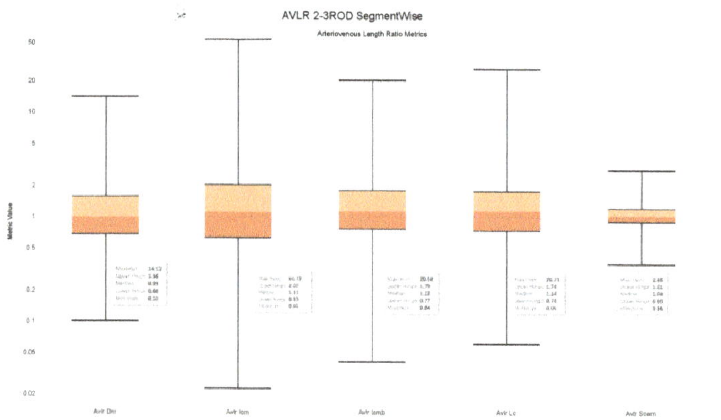

Figure 18. AVLR 2-3ROD Segmentwise.

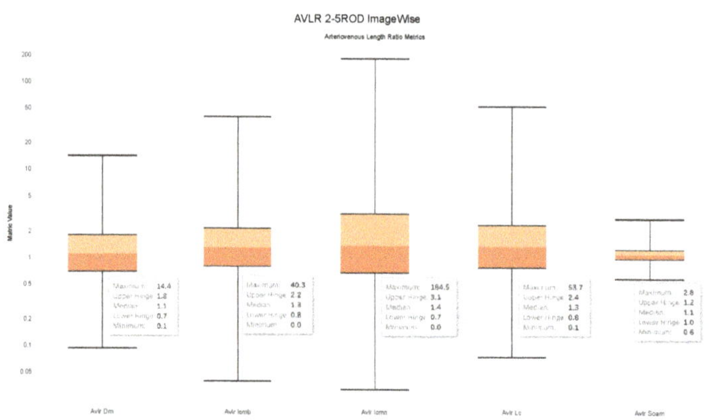

Figure 19. AVLR 2-5ROD Imagewise.

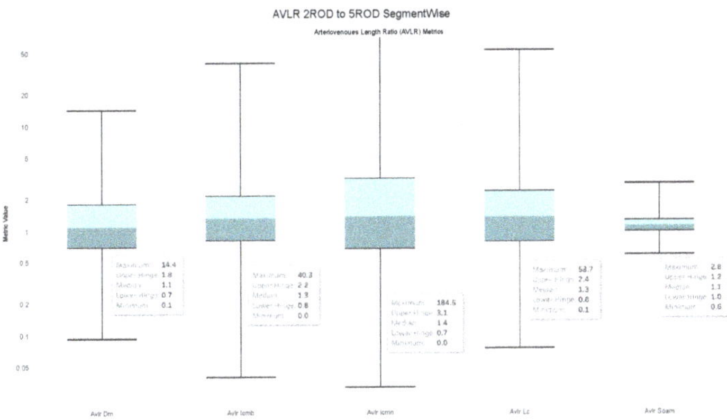

Figure 20. AVLR 2-5ROD Segmentwise.

Image Level Arterioveinous Length Ratios Analysis

The above observation reaffirms our statistical findings. To delve deeper into this relationship, we compute the image-based ratios across the 504 images in the AV-Classification dataset. We calculate them for each image as given in the equations. Additionally, we compute the ratio of the average tortuosity metric of retinal arteries over the average tortuosity metric of the retinal veins for each of the following length metrics: SOAM, ICM, ICMb, DM, and LC. Figures 17–20 show the box plots of the vessel length ratios image-wise and segment-wise using the equations in Section 3.6. Those ratios attribute each image in the AV-Classification dataset to the mean length of the arteries compared to the mean length of the veins in this specific retina. This could identify the healthiness of the vessels, whether their length varies or not, and how severe the deviation that can be noticed in each fundus image of the AV-Classification dataset is from those two ratios.

- The box plots shown in Figures 17–20 demonstrate that, across the 504 photos in the AV-Classification dataset, all metrics have a normal distribution. This leads us to a number of findings:
- The ratios for veins and arteries often show a comparable length pattern, indicating that the morphological lengthiness of veins and arteries is almost the same, with a ratio approaching 1.
- The center value of the box plots, or the median, slightly surpasses 1, indicating that the mean length of a vein is shorter than the average length of the artery.
- The ratio moves toward the top of the box plot when the artery-length mean is higher than that of the veins. As a consequence, when the length difference grows, the retinal image point rises in the box plot to approximately 50% of the normal distribution data.
- On the other hand, the ratio veers toward the bottom of the box plot when the average length of both arteries is smaller than the average length of both veins. The retinal image point in the box plot moves below the median as the length difference decreases. This tendency is seen in around 50% of the data from the normal distribution plot.
- In order to evaluate the difference between the lengths of the arteries and veins, the arc and chord lengths may be substituted because of the normal distribution of both ratios. In the meanwhile, for certain pictures in the AV-Classification dataset, the box plot in Figure 20 shows that the arteriovenous tortuosity ratios in (4)–(9) are skewed above the mean and median. This observation leads us to the following conclusions:
- The arteriovenous SOAM ratio illustrates how the curvature angle influences the calculation and limits the findings within the constrained range of the y-axis (0 to 360) by displaying a regular distribution with a mean and median of 1. However, any possible distinctions between veins and arteries may not be readily apparent because of this narrow range.

- The lengths of the arteries and veins lengthen and the ratio becomes closer to one when they have the same degree of AVLR. While the mean and median of the other AVLR ratios range from 0.99 to 1.4, the AV SOAM ratio has both a mean and a median of 1. The variations in the artery and vein diameters and their visibility on the retinal surface in each picture may be ascribed to this deviation.
- The ratio moves to the top part of the box plot if the average length of veins is shorter than the average length of arteries. This implies that the tortuosity of arteries is much greater than that of veins in the retinal image.
- Conversely, in the event that the average length of the arteries is lower than the mean lengths of the veins, the ratio will shift towards the lower end of the box plot. Vein tortuosity does not deviate substantially from the median in the box plots shown in Figures 17–20, suggesting that fewer cases of vein tortuosity arise in veins alone.

Based on the examination of the aforementioned arteriovenous length, it can be inferred that the observed results align with the statistically calculated outcomes, indicating that the measured lengths in the arteries are equivalent to those in the veins. However, the motivation for this study originates from a finding made during the creation of the AV-Classification dataset.

- The retinal images in the AV-Classification dataset display inconsistencies in the length and tortuosity of arteries and veins.
- Specifically, some images show tortuous arteries with normal veins, while others have tortuous veins but normal arteries.
- Additionally, there are images where both artery and vein lengths are either tortuous or normal.

Although numerous studies have focused on evaluating the tortuosity of retinal vessels, there is currently no formula available to specifically identify abnormalities in either veins, arteries or both. Such phenomena present a novel avenue for future research in ophthalmology, as it marks the first exploration of this aspect within the field. Furthermore, it is worth noting that existing tortuosity calculation methods found in the literature primarily assess overall tortuosity, whereas this study delves into the distinction in arteriovenous length behavior. The calculations for the arteriovenous length ratio (AVLR) were conducted for all segments of arteries and veins in each of the 504 retinal images from the AV-Classification dataset. Subsequently, formulae were developed to describe these phenomena, and statistical analysis demonstrated that, From a geometric standpoint, there are parallels in the tortuousness of veins and arteries. The quantification discussed by the authors is regarded as a distinctive and encouraging avenue for further exploration within the domain of retinal image pathology diagnosis.

4.7. AVLR New Dataset

The retinal vessel morphometry (RVM) research in [15] has been further enhanced by incorporating new labels for AVLR metrics, constituting an extension named the AV-Classification dataset. This extension encompasses vessel segment-wise and image-wise AVLR metrics quantified using the methodology proposed in this paper.

Specifically, as illustrated in Figure 21, the dataset now includes the following new AVLR metrics computed for each optimized vessel segment:

- Arteriovenous chord length ratio;
- Arteriovenous arc length ratio;
- Arteriovenous distance metric ratio;
- Arteriovenous inflection count metric ratio;
- Arteriovenous inflection count metric binomial ratio;
- Arteriovenous sum of angles metric ratio;
- Arteriovenous norm of curvature ratio;
- Arteriovenous standard deviation of average curvature ratio;
- Arteriovenous centerline length ratio.

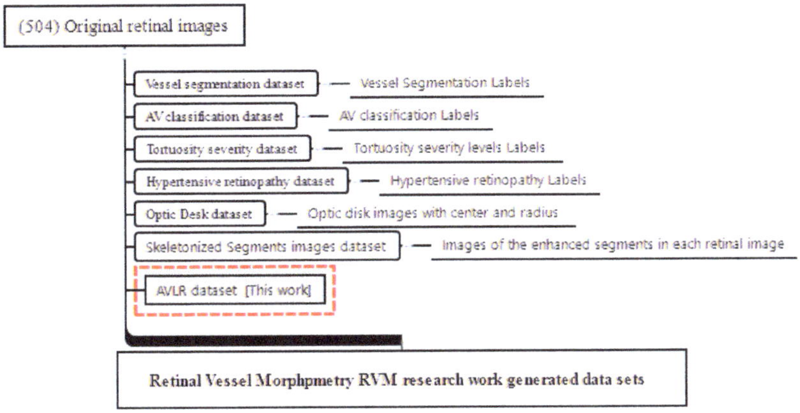

Figure 21. The AVLR dataset extends prior RVM work with new metrics quantified using the proposed methodology.

Additionally, image-level statistics are calculated by aggregating the segment-wise metrics for each image. The image-level statistics consist of summary values like count, mean, standard deviation, minimum, maximum, and median for all the above AVLR metrics.

In total, the enhanced AV-Classification dataset contains over 17,000 newly labeled AVLR values spanning across vessel segments and images. This expanded dataset with the additional AVLR annotations serves as a valuable resource for research and analysis of retinal vasculature morphology. The inclusion of the new AVLR metrics significantly augments the RVM dataset, providing a more comprehensive set of labels to quantify arteriovenous asymmetries and vascular abnormalities associated with various diseases.

The AV-Classification dataset will be made publicly available to the research community upon request, enabling further advancements in this domain. The dataset can facilitate the development of machine-learning models for automated AVLR computation from retinal images. In addition, the detailed AVLR labels can help identify morphological patterns and aid in screening vascular conditions like hypertensive retinopathy. Overall, the extended dataset allows a more in-depth analysis of retinal vessel morphology and its relationship to systemic diseases.

5. Conclusions

In conclusion, this paper proposes a novel method for evaluating retinal vasculature morphology and its diagnostic potential in hypertensive retinopathy (HR) and other eye-related diseases. The method introduces a new approach called the AVLR, which focuses on assessing the relative length of arteries and veins in the retinal vasculature. The methodology involves segmenting the retinal blood vessels, distinguishing between arteries and veins, and measuring the length of both within the area of interest encompassing the optic disc using an iterative approach. The effectiveness of the AVLR measurement was evaluated using the RVM dataset, demonstrating superior performance compared to existing techniques.

The major contributions of this research include the development of an automated system that integrates AVLR analysis and vessel tortuosity evaluation for the detection and grading of HR. This system provides a unique perspective on the morphological changes associated with HR and offers a valuable tool for the timely diagnosis of HR and other eye-related conditions. The methodology incorporates nine metrics that can accurately quantify the severity of AVLR, minimizing subjectivity and variability with manual assessments. Additionally, the comparative analysis of AVLR between the eyes enables a comprehensive evaluation of retinal vasculature in relation to HR and other conditions.

Finally, we enhanced prior retinal vessel morphometry research by adding new labels for metrics in an extended dataset called AV-Classification. This incorporates vessel segment-wise and image-wise AVLR metrics quantified using the proposed methodology.

The proposed method holds great promise for improving the efficiency and accuracy of HR diagnosis, facilitating timely treatment, and improving patient outcomes. Further research and validation on larger datasets and in clinical settings are warranted to fully establish the diagnostic potential of the AVLR and its integration into clinical decision support systems for ophthalmology.

Author Contributions: Conceptualization, S.A.B., M.T., G.A., D.G., I.E., M.A., N.M. and A.K.N.; Methodology, S.A.B., M.T., M.A.-.H., G.A., I.A.C., N.M. and A.K.N.; Software, S.A.B., D.G. and N.M.; Validation, S.A.B., M.T., M.A.-.H., D.G., I.E., I.A.C. and A.K.N.; Formal analysis, S.A.B., M.T., M.A.-.H., D.G., I.E., M.A., I.A.C., N.M. and A.K.N.; Investigation, S.A.B., D.G., I.E., M.A. and I.A.C.; Resources, S.A.B., M.A. and A.K.N.; Data curation, S.A.B., D.G., I.E., M.A., N.M. and A.K.N.; Writing—original draft, S.A.B., M.A.-.H. and G.A.; Writing—review & editing, S.A.B., M.T., M.A.-H., G.A. and D.G.; Visualization, S.A.B., M.A.-.H., G.A. and I.A.C.; Supervision, S.A.B.; Project administration, S.A.B. and I.E. The authors have contributed equally to this research. All authors have read and agreed to the final version of the manuscript.

Funding: This research received no external funding.

Institutional Review Board Statement: Not applicable.

Informed Consent Statement: Not applicable.

Data Availability Statement: Data are contained within article and our previous publication [16].

Acknowledgments: Ras Al-Khaimah public library under the umbrella of The UAE Ministry of Culture and Youth, and 'The American University of Ras Al Khaimah (AURAK)' both are appreciated and thanked by the authors for providing important resources to the researchers that enabled them to complete this study.

Conflicts of Interest: The authors declare no conflict of interest.

Abbreviations

The following abbreviations are used in this manuscript:

HR	Hypertensive Retinopathy
AVLR	Arteriovenous Length Ratio
AVR	Arteriovenous Ratio
ROI	Region of Interest
ROD	Radius of Optic disc
AV	Arteriovenous
DM	Distance Metric
SOAM	Sum of Angles Metric
ICM	Inflection Count Metric
ICMB	Inflection Count Metric Binomial
NC	Norm of Curvature
SDAC	Standard Deviation of Average Curvature
CL	Centerline Length
FCN	Fully Convolutional Neural Network
CNN	Convolutional Neural Network
OCT	Optical Coherence Tomography
RVM	Retinal Vessel Morphometry

References

1. Kanski, J.J.; Bowling, B. *Clinical Ophthalmology: A Systematic Approach*; Elsevier: Amsterdam, The Netherlands, 2011.
2. Dascalu, A.M.; Serban, D.; Tanasescu, D.; Vancea, G.; Cristea, B.M.; Stana, D.; Nicolae, V.A.; Serboiu, C.; Tribus, L.C.; Tudor, C.; et al. The Value of White Cell Inflammatory Biomarkers as Potential Predictors for Diabetic Retinopathy in Type 2 Diabetes Mellitus (T2DM). *Biomedicines* **2023**, *11*, 2106. [CrossRef] [PubMed]
3. Fraz, M.; Badar, M.; Malik, A.; Barman, S. Computational methods for exudates detection and macular edema estimation in retinal images: A survey. *Arch. Comput. Methods Eng.* **2019**, *26*, 1193–1220. [CrossRef]
4. Wegmann-Burns, M.; Gugger, M.; Goldblum, D. Hypertensive retinopathy. *Lancet* **2004**, *363*, 456. [CrossRef] [PubMed]
5. Badawi, S.A.; Fraz, M.M.; Shehzad, M.; Mahmood, I.; Javed, S.; Mosalam, E.; Nileshwar, A.K. Detection and grading of hypertensive retinopathy using vessels tortuosity and arteriovenous ratio. *J. Digit. Imaging* **2022**, *35*, 281–301. [CrossRef] [PubMed]
6. García-Sierra, R.; López-Lifante, V.M.; Isusquiza Garcia, E.; Heras, A.; Besada, I.; Verde Lopez, D.; Alzamora, M.T.; Forés, R.; Montero-Alia, P.; Ugarte Anduaga, J.; et al. Automated Systems for Calculating Arteriovenous Ratio in Retinographies: A Scoping Review. *Diagnostics* **2022**, *12*, 2865. [CrossRef] [PubMed]
7. Akbar, S.; Akram, M.U.; Sharif, M.; Tariq, A.; ullah Yasin, U. Arteriovenous ratio and papilledema based hybrid decision support system for detection and grading of hypertensive retinopathy. *Comput. Methods Programs Biomed.* **2018**, *154*, 123–141. [CrossRef] [PubMed]
8. Agurto, C.; Joshi, V.; Nemeth, S.; Soliz, P.; Barriga, S. Detection of hypertensive retinopathy using vessel measurements and textural features. In Proceedings of the 2014 36th Annual International Conference of the IEEE Engineering in Medicine and Biology Society, Chicago, IL, USA, 26–30 August 2014; pp. 5406–5409. [CrossRef]
9. Khitran, S.; Akram, M.U.; Usman, A.; Yasin, U. Automated system for the detection of hypertensive retinopathy. In Proceedings of the 2014 4th International Conference on Image Processing Theory, Tools and Applications (IPTA), Paris, France, 14–17 October 2014; pp. 1–6. [CrossRef]
10. Mapayi, T.; Tapamo, J.R.; Viriri, S.; Adio, A. Automatic Retinal Vessel Detection and Tortuosity Measurement. *Image Anal. Stereol.* **2016**, *35*, 117–135. [CrossRef]
11. Abbas, Q.; Qureshi, I.; Ibrahim, M.E.A. An Automatic Detection and Classification System of Five Stages for Hypertensive Retinopathy Using Semantic and Instance Segmentation in DenseNet Architecture. *Sensors* **2021**, *21*, 6936. [CrossRef] [PubMed]
12. Javaid, M.; Haleem, A.; Singh, R.P.; Suman, R.; Rab, S. Significance of machine learning in healthcare: Features, pillars and applications. *Int. J. Intel. Netw.* **2022**, *3*, 58–73. [CrossRef]
13. Foulsham, W.; Chien, J.; Lenis, T.L.; Papakostas, T.D. Optical Coherence Tomography Angiography: Clinical Utility and Future Directions. *J. Vitreoretin. Dis.* **2022**, *6*, 229–242. [CrossRef] [PubMed]
14. Gonzales, R.C.; Woods, R.E.; Eddins, S.L. *Digital Image Processing Using MATLAB*; Pearson Prentice Hall: Hoboken, NJ, USA, 2004.
15. Badawi, S.M. Segmentation, Classification, and Morphometric Analysis of Retinal Vasculature. Ph.D. Thesis, National University for Sciences and Technology, Islamabad, Pakistan, 2020.
16. Badawi, S.A.; Fraz, M.M. Multiloss function-based deep convolutional neural network for segmentation of retinal vasculature into arterioles and venules. *BioMed Res. Int.* **2019**, *2019*, 4747230. [CrossRef] [PubMed]
17. Badawi, S.A.; Fraz, M.M. Optimizing the trainable B-COSFIRE filter for retinal blood vessel segmentation. *PeerJ* **2018**, *6*, e5855. [CrossRef] [PubMed]
18. Badawi, S.A.; Takruri, M.; ElBadawi, I.; Chaudhry, I.A.; Mahar, N.U.; Nileshwar, A.K.; Mosalam, E. Enhancing Vessel Segment Extraction in Retinal Fundus Images Using Retinal Image Analysis and Six Sigma Process Capability Index. *Mathematics* **2023**, *11*, 3170. [CrossRef]

Journal of
Imaging

MDPI

Article

Retinal Microvasculature Image Analysis Using Optical Coherence Tomography Angiography in Patients with Post-COVID-19 Syndrome

Maha Noor [1], Orlaith McGrath [1], Ines Drira [1,2] and Tariq Aslam [1,3,*]

1 Department of Eye Research, Manchester Royal Eye Hospital, Manchester University NHS Foundation Trust, Oxford Road, Manchester M13 9WL, UK
2 Ophtalmologie Département, Centre Hospitalier Universitaire de Toulouse, 31300 Toulouse, France
3 Division of Pharmacy and Optometry, School of Health Sciences, Faculty of Biology, Medicine and Health, The University of Manchester, Oxford Road, Manchester M13 9PL, UK
* Correspondence: tariq.aslam@mft.nhs.uk

Abstract: Several optical coherence tomography angiography (OCT-A) studies have demonstrated retinal microvascular changes in patients post-SARS-CoV-2 infection, reflecting retinal-systemic microvasculature homology. Post-COVID-19 syndrome (PCS) entails persistent symptoms following SARS-CoV-2 infection. In this study, we investigated the retinal microvasculature in PCS patients using OCT-angiography and analysed the macular retinal nerve fibre layer (RNFL) and ganglion cell layer (GCL) thickness via spectral domain-OCT (SD-OCT). Conducted at the Manchester Royal Eye Hospital, UK, this cross-sectional study compared 40 PCS participants with 40 healthy controls, who underwent ophthalmic assessments, SD-OCT, and OCT-A imaging. OCT-A images from the superficial capillary plexus (SCP) were analysed using an in-house specialised software, OCT-A vascular image analysis (OCTAVIA), measuring the mean large vessel and capillary intensity, vessel density, ischaemia areas, and foveal avascular zone (FAZ) area and circularity. RNFL and GCL thickness was measured using the OCT machine's software. Retinal evaluations occurred at an average of 15.2 ± 6.9 months post SARS-CoV-2 infection in PCS participants. Our findings revealed no significant differences between the PCS and control groups in the OCT-A parameters or RNFL and GCL thicknesses, indicating that no long-term damage ensued in the vascular bed or retinal layers within our cohort, providing a degree of reassurance for PCS patients.

Keywords: post-COVID-19 syndrome (PCS); long COVID; optical coherence tomography angiography (OCT-A); SD-OCT; nerve fibre layer; ganglion cell layer; retina

1. Introduction

In March of 2020, the severe acute respiratory syndrome coronavirus 2 (SARS-CoV-2) gave rise to a global pandemic, incurring detrimental effects on the health, economy, and social infrastructures of populations worldwide. The effects of the SARS-CoV-2 infection, termed acute coronavirus disease 2019 (COVID-19), range from pneumonia and acute respiratory distress syndrome (ARDS) to thromboembolic disease, septic shock, and multi-organ failure [1]. In the aftermath of the pandemic, a post-viral sequela of SARS-CoV-2 infection emerged, referred to as post-COVID-19 syndrome (PCS) [2] or post-COVID-19 condition [3].

Post COVID-19 syndrome refers to the persistence of certain clinical symptoms more than 12 weeks after the initial COVID-19 infection, which cannot be explained by an alternative diagnosis [4]. Persistence of symptoms 4 or more weeks after the infection may be referred to as post-acute sequelae of SARS-CoV-2 infection (PASC) [5] or 'Long COVID'. Symptoms of PCS may include but are not limited to fatigue, dyspnoea, cough, autonomic symptoms (chest pain, palpitations, tachycardia), neurocognitive impairment i.e., 'brain fog', arthralgia, myalgia, headaches, anosmia, ageusia, gastrointestinal disturbances, sleep disturbances, hair loss, and psychiatric disorders such as depression and anxiety [5]. Immune dysregulation, autoimmunity, dysautonomia, viral persistence, re-activation of latent viral pathogens, neutrophil extracellular traps (NETs), dysregulation of the renin-angiotensin-aldosterone system (RAAS), coagulopathies (hypercoagulation, thrombosis), fibrin amyloid micro-clots, endothelial dysfunction, and impaired microvasculature, are some of the principle pathophysiological mechanisms underpinning post-COVID-19 syndrome described in the literature to date [5–34].

Studies have shown that increasing age, female sex, lower socioeconomic status, obesity, presence of co-morbidities, and smoking are associated with an increased risk of development of PCS [35–40]. It has been estimated that, globally, approximately 15% of COVID-19 patients experience persistent symptoms at 12 months following acute SARS-CoV-2 infection [41]. In the UK, 3.1% of the population self-reported suffering with long-term symptoms following their initial COVID-19 infection between September 2022 and January 2023 [42]. The worldwide influence of COVID-19 affects not only personal mental and physical well-being, but also social, economic, and productivity-related aspects in the healthcare, finance, and employment sectors. Decreased productivity, the requirement for medical assistance, and the growing need for diagnosis render post COVID-19 syndrome a deserving candidate for investment in the healthcare industry [43]. Therefore, it is important to investigate and research all characteristics of this condition.

SARS-CoV-2 enters cells via the angiotensin-converting enzyme-related carboxypeptidase (ACE2) receptor [44], which is also expressed on the surface of neuroretinal cells including Muller cells, retinal pigment epithelium, and pericytes of retinal and choroidal endothelial cells [45]. Whilst the most common ophthalmic manifestations of acute SARS-CoV-2 infection consist of conditions such as conjunctivitis and anterior uveitis [46–48], scleritis, episcleritis, inflammatory orbital diseases (dacryoadenitis, orbital cellulitis, mucormycosis), optic neuritis, papillophlebitis, cranial nerve palsies, choroiditis, retinitis, retinal vasculitis [48–51], and retinal artery and vein occlusion have also been reported [52]. Retinal examination in patients following COVID-19 infection has demonstrated a wide array of clinical signs ranging from cotton wool spots (CWS), retinal micro- and macro-haemorrhages, and venous tortuosity [51,53,54] indicative of acute vascular events and retinal ischaemia. One of the first studies pertaining to optical coherence tomography (OCT) imaging of the retina in patients with COVID-19 identified hyper-reflective plaques in the ganglion-cell-inner plexiform layer (GC-IPL) [54]. However, subsequent publications attributed this finding to normal variations in the retinal vasculature [55,56]. Increased peri-papillary retinal nerve fibre layer (RNFL) thickness [57,58], increased central macular thickness, and decreased ganglion cell layer and inner nuclear layer thickness have also been reported in patients following COVID-19 infection [59]. Therefore, SARS-CoV-2 infection can induce microstructural changes in the retina, which may persist long-term in patients with post-COVID-19 syndrome. Given that the retina and optic disc are considered to be intraorbital extensions of the central nervous system [60], alterations may prevail especially in those with neurocognitive symptoms.

The eye can be considered as a window into the body's microvascular system. Optical coherence tomography angiography (OCT-A) offers a non-invasive opportunity to analyse the retinal circulation in vivo, providing insight into the subject's systemic microvasculature by inference [61]. Several studies have investigated the retinal vasculature of patients infected with COVID-19 using OCT-A to date. A key finding of note is reduction in the macular vessel density from as early as 2 weeks following the infection up to 8 months afterward [62–75].

Considering the overwhelming amount of literature reporting an alteration in the retinal microvasculature in patients with a history of COVID-19 illness, it is pertinent to investigate whether these effects last long term, especially in patients with ongoing symptoms of post-COVID-19-syndrome. Furthermore, there is currently a paucity of literature examining the retinal vasculature of patients with PCS. Therefore, the primary aim of this study is to investigate the retinal microvasculature of patients with post-COVID-19 syndrome using OCT-angiography, in order to determine the long-term sequelae of SARS-CoV-2 infection on retinal tissues. Additionally, using spectral domain-OCT (SD-OCT) the thickness of the macular retinal nerve fibre layer and macular ganglion cell layer will also be examined, to determine if any anatomical alterations coincide with ongoing symptoms of PCS, particularly neurocognitive symptoms.

2. Materials and Methods

This was a prospective, cross-sectional observational study, conducted at the Manchester Royal Eye Hospital (MREH), UK. Ethical approval was obtained from the Health Research Authority (HRA) and Health and Care Research Wales (HCRW) along with the Office for Research Ethics Committees Northern Ireland (ORECNI). The study was conducted with adherence to the Declaration of Helsinki; written informed consent was obtained from all participants.

2.1. Participant Recruitment

Patients were recruited into two distinct groups for this comparative study. The first group comprised of patients over 18 years of age with an established clinical diagnosis of post-COVID-19 syndrome by the respiratory team at Manchester Royal Infirmary, UK. All included subjects either had a reverse transcription-polymerase chain reaction (RT PCR)—confirmed diagnosis of COVID-19 at an earlier stage or a clinical diagnosis of COVID-19 (as testing was not readily available in the early stages of the pandemic in the UK). Patients recruited within the post-COVID-19 syndrome cohort may have an initial mild, moderate, or severe initial illness with or without requirement for hospitalisation or outpatient treatment, allowing us to examine the retinal microvasculature in a wider range of PCS participants.

The second group, the controls, included patients over 18 years of age who did not have a recent history of COVID-19 infection or a diagnosis of post-COVID-19 syndrome. For both groups, we excluded patients with a history of diabetes, uncontrolled hypertension, stroke, haematological disorders, neurodegenerative diseases, high myopia or hypermetropia (above $\pm$ 6 dioptres), high astigmatism (>3 dioptres), significant media opacity compromising fundus imaging, or signs or previous history of choroidal atrophy, exudative age-related macular degeneration (AMD), central serous chorioretinopathy, glaucoma, acquired and hereditary optic neuropathy, hereditary retinal diseases, demyelinating disorders, and keratoconus. Both cohorts were age and sex matched. Recruitment commenced in April 2021 and extended to March 2023 due to delays caused by the COVID-19 pandemic.

Clinical history taking, visual acuity measurement, and OCT-A imaging was undertaken for each participant of the study. The clinical history taking comprised of details of the participants' acute COVID-19 illness, method of acute COVID-19 diagnosis (i.e., clinical or by PCR testing), disease course, ongoing symptoms of post-COVID-19 syndrome, vaccination history, relevant past medical history, smoking history, Body Mass Index (BMI) and HbA1c (if available), past ocular history, and ophthalmic prescription. The best corrected visual acuity was measured using Early Treatment Diabetic Retinopathy Study (ETDRS) charts at four metres, converted to the Logarithm of the Minimum Angle of Resolution (LogMAR).

2.2. OCT Imaging

Each patient then underwent wide-field (10 $\times$ 10 mm) macula and foveal (4 $\times$ 4 mm) OCT-A imaging using the Canon Xephilio OCTA-1 machine (Canon Medical Systems Europe B.V©, Amstelveen, The Netherlands). The field of investigation was centred on the foveal region. Scans had a 10-layer automated segmentation and a refresh rate of 70,000 A-scans/s. The depth of field of view was set to 10 $\times$ 10 mm and 4 $\times$ 4 mm with an axial sampling density of 464 $\times$ 464 px, with the number of repetitions set at two. For the purposes of this study, only the retinal superficial capillary plexus (SPC), which provided the most consistently high-quality images, was examined. Both eyes were imaged in each participant, however only the highest quality eye image was used for analysis. Inter-eye correlations and statistical complexities highlighted by Murdoch et al. would allow for us to include both eyes into the study, albeit with more complex and less easily recognised and interpreted techniques [76]. In this study however, image quality for accurate OCT-A measures, as noted by Czako et al., was of paramount importance [77]. Notably, we were aware that many PCS participants suffered with dry eyes, fatigue, and dyspnoea, and we opted to include the highest quality imaged eye only for each patient in this study protocol with standard statistical techniques to optimise overall image analysis validity. Lubricants were offered to all patients to mitigate effects of any dry eye disease. Pharmacological mydriasis was attained (tropicamide 1%) in cases where the quality of imaging was affected by lack of pupillary dilatation. The image acquisition technique was regimented, in that all patients were instructed to focus on the cross shape in the OCT-A machine to ensure standardisation of the macular image procured. Stability of the head was ensured, and all images were captured in dim lighting.

A two-fold strategy was employed to evaluate the quality of the images obtained. Initially, the Canon Quality Index from the OCT-A machine was utilised, accepting only images with an index of $\geq$7. Additionally, imagers with clinical expertise and a senior consultant conducted real-time evaluations of the OCT-A and OCT images to identify significant segmentation errors, and shadow or motion artefacts, leading to image exclusion if detected. The OCT-A images obtained from both participant groups (one eye per patient) were then analysed using an in-house specially designed image-processing software, OCT-A Vascular Image Analysis (OCTAVIA), which carried out an automated analysis and uploaded the specified measurements to a central Microsoft® Excel® 2021 spreadsheet.

Spectral domain OCT of the macula was also performed to analyse the average thicknesses of the macular retinal nerve fibre layer (mRNFL) and macular ganglion cell layer (mGCL) in microns. The OCT machine's internal software segmented the 10×10 macular image into the nine zones specified by the Early Treatment Diabetic Retinopathy Study (ETDRS) [78], giving an average value of RNFL or GCL thickness in each zone, shown in Figure 1. These values were exported into a Microsoft® Excel® 2021 spreadsheet, and a mean value for the thickness of the outer and inner segments was calculated. The final study parameters included the mean thickness of the outer segment, inner segment, and foveal (central) region of the mRNFL and mGCL.

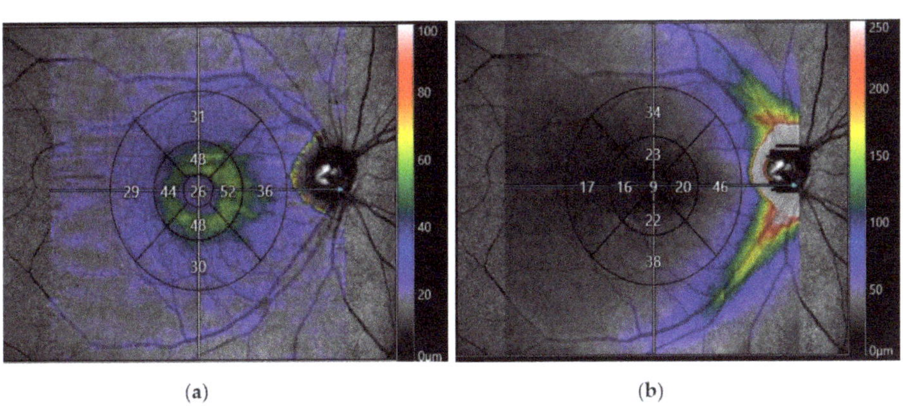

Figure 1. Spectral domain-optical coherence tomography (SD-OCT) of the macula obtained from Canon Xephilio OCT-A1 Machine (Canon Medical Systems Europe B.V©, Amstelveen, Netherlands) displaying a 10×10 mm macular image from a participant with post-COVID-19 syndrome segmented into nine EDTRS zones. The segments consist of superior outer, superior inner, nasal outer, nasal inner, inferior outer, inferior inner, temporal outer, temporal inner, and foveal (central) zones. (**a**) Displays the average thickness of the macular retinal nerve fibre layer (mRNFL) in nine EDTRS zones. (**b**) Displays the average thickness of the macular ganglion cell layer (mGCL) in nine EDTRS zones.

2.3. OCT-Angiography Image Processing Algorithm

The OCTAVIA algorithm was programmed using MATLAB® 2021 by the corresponding author (TA), developed from previously published work on small field OCT-A imaging in diabetic retinopathy (DR) [79]. Additional evidence for its reliability and validity are provided in Appendix A, Appendix A.1 Reliability and validity of the software.

OCT-angiography has demonstrated a range of retinal vascular changes, including enlargement of the foveal avascular zone (FAZ) and reduced macular vessel density, in diabetic retinopathy [80,81] but also specifically in patients with recent SARS-CoV-2 infection [62–75]. For our study we chose outcome measures to reflect a comprehensive but relevant assessment based upon previous research and clinical experience. The final measured parameters are listed in Table 1.

Table 1. Parameters evaluated in the analysis of optical coherence tomography-angiography (OCT-A) images.

10 × 10 mm Image	4 × 4 mm Image
Mean large vessel intensity	Mean capillary intensity
Mean capillary intensity	Percentage capillary network (vessel density)
Percentage capillary network (vessel density)	Area of the foveal avascular zone (FAZ)
Total area of ischaemia	Circularity of the foveal avascular zone (FAZ)

Large vessel and capillary intensity refer to the amount of blood flow through the large and capillary vessels, respectively. The percentage capillary network or vessel density is an index of vascularity, indicating the retinal area occupied by vessels divided by the total retinal area. Foveal avascular zone (FAZ) refers to the foveola and immediate parafoveal retina which lacks capillaries, relying on blood supply from the choriocapillaris. In addition to enlargement of the area of the FAZ, distortion of its circularity has also been observed in DR [82], leading us to examine this parameter to investigate the retinal microvasculature in detail.

To perform image analysis, OCTAVIA has two distinct processes, using both 4 × 4 and 10 × 10 images, shown in Figure 2. The software firstly inputs the subject's 10 × 10 mm macular image, followed by the OCT-A machine's own in-built proprietary binary interpretation of the same image. This process is repeated for the 4 × 4 mm foveal image. The 4 × 4 mm image allows for greater detail of the foveal avascular zone (FAZ) aiding more accurate measurement of its parameters. The proprietary binary images provide a template for accurately deriving the SCP vasculature. Morphological processing techniques are utilised to distinguish between larger and smaller vessels and thresholds applied onto processed images to identify low-intensity i.e., ischaemic areas. The large and capillary vessel intensity is measured by measuring the intensity of the pixels within the skeletonised vessels. During the initial software development phase, it became apparent that small vitreous opacities can cause darkened patches on the OCT-A images. To prevent the software misinterpreting areas of such low signal, including the optic disc, as ischaemia, the functionality to manually crop out these observed areas and the optic disc was integrated into an initial pre-processing step. The final data are directly outputted to a central Microsoft® Excel® spreadsheet by the software.

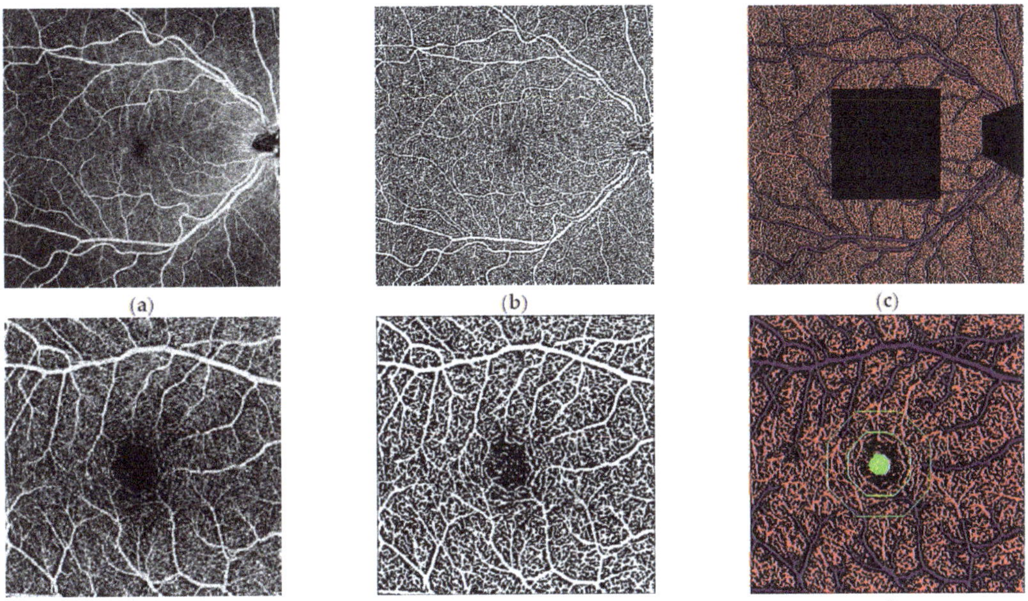

(a) (b) (c) (d) (e) (f)

Figure 2. Analysis of the macular 10 × 10 mm and 4 × 4 mm optical coherence tomography-angiography (OCT-A) images performed by our inhouse software. (**a**) 10 × 10 mm macular OCT-Angiography image of the right eye. (**b**) Binarisation of the 10 × 10 mm macular OCT-A image as a processing step. (**c**) Final segmentation of the image following removal of optic disc and the central 4 × 4 mm area which was analysed in separate dedicated 4 × 4 mm images (**d**) 4 × 4 mm macular OCT-Angiography image of the right eye. (**e**) Binarisation of the 4 × 4 mm macular OCT-A image. (**f**) Final segmentation of the 4 × 4 mm image with parafoveal and perifoveal zones highlighted.

2.4. Statistical Methodology

The sample size for this study was calculated based on data collected in a previous study on diabetic retinopathy and OCT-angiography [79] and is provided in full in Appendix A, Appendix A.2 Sample size calculation. We opted to calculate the sample size using data from a cohort of patients with diabetic retinopathy because the alterations observed in OCT-A studies in patients with diabetic retinopathy are comparable to those seen in the early stages of COVID-19 infection, such as the enlargement of the FAZ [80]. A minimum of 31 participants in each group were calculated. Therefore, we aimed to have 40 participants in each of our study cohorts (PCS and controls).

Statistical analyses were performed using jamovi (version 2.3) [83]. Normality was assessed using the Shapiro–Wilk test, histograms, and Q-Q plots. Quantitative variables following a normal distribution were studied with Student *t*-tests, while those without a normal distribution were analysed using the Mann–Whitney U-test. The adjusted p value was set at <0.00357 after Bonferroni correction for 14 study parameters. A linear regression was also undertaken, evaluating the effect of dependent variables including age, gender, and length of time since initial COVID-19 infection on a key outcome variable mean capillary intensity (10 × 10 mm OCT-A image).

3. Results

3.1. Demographic Distribution

A total of 80 eyes of 80 patients were included in this study (44 right eyes and 36 left eyes). There were 40 patients included in the group with post-COVID-19 syndrome and 40 controls. The PCS group was comprised of 31 females and 9 males, with an average age of 47.8 ± 10.4 years. Ethnic distribution consisted of thirty-five Caucasians, two Asians, one Black Caribbean and one Mixed Caucasian and Black Caribbean participant. Clinical assessment including clinical history, visual acuity measurement, OCT and OCT-A imaging was conducted at an average of 15.2 ± 6.9 months (range 3–32 months) after the initial SARS-CoV-2 infection. The control group was comprised of twenty-seven females and thirteen males, with an average age of 44.0 ± 14.6 years. Ethnic distribution consisted of thirty-four Caucasians, five Asians, and one Black African participant. No significant differences in age (p = 0.107) or sex (p = 0.317) distribution were found between the two groups. The mean LogMAR visual acuity was −0.0045 ± 0.168 in the PCS cohort, and 0.01652 ± 0.137 in the control subjects (p = 0.302). The average quality of the OCT-angiography 10 × 10 mm images was 7.40 ± 0.67 in the PCS cohort and 7.55 ± 0.64 in the controls (p = 0.465). Average quality of 4 × 4 mm images was 7.68 ± 0.76 in the PCS group and 7.83 ± 0.90 in the controls (p = 0.465). The quality of the SD-OCT macula images was 8.18 ± 0.82 in the PCS group and 8.24 ± 0.65 in the controls (0.779). Further details can be found in Table 2.

Table 2. Demographic details and analysis of participants in the post COVID-19 syndrome (PCS) and control cohorts.

	Total	PCS Group		Control Group		Statistical Test	Significance $p < 0.00357$
No. of Patients	80	40		40			
Female	58	31		27		Chi Squared test	0.317
Male	22	9		13			
		Mean	**SD**	**Mean**	**SD**		
Age		47.80	10.40	44.00	14.60	Mann-Whitney U test	0.107
LogMAR Visual Acuity		−0.0045	0.168	+0.0165	0.137	Mann-Whitney U test	0.302
OCT-A 10 × 10 mm Quality		7.40	0.67	7.55	0.64	Mann-Whitney U test	0.368
OCT-A 4 × 4 mm Quality		7.68	0.76	7.83	0.90	Mann-Whitney U test	0.465
SD-OCT Macula Quality		8.18	0.82	8.24	0.65	Mann-Whitney U test	0.779

Post COVID-19 syndrome, *PCS*; Standard Deviation, *SD*; Logarithm of the Minimum Angle of Resolution, *LogMAR*; Optical coherence tomography-angiography, *OCT-A*; Spectral-domain optical coherence tomography, *SD-OCT*. Note: H_a $\mu_{Control} \neq \mu_{PCS}$—The two-tailed alternative hypothesis for this study was that there is a significant difference in the measured parameters between the participants with post COVID-19 syndrome (PCS) and healthy control subjects.

3.2. Clinical History

Participants in the PCS cohort with notable past medical histories included six obese participants, eight asthmatic individuals, six with obstructive sleep apnoea, and one with fibromyalgia. Neither the controls nor any of the PCS cohort participants had a history of diabetes or hypertension. Only one patient in the cohort had a history of prior hospitalisation for COVID-19 pneumonia, with the remainder having had a mild COVID-19 infection which did not require any inpatient or outpatient treatment.

The most prevalent symptoms of post-COVID-19 syndrome identified during clinical history taking were fatigue (30/40), dyspnoea (23/40) of which 4/23 reported exertional dyspnoea, cognitive dysfunction termed "brain fog" (16/40), and palpitations (15/40). Four patients had been diagnosed with paroxysmal orthostatic tachycardia syndrome (POTS). Interestingly, 3/40 patients reported intermittent visual disturbance and 11/40 expressed presence of dry eyes. Table 3 provides further details on the PCS cohort's clinical symptoms, categorised by physiological systems, adapted from a comprehensive review of PCS by NICE guidelines [84].

Table 3. Categories of clinical symptoms of post-COVID-19 syndrome according to physiological systems.

Physiological System	Clinical Symptoms				
Systemic	**Fatigue**	**Dizziness**	**Fever**		
	30	5	0		
Cardiopulmonary	Dyspnoea	Chest Pain	Palpitations	Pericarditis	
	23	8	15	1	
Upper Respiratory	Blocked Nose	Cough	Sore Throat	Voice Changes	Laryngeal Disorders *
	0	4	0	1	2
Gastrointestinal	Nausea	Vomiting	Diarrhoea	Appetite Changes	Abdominal Pain	Weight Loss
	1	0	0	0	1	1
Musculoskeletal	Joint Pain	Muscle Pain	Worsened Mobility		
	3	3	0		
Neurological Or Neuromuscular	Headache	Hyposmia/ Anosmia	Hypogeusia/ Ageusia	Paraesthesia	
	10	3	3	1	
Psychological	Anxiety	Depression	Post-Traumatic Stress Disorder (PTSD)	Sleep Disturbances	
	1	2	0	2	
Neurocognitive	Cognitive Dysfunction i.e., Brain Fog (Reduced Memory And/Or Concentration)	Cognitive Impairment	Confusion		
	16	0	0		
Ophthalmic	Vision Disturbances	Dry Eyes			
	3	11			
Auditory	Reduced Hearing	Tinnitus			
	1	1			
Other	Hair Loss	Post-Menopausal Bleeding	Restless Legs		
	0	1	1		

* Sensation of Intermittent Laryngeal Obstruction; Coughing Hypersensitivity Related Inducible Laryngeal Obstruction.

3.3. OCT-Angiography Image Analysis

All 80 eyes were processed successfully through the OCTAVIA image analysis system, which provided output metrics as planned. No clinically observable significant macro- or micro-vascular abnormalities were detected in either participant groups by a retinal specialist (TA). The results indicated that there were no statistically significant differences between the PCS and control cohort in any of the SCP measures in the 10 × 10 images in terms of the mean large vessel intensity ($p = 0.588$), mean capillary intensity ($p = 0.099$), mean vessel densities ($p = 0.103$) and the total area of ischaemia ($p = 0.541$). In the 4 × 4 images also, no statistically significant differences were noted between the PCS and control cohort in relation to the mean vessel densities ($p = 0.895$), area of the FAZ ($p = 0.399$), and circularity of the FAZ ($p = 0.319$).

During the study, a required remote OCT-A software upgrade occurred. Although there was no visible effect on clinical examination of the image, we explored impact of this on images and there appeared to be a possible subtle change in values of the mean capillary intensities on 4 × 4 images but of no other parameters. Therefore, for the purposes of inter-cohort and intra-cohort analysis of the mean capillary intensity (4 × 4 mm images) only individuals recruited from December 2021 (PCS cohort $n = 20$, control cohort $n = 26$) have been included. The mean capillary intensity was measured as 139.25 ± 4.32 in the PCS cohort, and 140.51 ± 6.03 in the controls ($p = 0.350$). Further details are illustrated in Table 4.

Table 4. Analysis of optical coherence tomography-angiography (OCT-A) parameters in the post-COVID-19 syndrome (PCS) cohort compared with the control cohort.

					Shapiro-Wilk		Statistical Test	Statistic	df	$p < 0.00357$
	Cohort Category	N	Mean	SD	W	p				
Mean Large Vessel Intensity (10 × 10 mm)	PCS	40	225.133	2.978	0.967	0.298	Student's t	0.5435	78.0	0.588
	Control	40	225.480	2.739	0.965	0.243				
Mean Capillary Intensity (10 × 10 mm)	PCS	40	136.504	5.687	0.887	<0.001	Mann-Whitney U	628		0.099
	Control	40	134.035	3.850	0.986	0.890				
Percentage Capillary Network (Vessel Densities) (10 × 10 mm)	PCS	40	45.637	1.296	0.978	0.632	Mann-Whitney U	630		0.103
	Control	40	44.934	1.827	0.827	<0.001				
Total Area of Ischaemia (10 × 10 mm)	PCS	40	203.400	680.847	0.342	<0 .001	Mann-Whitney U	763		0.541
	Control	40	198.000	683.638	0.331	<0.001				
Percentage Capillary Network (Vessel Densities) (4 × 4 mm)	PCS	40	41.192	1.240	0.986	0.888	Student's t	−0.1327	78.0	0.895
	Control	40	41.147	1.746	0.950	0.073				
Mean Capillary Intensity (4 × 4 mm)	PCS	20	139.245	4.323	0.949	<0.359	Mann-Whitney U	217		0.350
	Control	26	140.512	6.028	0.795	<0.001				
Area of Foveal Avascular Zone (FAZ) (4 × 4 mm)	PCS	40	1917.725	405.880	0.573	<0.001	Mann-Whitney U	715		0.399
	Control	40	1925.050	389.784	0.589	<0.001				
Circularity of Foveal Avascular Zone (FAZ) (4 × 4 mm)	PCS	40	0.897	0.194	0.702	<0.001	Mann-Whitney U	699		0.319
	Control	40	0.869	0.202	0.775	<0 .001				

Post COVID-19 syndrome, *PCS*; Standard Deviation, *SD*; Shapiro Wilk test statistic, *W*; Degrees of Freedom, *df*. Note: H_a $\mu_{Control} \neq \mu_{PCS}$—The two-tailed alternative hypothesis for this study was that there is a significant difference in the measured parameters between the participants with post COVID-19 syndrome (PCS) and healthy control subjects.

3.4. OCT Analysis

3.4.1. Macular RNFL and GCL Thickness

An analysis of SD-OCT-macula images was performed in 39 participants with PCS and 34 controls following exclusion of images of inferior quality. Although increased thickness of the mean outer, inner, and foveal segments of the mRNFL was noted in the PCS cohort compared to the controls, the differences were not statistically significant. Furthermore, a reduction in the thickness of the outer and inner segment of mGCL was observed in the PCS cohort compared to the control cohort, however these findings were not statistically significant. Details provided in Table 5.

Table 5. Analysis of the thickness (microns) of the macular retinal nerve fibre layer (RNFL) and ganglion cell layer (GCL) in the post COVID-19 syndrome (PCS) group (n = 39) and control group (n = 34).

				Shapiro-Wilk Test		Statistical Test	Statistic	df	p < 0.00357
Mean Thickness (Microns)	**Cohort Category**	**Mean**	**SD**	**W**	p				
Mean Outer Segment mRNFL	**PCS**	36.11	5.23	0.930	0.018	Mann-Whitney U	617		0.615
	Control	35.31	4.66	0.978	0.698				
Mean Inner Segment mRNFL	**PCS**	21.71	2.30	0.959	0.166	Student's t	1.1007	71.0	0.275
	Control	21.18	1.78	0.945	0.085				
Foveal (Central) Segment mRNFL	**PCS**	8.67	1.90	0.913	0.005	Mann-Whitney U	654		0.920
	Control	8.65	2.52	0.928	0.027				
Mean Outer Segment mGCL	**PCS**	30.34	3.48	0.973	0.454	Mann-Whitney U	599		0.479
	Control	31.63	5.05	0.706	<0.001				
Mean Inner Segment mGCL	**PCS**	50.42	5.96	0.946	0.058	Mann-Whitney U	645		0.842
	Control	50.95	5.91	0.931	0.033				
Foveal (Central) Segment mGCL	**PCS**	19.85	5.94	0.865	<0.001	Mann-Whitney U	638		0.786
	Control	19.35	5.71	0.965	0.328				

Post COVID-19 syndrome, *PCS*; Macular Retinal Nerve Fibre Layer, *mRNFL*; Macular Ganglion Cell Layer, *mGCL*; Standard Deviation, *SD*; Shapiro Wilk test statistic, *W*; Degrees of Freedom, *df*. Note: $H_a\ \mu_{Control} \neq \mu_{PCS}$—The two-tailed alternative hypothesis for this study was that there is a significant difference in the measured parameters between the participants with post COVID-19 syndrome (PCS) and healthy control subjects.

3.4.2. Neurocognitive Symptoms and Macular RNFL and GCL Thickness

Within the PCS cohort, we evaluated the thickness of the macular RNFL and GCL segments in patients with ongoing neurocognitive symptoms, encompassing cognitive dysfunction i.e., brain fog and headaches (n = 24) compared with PCS participants without these symptoms (n = 15). No statistically significant differences were noted in the thickness of the macular RNFL or GCL segments within these sub-groups, described in Table 6.

Table 6. Analysis of the thickness (microns) of the macular retinal nerve fibre layer (RNFL) and ganglion cell layer (GCL) in patients with post COVID-19 syndrome (PCS) and neurocognitive symptoms (n = 24) compared to post COVID-19 syndrome patients without neurocognitive symptoms (n = 15).

				Shapiro–Wilk		Statistical Test	Statistic	df	p < 0.00357
Mean Thickness (Microns)	**Neurocognitive Symptoms**	**Mean**	**SD**	**W**	p				
Mean Outer mRNFL	Y	36.15	5.62	0.861	0.003	Mann-Whitney U	167		0.707
	N	36.05	4.74	0.939	0.367				
Mean Inner mRNFL	Y	21.79	2.35	0.917	0.050	Student's t	0.2722	37.0	0.787
	N	21.58	2.28	0.957	0.647				
Foveal (Central) mRNFL	Y	8.88	1.73	0.906	0.029	Mann-Whitney U	157		0.510
	N	8.33	2.16	0.926	0.234				
Mean Outer mGCL	Y	30.61	3.10	0.929	0.091	Student's t	0.6193	37.0	0.540
	N	29.90	4.09	0.954	0.581				
Mean Inner mGCL	Y	50.80	5.47	0.935	0.127	Student's t	0.5056	37.0	0.616
	N	49.80	6.83	0.934	0.315				
Foveal (Central) mGCL	Y	20.54	5.16	0.870	0.005	Mann-Whitney U	129		0.140
	N	18.73	7.06	0.823	0.007				

Macular Retinal Nerve Fibre Layer, *mRNFL*; Macular Ganglion Cell Layer, *mGCL*; Neurocognitive symptoms present, *Y*; Neurocognitive symptoms not present, *N*; Standard Deviation, *SD*; Shapiro Wilk test statistic, *W*; Degrees of Freedom, *df*. Note: $H_a\ \mu_Y \neq \mu_N$—The two-tailed alternative hypothesis was that there is a significant difference in the measured parameters between the participants with post COVID-19 syndrome (PCS) with neurocognitive symptoms and those without neurocognitive symptoms.

3.5. Linear Regression

Linear regression analysis was undertaken to evaluate the effect of age, gender, and length of time since COVID-19 infection on mean capillary intensity (10 × 10 OCT-A image), a key measure of retinal microvasculature examined in our study. No significant relationships were observed with respect to the above independent variables on the mean capillary intensity (age, $p = 0.922$; gender, $p = 0.966$; length since initial infection, $p = 0.332$). This has been demonstrated below in Table 7.

Table 7. Linear regression analysis of independent variables (age, gender, length since initial infection) and dependent variable, mean capillary intensity, measured on 10 × 10 mm optical coherence tomography-angiography (OCT-A) Image.

Model	R		R^2	
1	0.164		0.0288	
Normality Test (Shapiro–Wilk)	**Statistic (W)**		*p*	
	0.928		0.014	
Independent Variables	**Estimate**	**SE**	**t**	*p*
Intercept [a]	139.061	5.3596	25.7593	<0 .001
Age	0.00954	0.0967	0.0986	0.922
Gender:				
F—M	0.09885	2.3094	0.0428	0.966
Length Since Initial COVID-19 Infection	−0.13636	0.1387	−0.9832	0.332

Correlation between the independent and the dependent variable, *R*; The proportion of variance in the dependent variable that can be explained by the independent variable, R^2; Normality of the dependent variable, *Shapiro–Wilk Statistic (W) and p value*; Reference Level, *Intercept* [a]; Standard Error, *SE*; The number of standard errors the estimated coefficient is away from the hypothesised value, *t*; Determines if there is a significant relationship between an independent and dependent variable in the model, *p*.

4. Discussion

Post-COVID-19 syndrome has been linked to a persistent impairment of the systemic microvasculature. This study explored the retinal microvasculature network as a potential window into the pathophysiology of post-COVID-19, considering the known homology of the retinal vascular bed with systemic diseases.

We used custom-designed image analysis algorithms to assess a range of features using the most modern retinal imaging techniques including OCT and narrow and wide-field OCT-A imaging, in patients with and without post-COVID-19 syndrome. Our study shows that there were no significant differences found in any of the comprehensive measures used between our populations of people with and without this syndrome. There were no defects or abnormalities detected in the OCT of retinal layers or OCT-A of retinal vasculature.

Most studies discussing OCT-A in relation to COVID-19 primarily concentrate on patients who were hospitalised and/or treated for COVID-19 during the early stages of recovery from the infection as opposed to those experiencing post-COVID-19 syndrome (PCS). As a result, any inferences about persistent changes in the retina may be limited in their generalisability. A prominent finding in these studies has been the reduction in the central vessel density (VD) in patients with COVID-19 infection as compared to control patients [62–75].

Further studies entailing a slightly longer length of time between initial infection and imaging comprise of a duration of 3 month [71], up to 4 months [72], 6 months [73,74], with the longest follow-up being at 8 months post-infection [75]. The key findings were of reduced VDs in the superficial [71–75], deep [71–73,75], and radial peripapillary plexi [73], GCL thinning [74], parafoveal RNFL thinning [73,74] and FAZ enlargement [72,74,75].

In contrast to the studies above, the patients in our study cohort had a longer length of time since initial SARS-CoV-2 infection (15.2 ± 6.9 months (range 3–32)). Our analyses did not demonstrate any significant reduction in vessel densities or intensities in the SCP nor any differences in the area of the FAZ. The results of our investigation are encouraging, therefore, as they may indicate that any alterations in the retinal vasculature of individuals with a recent COVID-19 disease may not necessarily be long-term.

Alternatively, our negative findings may be due to our particular cohort—Other studies have been performed on patients with a moderate-severe SARS-CoV-2 infection [62–65,67,70,71,73–75] whilst in our study almost all participants had experienced a milder form of the disease which did not require hospitalisation and/or outpatient treatment.

A similar narrative to OCT-A research can be seen when assessing literature on structural OCT changes. Mavi et al. found statistically significant changes with higher central foveal, mean outer nuclear layer, and mean peri-papillary RNFL thickness in the post COVID-19 patients compared to normal [85]. Ugurlu et al. also examined SD-OCTs of 129 patients with COVID-19 infection 29 to 45 days following a positive PCR test, with findings of a statistically thinner macular RNFL and GCL layer in COVID-19 patients with neurological symptoms during the acute infection compared to those with non-neurological symptoms, no symptoms/pauci-symptoms, and control subject [86]. Interestingly, Taskiran-Sag et al. studied 40 patients 113 $\pm$ 62 (SD) days after recovering from acute COVID-19 infection. Within the COVID-19 recovered cohort, significantly reduced GCL thickness were found in patients with symptoms of cognitive disturbance and headaches [87]. More recently, Kanra et al. examined 34 eyes of 20 patients with neurological symptoms 4.3 $\pm$ 2.7 (range, 1–12) months following the initial COVID-29 infection. Thinning of the macular RFNL, the GCL, and inner plexiform layer (IPL) segments were noted [88]. In our study, analysis of the SD-OCT of 39 PCS and 34 control subjects did not demonstrate any differences in the mRNFL and mGCL between PCS and control cohorts. Furthermore, no distinguishable results were noted within the PCS cohort in patients with and without neurocognitive symptoms. It is therefore possible that structural changes noted in the retinal layers post-COVID-19 infection may not persist long-term. A summary of studies pertaining to OCT-A and OCT studies in COVID-19 can be found in Appendix B, Tables A5 and A6.

We opted to analyse OCT-A images of the superficial retinal plexus, noting the clinical importance of this region and improved quality of imaging compared to intermediate or deep plexi as well as its predominance as a focus of analysis in other publications. However, Schlick et al. recently explored the retinal microvasculature of patients with post-COVID-19 syndrome using OCT-Angiography and found significant changes in the intermediate capillary plexus (ICP), as compared to the controls [89]. Future studies may benefit from attention to improved imaging of intermediate and deeper plexi to assess if this effect is seen longer term.

Overall, no long-term structural changes were noted in the retinal microvasculature pertaining to the SCP, and the thickness of the RNFL and GCL layers within our PCS cohort. Our PCS cohort comprised of patients with a predominantly mild initial COVID-19 illness, no underlying conditions known to affect the retinal microvasculature such as diabetes or hypertension, with an extended length of time since the initial infection. Therefore, our results should be interpreted with this in mind when comparing to other studies. Further examination with improved imaging of the intermediate and deep plexi and the choriocapillaris, as well as recruitment of patients with varying severity of initial disease, could further enhance our understanding of the long-term implications of COVID-19 on the retinal microvasculature in different patient groups.

SARS-CoV-2 enters cells by binding to angiotensin-converting enzyme 2 (ACE2), downregulating its activity and causing a disruption in the signalling effects of Angiotensin II and its receptor (Angiotensin II type 1 receptor, AT_1). This leads to an accumulation of Angiotensin II, resulting in vasoconstriction, inflammation, cellular differentiation and growth, endothelial dysfunction, formation of reactive oxidative species (ROS), and microvascular thrombosis [90]. ACE2 is expressed within multiple retinal tissues, including the vascular endothelium, making it susceptible to Ag II/AT1 signalling effects and the resulting activation of the caspase 1/inflammasome pathway, responsible for the release of inflammatory cytokines [91]. Additionally, both dysregulation of the renin-angiotensin-aldosterone system (RAAS) and inflammation have been elucidated in the aetiology of post-COVID-19 syndrome [5]. No protracted alterations in the retinal microvasculature and structural layer thickness were observed in our study group following COVID-19 infection. We postulate that retinal microvascular alterations noted in the acute period post-SARS-CoV-2 infection might be predominantly ascribed to pro-inflammatory mechanisms linked to Angiotensin II. During a profound COVID-19 infection, compounded by additional co-morbidities, this response may be amplified. This is supported by studies demonstrating increased severity and mortality of COVID-19 in patients with diabetes, hypertension, and cardiovascular disease [92]. Additionally, COVID-19 disease severity has been found to affect the presence of retinopathy as a higher incidence was reported in moderate to severe disease [51,91,93]. Therefore, it is possible that, in the presence of a diminished disease severity, no underlying comorbidities, and as the duration since the initial infection extends, these inflammatory changes subside, without the occurrence of long-term ischaemic damage.

Our study was limited predominantly by challenges in imaging and defining study populations. Due to the immobility of the ophthalmic imaging apparatus, participant recruitment faced limitations as individuals were required to visit our facility instead of us conducting assessments at their respective respiratory clinics. Challenges were also encountered in image acquisition of the participants, especially with PCS due to the debilitating symptoms encompassing the disease, including dyspnoea, fatigue, and dry eyes. Movement of the head up and down due to dyspnoea posed limitations in maintaining a still stature whilst imaging was undertaken. Dry eyes significantly increased the participants' blink frequency, and despite provision of lubricants, interfered with image acquisition. Fatigue resulted in easy tiring during imaging, reducing the number of repetitions which could be utilised to capture high quality images. These challenges were further compounded by a required imaging software update that led to potential changes in specific scans which were not visible clinically but excluded them from image analysis. In order to address these expected challenges in imaging, we dedicated some time at the beginning of the study to explore different imaging techniques with the OCT-A camera and developed a protocol that was standardised yet gave the best possible results for both cohorts of patients. We incorporated software routines for example to negate the effect of artefacts including vitreous aberrations that could obscure some regions of wide field OCTA imaging. Therefore, despite imaging challenges, we are confident that the images that were ultimately accepted into analysis in our study were all of adequate quality in both groups.

We defined patients as clearly as possible as those with a clinical diagnosis of post-COVID-19 syndrome, who were all recruited from specialist clinics designed to treat these patients. However, we recognised the heterogeneity of patients within this group. In addition, due to practical restrictions, our control sample included patients who have had COVID-19 infection but did not develop post COVID-19 symptoms. Ideally, we would have benefitted from comparing against patients who had not had a previous COVID-19 infection. Furthermore, in an ideal situation we would have compared the scans of patients with post-COVID-19 syndrome to scans they had prior to their infection. Again, this was not possible for practical reasons within the confines of this study.

This study is distinguishable within the literature due to a multitude of reasons. To our knowledge, it is the longest study examining the effects of SARS-CoV-2 on retinal tissues in patients up to 32 months following initial acute infection, particularly those who continue to suffer with symptoms of post-COVID-19 syndrome. Furthermore, our work is distinct as our in-house specially designed image analysis software, OCTAVIA, provides a comprehensive analysis of the data obtained from OCT-A imaging, with evidence for its reliability and validity. Parameters beyond the vessel densities are obtained, providing in-depth detail on the microvasculature of the retina, including vessel intensities, presence of ischaemia, and FAZ area and circularity measures.

5. Conclusions

In this study, we have demonstrated that there were no statistically significant differences in the retinal microvasculature of patients with post-COVID-19 syndrome compared to healthy cohorts. Furthermore, no significant structural differences were observed in the macular retinal nerve fibre layer (RNFL) and ganglion cell layer (GCL) of the study participants. The findings of this study indicate that despite an extensive investigation in patients with post-COVID-19 syndrome, there were no long-term structural signs of damage after detailed analysis of this accessible microvasculature bed that is known to have homology with systemic vasculature. This may serve as some positive reassurance for patients experiencing ongoing symptoms of PCS.

Author Contributions: Conceptualization, T.A., O.M. and M.N.; funding acquisition, T.A.; data curation, M.N., O.M. and I.D.; investigation, M.N., O.M. and I.D.; software design, T.A.; validation, M.N., O.M., I.D. and T.A.; formal analysis, M.N., O.M. and I.D.; methodology, M.N., O.M., I.D. and T.A.; project administration, M.N., O.M., I.D. and T.A.; resources, T.A.; supervision, T.A.; visualization, M.N., O.M., I.D. and T.A.; writing—original draft preparation, M.N.; writing—review and editing, T.A. All authors have read and agreed to the published version of the manuscript.

Funding: The position of the most recent research fellow (M.N.) was funded by the University of Manchester. The position of the previous research fellow (O.M.G.) was partly funded by Canon Medical EU, Zilverstraat, The Netherlands. The funders had no role in the design of the study; in the collection, analyses, or interpretation of data; in the writing of the manuscript; or in the decision to publish the results.

Institutional Review Board Statement: The study was conducted in accordance with the Declaration of Helsinki and approved by the Ethics Committees Health Research Authority (HRA) and Health and Care Research Wales (HCRW) along with the Office for Research Ethics Committees Northern Ireland (ORECNI). The IRAS project ID is 291767 and the reference number is 21/NI/0045.

Informed Consent Statement: Informed consent was obtained from all subjects involved in the study. Written informed consent has been obtained from the patient(s) to publish this paper.

Data Availability Statement: All data generated or analysed during this study are included in this published article. The datasets generated during and/or analysed during the current study are available from the corresponding author on reasonable request.

Acknowledgments: We would like to express our gratitude to the participants both with post-COVID-19 syndrome and without for their invaluable contribution to this study. We would also like to extend our appreciation to the respiratory team at the Manchester Royal Infirmary, UK, including Respiratory and Acute Medicine consultant John Bright and specialist respiratory nurses for their support in the recruitment of patients with post-COVID-19 syndrome.

Conflicts of Interest: Dr Maha Noor, Dr Orlaith McGrath, and Dr Ines Drira declare that they have no personal, financial, commercial, or academic conflict of interest. Professor Tariq M. Aslam has received funding and educational grants from Bayer, Novartis, Roche, Allergan, Laboratoires, Thea, Oraya, and Bausch and Lomb, Topcon, Heidelberg, Canon. The funders had no role in the design of the study; in the collection, analyses, or interpretation of data; in the writing of the manuscript; or in the decision to publish the results.

Appendix A

Appendix A.1. Reliability and Validity of the Software

To verify the utility of the OCTAVIA software, we appraised the reliability by assessing its automated large vessel intensity measurement in the 10 × 10 mm images of 20 participants on two separate occasions by the same examiner (T.A., a medical retina consultant). This included the manual pre-processing steps involving removal of artefacts. Data is shown below in Table A1.

A Pearson correlation coefficient between the two sets of measurements of 0.9963 was found, demonstrating excellent reliability. A Bland-Altman plot to graphically examine the repeatability was also plotted, shown in Figure A1, which also indicated high reliability.

Using jamovi, additional reliability statistics were obtained which were also found to be supportive of our results (Cronbach's α score of 0.998 and McDonalds's ω score of 0.997).

Table A1. Measurement of large vessel intensities in the 10 × 10 mm OCT-A images of 20 participants, calculated twice to evaluate reliability.

	OCTAVIA1	OCTAVIA2	Mean	Difference
1	200.2152	200.8782	200.546715	−0.66295
2	214.7086	214.2896	214.4990603	0.418981
3	214.8738	215.4842	215.1789925	−0.61038
4	226.5071	226.0975	226.3023362	0.409616
5	224.3287	225.3507	224.8396983	−1.02203
6	213.7813	213.7845	213.7829488	−0.0032
7	223.4939	223.8539	223.6739096	−0.36005
8	216.0177	216.0692	216.0434456	−0.05153
9	215.3221	215.9063	215.6141776	−0.58424
10	217.7659	218.0761	217.9210217	−0.31021
11	212.5384	213.0391	212.7887705	−0.50068
12	216.6652	216.777	216.7211303	−0.11176
13	216.6811	216.6882	216.6846339	−0.00713
14	213.7026	213.9167	213.8096195	−0.2141
15	218.6594	217.7792	218.2192833	0.880265
16	220.9376	221.0125	220.9750348	−0.07491
17	217.0157	216.9396	216.9776428	0.076176
18	217.3771	217.8395	217.608298	−0.46247
19	213.694	213.1311	213.4125719	0.562955
20	218.2591	218.2816	218.2703393	−0.02244

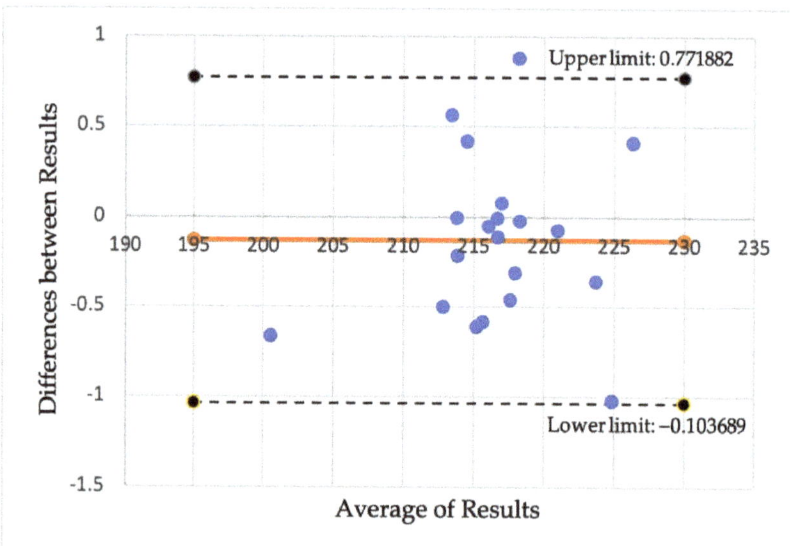

Figure A1. A Bland-Altmann plot demonstrating the difference between the results measuring large vessel intensities using the OCTAVIA software on two occasions by the same assessor.

To measure the validity of the OCTAVIA software, Image J was utilised to manually measure the area of the FAZ, comparing this with the measure outputted from the automated OCTAVIA software in the same 20 participants. Data is shown below in Table A2. This was demonstrated in a scatterplot, Figure A2, illustrating that the line of best fit for the relationship between the manual and automatic measurements was not significantly different from the 1:1 line (where a perfect match of both measurements would lie).

Table A2. Measurement of the area of the foveal avascular zone (FAZ) using the OCTAVIA software vs manual assessment utilising image J.

	OCTAVIA	Image J
1	0.005788	0.005673325
2	0.0212	0.0227944
3	0.014488	0.01531618
4	0.026428	0.027984508
5	0.017475	0.01882717
6	0.006208	0.00685339
7	0.019427	0.02044105
8	0.007463	0.00819081
9	0.008405	0.00898651
10	0.017428	0.0184428
11	0.007847	0.0070062
12	0.007238	0.007172576
13	0.022608	0.02153121
14	0.004226	0.004765234
15	0.00386	0.00383916
16	0.009413	0.009449548
17	0.004861	0.004664085
18	0.011756	0.010274797
19	0.005088	0.005610388
20	0.011491	0.011762485

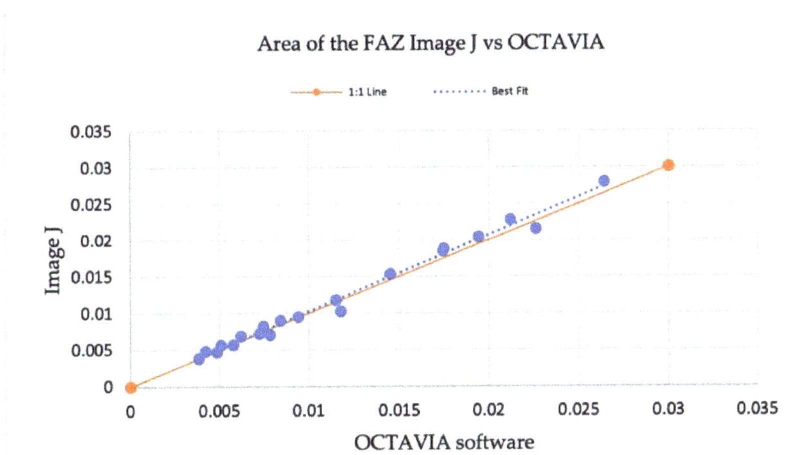

Figure A2. A scatterplot demonstrating the measurements of the area of the foveal avascular zone (FAZ) using Image J and the OCTAVIA software, outlining the relationship between a line of best fit compared to the 1:1 line.

An additional evaluation of validity was performed by comparing the measurement of mean large vessel intensities in 10 × 10 mm OCT-A images outputted by the OCTAVIA software with manual measurements made using large vessel data points in MATLAB®® 2021. In the 10 × 10 mm OCT-A images of 20 study participants, three large vessels were selected, and data corresponding with vessel intensity was obtained by selecting two points at the centre of the vessel and two near the margin in each vessel. The data was noted in a Microsoft® Excel® spreadsheet and a total average was calculated to compare against

OCTAVIA measurements, shown in Table A3. The results are also demonstrated in a scatter plot, Figure A3, illustrating no significant differences between the line of best fit and 1:1 line.

Table A3. Data points corresponding with vessel intensities measured in three large vessels in the 10 × 10 mm OCT-A images of 20 participants using MATLAB, compared with measurements from the OCTAVIA algorithm.

	Large Vessel 1				Large Vessel 2				Large Vessel 3				Manual Average	OCTAVIA
	Centre	Centre	Margin	Margin	Centre	Centre	Margin	Margin	Centre	Centre	Margin	Margin		
1	1	0.929412	0.709804	0.788235	0.92549	1	0.815686	0.792157	0.941176	1	0.85098	0.839216	0.882679667	0.885820538
2	0.913725	1	0.878431	0.768627	0.933333	0.835294	0.898039	0.890196	0.92549	1	0.85098	0.843137	0.894771	0.893043548
3	0.894118	0.945098	0.745098	0.996078	0.933333	0.733333	0.803922	0.913725	0.901961	1	0.784314	0.964706	0.8846405	0.90166909
4	1	0.929412	0.72549	0.819608	0.866667	0.945098	0.8	0.941176	0.901961	1	0.854902	0.85098	0.8862745	0.889066606
5	1	0.933333	0.882353	0.831373	1	0.988235	0.823529	0.843137	0.878431	0.976471	0.803922	0.831372	0.899346333	0.893814466
6	1	0.964706	0.87451	0.752941	0.909804	0.917647	0.870588	0.843137	0.968627	0.929412	0.811765	0.764706	0.883986917	0.883226069
7	0.984314	0.835294	0.819608	0.882353	1	0.988235	0.847059	0.784314	1	0.972549	0.768627	0.784314	0.888888917	0.887244315
8	0.980392	1	0.796078	0.839216	1	0.905882	0.870588	0.709804	0.976471	0.811765	0.85098	0.741176	0.873529333	0.875732357
9	0.92549	1	0.764706	0.839216	1	0.886275	0.7333333	0.827451	0.972549	1	0.827451	0.752941	0.877451025	0.864223212
10	0.933333	0.898039	0.831373	0.879588	0.992157	0.964706	0.858824	0.803922	0.976471	0.972549	0.815686	0.752941	0.88996575	0.885640208
11	0.984314	0.960784	0.815686	0.776471	0.85098	1	0.831373	0.85098	0.905882	0.988235	0.713725	0.768627	0.870588083	0.868688789
12	1	0.945098	0.823529	0.827451	0.996078	0.992157	0.745098	0.784314	0.921569	0.992157	0.784314	0.847059	0.888235333	0.871880492
13	0.917647	0.968627	0.894118	0.780392	1	0.905882	0.898039	0.760784	0.811765	1	0.866667	0.72549	0.877450917	0.874451774
14	0.905882	0.886275	0.878431	0.870588	1	1	0.8	0.780392	0.905882	1	0.807843	0.717647	0.879411667	0.874872568
15	1	0.976471	0.878431	0.835294	0.976471	0.882353	0.760784	0.87451	0.968627	0.968627	0.796078	0.807843	0.89379075	0.893920821
16	0.87451	0.937255	0.835294	0.709804	0.937255	0.929412	0.847059	0.862745	1	0.972549	0.886275	0.784314	0.881372667	0.882249605
17	0.980392	1	0.878431	0.839216	1	0.94902	0.827451	0.752941	0.815686	0.894118	0.894118	0.823529	0.8879085	0.8875187
18	1	0.913725	0.85098	0.843137	0.968627	0.776471	0.87451	0.862745	0.956863	0.917647	0.737255	0.72549	0.868954167	0.862268732
19	1	1	0.815686	0.94902	0.996078	0.92549	0.764706	0.764706	1	0.866667	0.796078	0.917647	0.899673167	0.90460477
20	0.996078	1	0.890196	0.792157	0.882353	1	0.835294	0.745098	0.992157	1	0.768627	0.831373	0.894444417	0.897619232

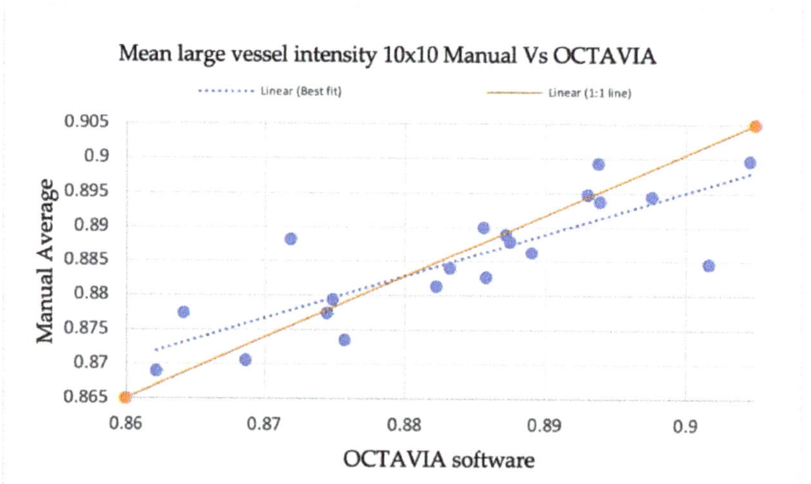

Figure A3. A scatterplot demonstrating manual measurements of large vessel intensities in 10 × 10 OCT-A images compared with OCTAVIA software, outlining the relationship between the line of best fit and 1:1 line.

Appendix A.2. Sample Size Calculation

The sample size for this study was calculated based on data collected in a previous study on diabetic retinopathy and OCT-Angiography [79]. The method was as follows:

Typically, we set:

$$
\begin{aligned}
\alpha &= 0.05 \text{ so } Z_{\alpha/2} = 1.960 \\
\beta &= 0.20 \text{ so } Z_{\beta/2} = 0.8416 \\
&\text{which gives :} \\
n_g &\doteq 2(z_{\alpha/2} + z_\beta)^2 \left(\frac{\sigma}{\mu_1 - \mu_2}\right)^2 \\
&\doteq 2 \times (1.960 + 0.8416)^2 \left(\frac{\sigma}{\mu_1 - \mu_2}\right)^2 \\
&\doteq 2 \times 7.849 \left(\frac{\sigma}{\mu_1 - \mu_2}\right)^2
\end{aligned}
$$

σ is the population standard deviation. $\mu_1 - \mu_2$ is the smallest difference with scientific significance. The sample size for this study is based on the sample size for the key metric, mean capillary intensity. The mean capillary intensity of a normal patient from the study was 98.69, shown in Table A4, with σ = 6.23. The difference between mean capillary intensities in Normal vs Diabetic patients with no retinopathy suggested a clinically important difference in this metric, $\mu_1 - \mu_2 = 98.69 - 94.22 = 4.47$. Therefore: $2 \times 7.849 \left(\frac{6.23}{4.47}\right)^2 = 30.49 \approx 31$. A minimum of 31 in each group was needed, hence, we aimed to have 40 participants in each of our study cohorts (PCS and controls).

Table A4. Table excerpt from Aslam et al.'s study of optical coherence tomography angiography in diabetic retinopathy [79] demonstrating the strategy utilised in the calculation of sample size for this study.

Variable	Statistic	Normal, N (*n* = 49)	Diabetic No Retinopathy, DnR (*n* = 50)	Diabetic with Retinopathy, DR (*n* = 53)
Sex		Male 32, Female 17	Male 36, Female 14	Male 43, Female 10
Age	Mean (SD) Median	57.14 (13.56) 57.0	61.06 (12.77) 61.5	58.38 (13.06) 60
Best Corrected Visual Acuity (BCVA)	Mean (SD) Median	85 (19.66) 95	96.36 (7.39) 95	86.17 (13.3 90
Mean Vessel Intensity	Mean (SD) Median	180.65 (6.43) 182.0	181.38 (6.04) 183.5	179.28 (7.45) 181.0
Mean Capillary Intensity	Mean (SD) Median	98.69 (6.23) 99	94.22 (5.41) 95	93.47 (6.03) 94

Standard Deviation, *SD*.

Appendix B

A summary of studies examining OCT-A and SD-OCT images in patients following COVID-19 Infection can be found in Tables A5 and A6.

Table A5. A summary of studies pertaining to OCT-Angiography and SD-OCT macula findings in patients following COVID-19 infection.

Study	Time Following Initial SARS-CoV-2 Infection	Severity of Infection	Size (Case, Control)	Key Findings (Statistically Significant)
OCT-Angiography: Studies Demonstrating Reduction in Vessel Densities in Patients with a History of COVID-19 Infection				
Zapata et al., 2022 [62]	≤90 days (3 months)	Mild, Moderate, Severe	24, 24, 21, 27	Reduced VDs in the SCP of patients with moderate and severe disease, compared to mild disease and control subjects.
Turker et al., 2021 [63]	7 days following hospital discharge	Moderate	27, 27	Reduced VD in SCP and DCP regions. No difference in area of the FAZ.
Abrishami et al., 2021 [64]	≥14 days following recovery	Moderate, Severe	31, 23	Reduced VD in the SCP and DCP. No difference in area of the FAZ.
Gonzalez-Zamora et al., 2021 [65]	14 days following hospital discharge	Severe	25, 25	Reduced VD in SCP and DCP regions. Enlargement of FAZ area.

Table A5. *Cont.*

Study	Time Following Initial SARS-CoV-2 Infection	Severity of Infection	Size (Case, Control)	Key Findings (Statistically Significant)
Hazar et al., 2021 [66]	≈30 days (1 month) following hospital discharge	Mild, Moderate	50, 55	Reduced VD in SCP and DCP regions. No difference in area of the FAZ.
Guemes Villahoz et al., 2021 [67]	88 days following initial diagnosis	Moderate, Severe	66 (19, 47), 29	Reduction in VDs in the SCP and reduced perfusion density in the fovea. No difference in the area of the FAZ.
Rodman et al., 2021 [68]	-	Mild	18, 18	Reduced VDs in regions of the SCP.
Yilmaz Cebi et al., 2022 [69]	67–86 days	Mild, Moderate	52, 42	Reduced VDs in SCP and DCP.
Cetinkaya et al., 2022 [70]	≈30 days following hospital discharge	Moderate	45, 45	Reduced VDs in SCP, DCP, and CC.
Abrishami et al., 2022 [71]	7 days, 1 month, 3 months	Moderate, Severe	18 (follow-up study)	Reduced VDs in the SCP and DCP, no difference in the area of the FAZ.
Nageeb Louz et al., 2022 [72]	30–120 days (1–4 months)	Mild, Moderate, Severe	45, 45	Reduced VDs in the SCP, DCP, and enlargement of FAZ.
Cennamo et al., 2021 [73]	180 days (6 months)	Moderate	40, 40	Reduced VDs in SCP, RPC, DCP. RNFL thickness reduced.
Bilbao-Malave V et al., 2021 [74]	6 months from hospital discharge	Severe	17 (follow-up study)	Reduced VDs in SCP and DCP, enlargement of FAZ, Thinner GCL and RNFL.
Banderas Garcia S et al., 2022 [75]	8 months after initial infection	Mild, Moderate, Severe	75, 19	Reduced VDs in the SCP and DCP of patients with moderate and severe disease, compared to mild disease and control subjects. Enlargement of FAZ in patients with more severe disease.
OCT-Angiography: Studies demonstrating no difference in vessel densities in patients with a history of COVID-19 infection				
Savastano et al., 2021 [93]	1 month following hospital discharge	Moderate	70, 22	No differences in VD and VP in the SCP and DCP.
Szkodny et al., 2021 [94]	1–4 months following infection	Mild, Moderate	156, 98	No differences in the VDs of the SVP, area of the FAZ, macular RNFL thickness, and central macular thickness.
Kal M et al., 2022 [95]	8 weeks following hospital discharge	Severe	63, 45	No difference in the VDs in SCP or DCP between the two groups. Reduced VD in the foveal CC. Enlargement of area of FAZ.
Chiosi F et al., 2022 [96]	1 month following recovery from infection	Mild, Moderate, Severe	142, 60	No difference in the VDs in SCP. Reduced VD in the DCP and CC.
OCT-Angiography: Studies demonstrating increase in vessel densities in patients with a history of COVID-19 infection				
Naderi Beni A et al., 2022 [58]	40–95 days following initial infection	Moderate	51, 37	Increased VDs in the SCP and DCP. Increased peripapillary RNFL thickness.
OCT macula structural retinal findings in patients with a history of COVID-19 infection				
Sim et al., 2021 [52]	16.1 ± 3.6 days	Mild	108, 0	Microhaemorrhages, retinal vascular tortuosity, cotton wool spots, hyper-reflective plaques in the GCL-IPL.
Marinho et al., 2020 [54]	11–33 days	Mild to Moderate	12	Hyper-reflective plaques in the GCL-IPL, cotton wool spots, microhaemorrhages.
Burgos-Blasco et al., 2022 [57]	4 weeks following recovery	Mild, Moderate, Severe	90, 70	Increased peri papillary RNFL and macular GCL thickness (more significant in patients with anosmia and ageusia) and decreased macular RNFL thickness.
Oren et al., 2021 [59]	14–30 days	Mild	35, 25	Increased central macular thickness, decreased GCL and INL thickness.

Table A5. *Cont.*

Study	Time Following Initial SARS-CoV-2 Infection	Severity of Infection	Size (Case, Control)	Key Findings (Statistically Significant)
Mavi et al., 2022 [85]	2–8 weeks	Moderate	63 (30 hospitalised), 59	Increased central foveal, mean outer nuclear layer, mean peri papillary RNFL thickness.
Ugurlu A et al., 2022 [86]	29–45 days	Moderate, Severe	129, 130	No difference between COVID-19 and controls. Thinner macular RNFL and GCL in patients with neurological symptoms within the COVID-19 cohort. Reduced VD in SCP, DCP, RPCP, enlargement of FAZ area, reduction of FAZ circularity in symptomatic COVID-10 patients.
Taskiran-Sag et al., 2022 [87]	113 ± 62 days following recovery from infection	Mild, Moderate	40, 40	No difference between GCL thickness between COVID-19 and controls. Thinner macular GCL in patients with neuro-cognitive symptoms.

Choriocapillaris, CC; Cotton wool spots, *CWS*; Deep capillary plexus; *DCP*; Foveal avascular zone, *FAZ*; Ganglion cell layer; *GCL*; Inner nuclear layer, *INL*; Inner plexiform layer, *IPL*; Retinal nerve fibre layer, *RNFL*; Radial peri papillary capillary plexus, *RPCP*; radial peripapillary capillaries, *RPC*; Superficial capillary plexus, *SCP*; Vessel density, *VD*; Vessel perfusion, *VP*.

Table A6. A Summary of studies pertaining to OCT-Angiography and SD-OCT macula findings in patients with post-COVID-19 syndrome.

Study	Time Following Initial SARS-CoV-2 Infection	Severity of Infection	Size (Case, Control)	Key Findings (Statistically Significant)
Kanra AY et al., 2022 [88]	4.3 ± 2.7 months (1–12)	Mild to Moderate	20, 23	Thinning in segments of the macular RNFL, GCL, and IPL.
Schlick et al., 2022 [89]	231 ± 111 days (7.59 ± 3.65 months)	-	173, 28	Reduced VDs in ICP, no difference in SVP, DCP. Females with PCS had lower VDs in SVP than males. PCS participants with CF had lower VDs in SVP than those without.

Ganglion cell layer, *GCL*; Inner Plexiform Layer, *IPL*; Intermediate capillary plexus, *ICP*; Post COVID-19 syndrome, *PCS*; Retinal nerve fibre layer, *RNFL*; Vessel density, *VD*.

References

1. Wiersinga, W.J.; Rhodes, A.; Cheng, A.C.; Peacock, S.J.; Prescott, H.C. Pathophysiology, Transmission, Diagnosis, and Treatment of Coronavirus Disease 2019 (COVID-19): A Review. *JAMA* **2020**, *324*, 782–793. [CrossRef] [PubMed]
2. National Institute of Health and Care Excellence (NICE). *COVID-19 Rapid Guideline: Managing the Long-Term Effects of COVID-19*; NICE: Sutton-in-Ashfield, UK, 2021. Available online: https://www.nice.org.uk/guidance/ng188 (accessed on 3 March 2023).
3. Soriano, J.B.; Murthy, S.; Marshall, J.C.; Relan, P.; Diaz, J.V. A clinical case definition of post-COVID-19 condition by a Delphi consensus. *Lancet Infect. Dis.* **2022**, *22*, e102–e107. [CrossRef] [PubMed]
4. Nalbandian, A.; Sehgal, K.; Gupta, A.; Madhavan, M. Post-acute COVID-19 syndrome. *Nat. Med.* **2021**, *27*, 601–615. [CrossRef] [PubMed]
5. Sherif, Z.A.; Gomez, C.R.; Connors, T.J.; Henrich, T.J.; Reeves, W.B. Pathogenic mechanisms of post-acute sequelae of SARS-CoV-2 infection (PASC). *eLife* **2023**, *12*, e86002. [CrossRef]
6. Ahamed, J.; Laurence, J. Long COVID endotheliopathy: Hypothesized mechanisms and potential therapeutic approaches. *J. Clin. Investig.* **2022**, *132*, e161167. [CrossRef] [PubMed]
7. Al-Ramadan, A.; Rabab'h, O.; Shah, J.; Gharaibeh, A. Acute and Post-Acute Neurological Complications of COVID-19. *Neurol. Int.* **2021**, *13*, 102–119. [CrossRef]
8. Amenta, E.M.; Spallone, A.; Rodriguez-Barradas, M.C.; El Sahly, H.M.; Atmar, R.L.; A Kulkarni, P. Postacute COVID-19: An Overview and Approach to Classification. *Open Forum Infect. Dis.* **2020**, *7*, ofaa509. [CrossRef]
9. Camargo-Martínez, W.; Lozada-Martínez, I.; Escobar-Collazos, A.; Navarro-Coronado, A.; Moscote-Salazar, L.; Pacheco-Hernández, A.; Janjua, T.; Bosque-Varela, P. Post-COVID 19 neurological syndrome: Implications for sequelae's treatment. *J. Clin. Neurosci.* **2021**, *88*, 219–225. [CrossRef]
10. Choutka, J.; Jansari, V.; Hornig, M.; Iwasaki, A. Unexplained post-acute infection syndromes. *Nat. Med.* **2022**, *28*, 911–923. [CrossRef]

11. Carod-Artal, F.J. Post-COVID-19 syndrome: Epidemiology, diagnostic criteria and pathogenic mechanisms involved. *Rev. Nuerol.* **2021**, *72*, 384–396.
12. Cortés-Telles, A.; López-Romero, S.; Figueroa-Hurtado, E.; Pou-Aguilar, Y.N.; Wong, A.W.; Milne, K.M.; Ryerson, C.J.; Guenette, J.A. Pulmonary function and functional capacity in COVID-19 survivors with persistent dyspnoea. *Respir. Physiol. Neurobiol.* **2021**, *288*, 103644. [CrossRef] [PubMed]
13. Doykov, I.; Hällqvist, J.; Gilmour, K.C.; Grandjean, L.; Mills, K.; Heywood, W.E. 'The long tail of COVID-19'—The detection of a prolonged inflammatory response after a SARS-CoV-2 infection in asymptomatic and mildly affected patients. *F1000 Res.* **2021**, *9*, 1349. [CrossRef] [PubMed]
14. Gottschalk, C.G.; Peterson, D.; Armstrong, J.; Knox, K.; Roy, A. Potential molecular mechanisms of chronic fatigue in long haul COVID and other viral diseases. *Infect. Agent Cancer* **2023**, *18*, 7. [CrossRef]
15. Fernández-De-Las-Peñas, C.; Palacios-Ceña, D.; Gómez-Mayordomo, V.; Cuadrado, M.L.; Florencio, L.L. Defining Post-COVID Symptoms (Post-Acute COVID, Long COVID, Persistent Post-COVID): An Integrative Classification. *Int. J. Environ. Res. Public Health* **2021**, *18*, 2621. [CrossRef] [PubMed]
16. Andrade, B.S.; Siqueira, S.; Soares, W.R.d.A.; Rangel, F.d.S.; Santos, N.O.; Freitas, A.d.S.; da Silveira, P.R.; Tiwari, S.; Alzahrani, K.J.; Góes-Neto, A.; et al. Long-COVID and Post-COVID Health Complications: An Up-to-Date Review on Clinical Conditions and Their Possible Molecular Mechanisms. *Viruses* **2021**, *13*, 700. [CrossRef] [PubMed]
17. Lee, J.C.; Nallani, R.; Cass, L.; Bhalla, V.; Chiu, A.G.; Villwock, J.A. A Systematic Review of the Neuropathologic Findings of Post-Viral Olfactory Dysfunction: Implications and Novel Insight for the COVID-19 Pandemic. *Am. J. Rhinol. Allergy* **2020**, *35*, 323–333. [CrossRef]
18. Mantovani, A.; Morrone, M.C.; Patrono, C.; Santoro, M.G.; Schiaffino, S.; Remuzzi, G.; Bussolati, G.; Cappuccinelli, P. Long Covid: Where we stand and challenges ahead. *Cell Death Differ.* **2022**, *29*, 1891–1900. [CrossRef]
19. McFarland, A.J.; Yousuf, M.S.; Shiers, S.; Price, T.J. Neurobiology of SARS-CoV-2 interactions with the peripheral nervous system: Implications for COVID-19 and pain. *Pain Rep.* **2021**, *6*, e885. Available online: https://journals.lww.com/painrpts/Fulltext/2021/01000/Neurobiology_of_SARS_CoV_2_interactions_with_the.1.aspx (accessed on 15 March 2023). [CrossRef]
20. Oronsky, B.; Larson, C.; Hammond, T.C.; Oronsky, A.; Kesari, S.; Lybeck, M.; Reid, T.R. A Review of Persistent Post-COVID Syndrome (PPCS). *Clin. Rev. Allergy Immunol.* **2023**, *64*, 66–74. [CrossRef]
21. Song, W.-J.; Hui, C.K.M.; Hull, J.H.; Birring, S.S.; McGarvey, L.; Mazzone, S.B.; Chung, K.F. Confronting COVID-19-associated cough and the post-COVID syndrome: Role of viral neurotropism, neuroinflammation, and neuroimmune responses. *Lancet Respir. Med.* **2021**, *9*, 533–544. [CrossRef]
22. Toniolo, S.; Scarioni, M.; Di Lorenzo, F.; Hort, J.; Georges, J.; Tomic, S.; Nobili, F.; Frederiksen, K.S.; the Management Group of the EAN Dementia and Cognitive Disorders Scientific Panel. Dementia and COVID-19, a Bidirectional Liaison: Risk Factors, Biomarkers, and Optimal Health Care. *J. Alzheimer's Dis.* **2021**, *82*, 883–898. [CrossRef] [PubMed]
23. Touyz, R.M.; Boyd, M.O.; Guzik, T.; Padmanabhan, S.; McCallum, L.; Delles, C.; Mark, P.B.; Petrie, J.R.; Rios, F.; Montezano, A.C.; et al. Cardiovascular and Renal Risk Factors and Complications Associated with COVID-19. *CJC Open* **2021**, *3*, 1257–1272. [CrossRef] [PubMed]
24. Townsend, L.; Fogarty, H.; Dyer, A.; Martin-Loeches, I.; Bannan, C.; Nadarajan, P.; Bergin, C.; O'Farrelly, C.; Conlon, N.; Bourke, N.M.; et al. Prolonged elevation of D-dimer levels in convalescent COVID-19 patients is independent of the acute phase response. *J. Thromb. Haemost.* **2021**, *19*, 1064–1070. [CrossRef] [PubMed]
25. Wong, S.W.; Fan, B.E.; Huang, W.; Chia, Y.W. ST-segment elevation myocardial infarction in post-COVID-19 patients: A case series. *Ann. Acad. Med. Singap.* **2021**, *50*, 425–430. [CrossRef] [PubMed]
26. Wirth, K.J.; Scheibenbogen, C. Pathophysiology of skeletal muscle disturbances in Myalgic Encephalomyelitis/Chronic Fatigue Syndrome (ME/CFS). *J. Transl. Med.* **2021**, *19*, 162. [CrossRef]
27. Kell, D.B.; Laubscher, G.J.; Pretorius, E. A central role for amyloid fibrin microclots in long COVID/PASC: Origins and therapeutic implications. *Biochem. J.* **2022**, *479*, 537–559. [CrossRef]
28. Szewczykowski, C.; Mardin, C.; Lucio, M.; Wallukat, G.; Hoffmanns, J.; Schröder, T.; Raith, F.; Rogge, L.; Heltmann, F.; Moritz, M.; et al. Long COVID: Association of Functional Autoantibodies against G-Protein-Coupled Receptors with an Impaired Retinal Microcirculation. *Int. J. Mol. Sci.* **2022**, *23*, 7209. [CrossRef]
29. Fernández-de-las-Peñas, C.; Pellicer-Valero, O.J.; Navarro-Pardo, E.; Palacios-Ceña, D.; Florencio, L.L.; Guijarro, C.; Martín-Guerrero, J.D. Symptoms Experienced at the Acute Phase of SARS-CoV-2 Infection as Risk Factor of Long-term Post-COVID Symptoms: The LONG-COVID-EXP-CM Multicenter Study. *Int. J. Infect. Dis.* **2022**, *116*, 241–244. [CrossRef]
30. Gaebler, C.; Wang, Z.; Lorenzi, J.C.C.; Muecksch, F.; Finkin, S.; Tokuyama, M.; Cho, A.; Jankovic, M.; Schaefer-Babajew, D.; Oliveira, T.Y.; et al. Evolution of antibody immunity to SARS-CoV-2. *Nature* **2021**, *591*, 639–644. [CrossRef]
31. Hopkinson, N.S.; Jenkins, G.; Hart, N. COVID-19 and what comes after? *Thorax* **2021**, *76*, 324. [CrossRef]
32. Gupta, A.; Madhavan, M.V.; Sehgal, K.; Nair, N.; Mahajan, S.; Sehrawat, T.S.; Bikdeli, B.; Ahluwalia, N.; Ausiello, J.C.; Wan, E.Y.; et al. Extrapulmonary manifestations of COVID-19. *Nat. Med.* **2020**, *26*, 1017–1032. [CrossRef] [PubMed]
33. Fogarty, H.; Townsend, L.; Morrin, H.; Ahmad, A.; Comerford, C.; Karampini, E.; Englert, H.; Byrne, M.; Bergin, C.; O'sullivan, J.M.; et al. Persistent endotheliopathy in the pathogenesis of long COVID syndrome. *J. Thromb. Haemost.* **2021**, *19*, 2546–2553. [CrossRef] [PubMed]

34. Hohberger, B.; Harrer, T.; Mardin, C.; Kruse, F.; Hoffmanns, J.; Rogge, L.; Heltmann, F.; Moritz, M.; Szewczykowski, C.; Schottenhamml, J.; et al. Case Report: Neutralization of Autoantibodies Targeting G-Protein-Coupled Receptors Improves Capillary Impairment and Fatigue Symptoms After COVID-19 Infection. *Front. Med.* **2021**, *8*, 754667. Available online: https://www.frontiersin.org/articles/10.3389/fmed.2021.754667 (accessed on 15 March 2023). [CrossRef] [PubMed]
35. Robertson, M.M.; Qasmieh, S.A.; Kulkarni, S.G.; Teasdale, C.A.; Jones, H.E.; McNairy, M.; Borrell, L.N.; Nash, D. The Epidemiology of Long COVID in US Adults. *Clin. Infect. Dis.* **2022**, *76*, 1636–1645. [CrossRef]
36. Thompson, E.J.; Williams, D.M.; Walker, A.J.; Mitchell, R.E.; Niedzwiedz, C.L.; Yang, T.C.; Huggins, C.F.; Kwong, A.S.F.; Silverwood, R.J.; Di Gessa, G.; et al. Long COVID burden and risk factors in 10 UK longitudinal studies and electronic health records. *Nat. Commun.* **2022**, *13*, 3528. [CrossRef]
37. Subramanian, A.; Nirantharakumar, K.; Hughes, S.; Myles, P.; Williams, T.; Gokhale, K.M.; Taverner, T.; Chandan, J.S.; Brown, K.; Simms-Williams, N.; et al. Symptoms and risk factors for long COVID in non-hospitalized adults. *Nat. Med.* **2022**, *28*, 1706–1714. [CrossRef]
38. Chen, C.; Haupert, S.R.; Zimmermann, L.; Shi, X.; Fritsche, L.G.; Mukherjee, B. Global Prevalence of Post-Coronavirus Disease 2019 (COVID-19) Condition or Long COVID: A Meta-Analysis and Systematic Review. *J. Infect. Dis.* **2022**, *226*, 1593–1607. [CrossRef]
39. Su, Y.; Yuan, D.; Chen, D.G.; Ng, R.H.; Wang, K.; Choi, J.; Li, S.; Hong, S.; Zhang, R.; Xie, J.; et al. Multiple early factors anticipate post-acute COVID-19 sequelae. *Cell* **2022**, *185*, 881–895.e20. [CrossRef]
40. Desgranges, F.; Tadini, E.; Munting, A.; Regina, J.; Filippidis, P.; Viala, B.; Karachalias, E.; Suttels, V.; Haefliger, D.; Kampouri, E.; et al. Post-COVID-19 Syndrome in Outpatients: A Cohort Study. *J. Gen. Intern. Med.* **2022**, *37*, 1943–1952. [CrossRef]
41. Hanson, S.W.; Abbafati, C.; Aerts, J. A global systematic analysis of the occurrence, severity, and recovery pattern of long COVID in 2020 and 2021. *medRxiv* **2022**. [CrossRef]
42. Office for National Statistics (ONS). *Prevalence of Ongoing Symptoms Following Coronavirus (COVID-19) Infection in the UK: 2 February 2023*; Office for National Statistics (ONS) Website, Statistical Bulletin: London, UK, 2023. Available online: https://www.ons.gov.uk/peoplepopulationandcommunity/healthandsocialcare/conditionsanddiseases/bulletins/prevalenceofongoingsymptomsfollowingcoronaviruscovid19infectionintheuk/2february2023 (accessed on 15 March 2023).
43. Davis, H.E.; Assaf, G.S.; McCorkell, L.; Wei, H.; Low, R.J.; Re'Em, Y.; Redfield, S.; Austin, J.P.; Akrami, A. Characterizing long COVID in an international cohort: 7 months of symptoms and their impact. *EClinicalMedicine* **2021**, *38*, 101019. [CrossRef] [PubMed]
44. Ni, W.; Yang, X.; Yang, D.; Bao, J.; Li, R.; Xiao, Y.; Hou, C.; Wang, H.; Liu, J.; Yang, D.; et al. Role of angiotensin-converting enzyme 2 (ACE2) in COVID-19. *Crit. Care* **2020**, *24*, 422. [CrossRef] [PubMed]
45. Gheblawi, M.; Wang, K.; Viveiros, A.; Nguyen, Q.; Zhong, J.C.; Turner, A.J.; Raizada, M.K.; Grant, M.B.; Oudit, G.Y. Angiotensin-Converting Enzyme 2: SARS-CoV-2 Receptor and Regulator of the Renin-Angiotensin System. *Circ. Res.* **2020**, *126*, 1456–1474. [CrossRef] [PubMed]
46. Lomi, N.; Sindhuja, K.; Asif, M.; Tandon, R. Clinical profile and prevalence of conjunctivitis in mild COVID-19 patients in a tertiary care COVID-19 hospital: A retrospective cross-sectional study. *Indian J. Ophthalmol.* **2020**, *68*, 1546–1550. [CrossRef]
47. Costa, F.; Bonifácio, L.P.; Bellissimo-Rodrigues, F.; Rocha, E.M.; Jorge, R.; Bollela, V.R.; Antunes-Foschini, R. Ocular findings among patients surviving COVID-19. *Sci. Rep.* **2021**, *11*, 11085. [CrossRef]
48. Honavar, S.; Sen, M.; Sharma, N.; Sachdev, M. COVID-19 and Eye: A Review of Ophthalmic Manifestations of COVID-19. *Indian J. Ophthalmol.* **2021**, *69*, 488–509. [CrossRef]
49. Feizi, S.M.; Meshksar, A.; Naderi, A.; Esfandiari, H. Anterior Scleritis Manifesting After Coronavirus Disease 2019: A Report of Two Cases. *Cornea* **2021**, *40*, 1204–1206. [CrossRef]
50. Bertoli, F.; Veritti, D.; Danese, C.; Samassa, F.; Sarao, V.; Rassu, N.; Gambato, T.; Lanzetta, P. Ocular Findings in COVID-19 Patients: A Review of Direct Manifestations and Indirect Effects on the Eye. *J. Ophthalmol.* **2020**, *2020*, 4827304. [CrossRef]
51. Teo, K.Y.; Invernizzi, A.; Staurenghi, G.; Cheung, C.M.G. COVID-19-Related Retinal Micro-vasculopathy—A Review of Current Evidence. *Arch. Ophthalmol.* **2021**, *235*, 98–110. [CrossRef]
52. Sen, S.; Kannan, N.B.; Kumar, J.; Rajan, R.P.; Kumar, K.; Baliga, G.; Reddy, H.; Upadhyay, A.; Ramasamy, K. Retinal manifestations in patients with SARS-CoV-2 infection and pathogenetic implications: A systematic review. *Int. Ophthalmol.* **2022**, *42*, 323–336. [CrossRef]
53. Sim, R.; Cheung, G.; Ting, D.; Wong, E.; Wong, T.Y.; Yeo, I.; Wong, C.W. Retinal microvascular signs in COVID-19. *Br. J. Ophthalmol.* **2022**, *106*, 1308–1312. [CrossRef] [PubMed]
54. Marinho, P.M.; Marcos, A.A.A.; Romano, A.C.; Nascimento, H.; Belfort, R. Retinal findings in patients with COVID-19. *Lancet* **2020**, *395*, 1610. [CrossRef] [PubMed]
55. Vavvas, D.G.; Sarraf, D.; Sadda, S.R.; Eliott, D.; Ehlers, J.P.; Waheed, N.K.; Morizane, Y.; Sakamoto, T.; Tsilimbaris, M.; Miller, J.B. Concerns about the interpretation of OCT and fundus findings in COVID-19 patients in recent Lancet publication. *Eye* **2020**, *34*, 2153–2154. [CrossRef] [PubMed]
56. Ouyang, P.; Zhang, X.; Peng, Y.; Jiang, B. Seeking clarity on retinal findings in patients with COVID-19. *Lancet* **2020**, *396*, e35. [CrossRef]

57. Burgos-Blasco, B.; Güemes-Villahoz, N.; Vidal-Villegas, B.; Martinez-De-La-Casa, J.M.; Donate-Lopez, J.; Martín-Sánchez, F.J.; González-Armengol, J.J.; Porta-Etessam, J.; Martin, J.L.R.; Garcia-Feijoo, J. Optic nerve and macular optical coherence tomography in recovered COVID-19 patients. *Eur. J. Ophthalmol.* **2022**, *32*, 628–636. [CrossRef]
58. Naderi Beni, A.; Dehghani, A.; Kianersi, F.; Ghanbari, H.; Habibidastenae, Z.; Memarzadeh, S.E.; Beni, Z.N. Retinal findings of COVID-19 patients using ocular coherence tomography angiography two to three months after infection: Ocular appearance recovered COVID-19 patient. *Photodiagnos. Photodyn. Ther.* **2022**, *38*, 102726. [CrossRef]
59. Oren, B.; Aydemır, G.A.; Aydemır, E.; Atesoglu, H.I.; Goker, Y.S.; Kızıltoprak, H.; Ozcelık, K.C. Quantitative assessment of retinal changes in COVID-19 patients. *Clin. Exp. Optom.* **2021**, *104*, 717–722. [CrossRef]
60. Salvi, L.; Plateroti, P.; Balducci, S.; Bollanti, L.; Conti, F.G.; Vitale, M.; Recupero, S.M.; Enrici, M.M.; Fenicia, V.; Pugliese, G. Abnormalities of retinal ganglion cell complex at optical coherence tomography in patients with type 2 diabetes: A sign of diabetic polyneuropathy, not retinopathy. *J. Diabetes Complicat.* **2016**, *30*, 469–476. [CrossRef]
61. Patton, N.; Aslam, T.; MacGillivray, T.; Pattie, A.; Deary, I.J.; Dhillon, B. Retinal vascular image analysis as a potential screening tool for cerebrovascular disease: A rationale based on homology between cerebral and retinal microvasculatures. *J. Anat.* **2005**, *206*, 319–348. [CrossRef]
62. Zapata, M.Á.; Banderas García, S.; Sánchez-Moltalvá, A.; Falcó, A.; Otero-Romero, S.; Arcos, G.; Velazquez-Villoria, D.; García-Arumí, J. Retinal microvascular abnor-malities in patients after COVID-19 depending on disease severity. *Br. J. Ophthalmol.* **2022**, *106*, 559. [CrossRef]
63. Turker, I.C.; Dogan, C.U.; Guven, D.; Kutucu, O.K.; Gul, C. Optical coherence tomography angiography findings in patients with COVID-19. *Can. J. Ophthalmol.* **2021**, *56*, 83–87. [CrossRef] [PubMed]
64. Abrishami, M.; Emamverdian, Z.; Shoeibi, N.; Omidtabrizi, A.; Daneshvar, R.; Rezvani, T.S.; Saeedian, N.; Eslami, S.; Mazloumi, M.; Sadda, S.; et al. Optical coherence tomography angiography analysis of the retina in patients recovered from COVID-19: A case-control study. *Can. J. Ophthalmol.* **2021**, *56*, 24–30. [CrossRef] [PubMed]
65. González-Zamora, J.; Bilbao-Malavé, V.; Gándara, E.; Casablanca-Piñera, A.; Boquera-Ventosa, C.; Landecho, M.F.; Zarranz-Ventura, J.; García-Layana, A. Retinal Microvascular Impairment in COVID-19 Bilateral Pneumonia Assessed by Optical Coherence Tomography Angiography. *Biomedicines* **2021**, *9*, 247. [CrossRef] [PubMed]
66. Hazar, L.; Karahan, M.; Vural, E.; Ava, S.; Erdem, S.; Dursun, M.E.; Keklikçi, U. Macular vessel density in patients recovered from COVID 19. *Photodiagnos. Photodyn. Ther.* **2021**, *34*, 102267. [CrossRef]
67. Guemes-Villahoz, N.; Burgos-Blasco, B.; Vidal-Villegas, B.; Donate-López, J.; de la Muela, M.H.; López-Guajardo, L.; Martín-Sánchez, F.J.; García-Feijoó, J. Reduced macular vessel density in COVID-19 patients with and without associated thrombotic events using optical coherence tomography angiography. *Graefe's Arch. Clin. Exp. Ophthalmol.* **2021**, *259*, 2243–2249. [CrossRef]
68. Rodman, J.; Ferraz, M.; Baran, A.; Zhang, B. Optical coherence tomography angiography of retinal vasculature in recovered COVID-19 patients compared to age and ethnic matched controls. *Clin. Exp. Optom.* **2022**, *105*, 842–847. [CrossRef]
69. Yılmaz Çebi, A.; Kılıçarslan, O.; Uçar, D. Evaluation of Retinal Microvascular Impairment after COVID-19 and Its Clinical Correlates Using Optical Coherence Tomography Angiography. *Turk. J. Ophthalmol.* **2022**, *52*, 324–330. [CrossRef]
70. Cetinkaya, T.; Kurt, M.M.; Akpolat, C. Analysis of swept-source optical coherence tomography angiography measurement al-terations in adult patients recovered from COVID-19. *Clin. Exp. Optom.* **2022**, *105*, 848–852. [CrossRef]
71. Abrishami, M.; Hassanpour, K.; Hosseini, S.; Emamverdian, Z.; Ansari-Astaneh, M.R.; Zamani, G.; Gharib, B.; Abrishami, M. Macular vessel density re-duction in patients recovered from COVID-19: A longitudinal optical coherence tomography angiography study. *Graefe's Arch. Clin. Exp. Ophthalmol.* **2022**, *260*, 771–779. [CrossRef]
72. Nageeb Louz, R.E.; Salah Eddin, M.A.; Macky, T.A.; Tolba, D.A.A. Post COVID-19 Retinal Evaluation Using Optical Coherence Tomography Angiography: A Case Control Study. *Ocul. Immunol. Inflamm.* **2022**, *31*, 1175–1183. [CrossRef]
73. Cennamo, G.; Reibaldi, M.; Montorio, D.; D'Andrea, L.; Fallico, M.; Triassi, M. Optical Coherence Tomography Angiography Features in Post-COVID-19 Pneumonia Patients: A Pilot Study. *Am. J. Ophthalmol.* **2021**, *227*, 182–190. [CrossRef] [PubMed]
74. Bilbao-Malavé, V.; González-Zamora, J.; de Viteri, M.S.; de la Puente, M.; Gándara, E.; Casablanca-Piñera, A.; Boquera-Ventosa, C.; Zarranz-Ventura, J.; Landecho, M.F.; García-Layana, A. Persistent Retinal Microvascular Impairment in COVID-19 Bilateral Pneumonia at 6-Months Follow-Up Assessed by Optical Coherence Tomography Angiography. *Biomedicines* **2021**, *9*, 502. [CrossRef] [PubMed]
75. García, S.B.; Aragón, D.; Azarfane, B.; Trejo, F.; Garrell-Salat, X.; Sánchez-Montalvá, A.; Otero-Romero, S.; Garcia-Arumi, J.; Zapata, M.A. Persistent reduction of retinal microvascular vessel density in patients with moderate and severe COVID-19 disease. *BMJ Open Ophthalmol.* **2022**, *7*, e000867. [CrossRef] [PubMed]
76. E Murdoch, I.; Morris, S.S.; Cousens, S.N. People and eyes: Statistical approaches in ophthalmology. *Br. J. Ophthalmol.* **1998**, *82*, 971–973. [CrossRef] [PubMed]
77. Czakó, C.; István, L.; Benyó, F.; Élő, Á.; Erdei, G.; Horváth, H.; Nagy, Z.Z.; Kovács, I. The Impact of Deterministic Signal Loss on OCT Angiography Measurements. *Transl. Vis. Vision Sci. Technol.* **2020**, *9*, 10. [CrossRef]
78. Early Treatment Diabetic Retinopathy Study Research Group. Early Photocoagulation for Diabetic Retinopathy: ETDRS Report Number 9. *Ophthalmology* **1991**, *98*, 766–785. [CrossRef]
79. Aslam, T.M.; Hoyle, D.C.; Puri, V.; Bento, G. Differentiation of Diabetic Status Using Statistical and Machine Learning Techniques on Optical Coherence Tomography Angiography Images. *Transl. Vis. Sci. Technol.* **2020**, *9*, 2. [CrossRef]

80. Thompson, I.A.; Durrani, A.K.; Patel, S. Optical coherence tomography angiography characteristics in diabetic patients without clinical diabetic retinopathy. *Eye* **2019**, *33*, 648–652. [CrossRef]
81. Mastropasqua, R.; Toto, L.; Mastropasqua, A.; Aloia, R.; De Nicola, C.; A Mattei, P.; Di Marzio, G.; Di Nicola, M.; Di Antonio, L. Foveal avascular zone area and parafoveal vessel density measurements in different stages of diabetic retinopathy by optical coherence tomography angiography. *Int. J. Ophthalmol.* **2017**, *10*, 1545–1551. [CrossRef]
82. Krawitz, B.D.; Mo, S.; Geyman, L.S.; Agemy, S.A.; Scripsema, N.K.; Garcia, P.M.; Chui, T.Y.; Rosen, R.B. Acircularity index and axis ratio of the foveal avascular zone in diabetic eyes and healthy controls measured by optical coherence tomography angiography. *Vis. Res.* **2017**, *139*, 177–186. [CrossRef]
83. The Jamovi Project [Computer Software]. 2021. Available online: http://www.jamovi.org (accessed on 1 January 2020).
84. National Institute of Health and Care Excellence (NICE). *COVID-19 Rapid Evidence Review: Risk Factors for Long-Term Effects of COVID-19*; NICE: Sutton-in-Ashfield, UK, 2021. Available online: https://files.magicapp.org/guideline/08d10c67-1331-4146-9471-b15d1d93e707/files/Evidence_review_for_risk_factors_FINAL_r400881.pdf (accessed on 15 March 2023).
85. Mavi Yildiz, A.; Ucan Gunduz, G.; Yalcinbayir, O.; Acet Ozturk, N.A.; Avci, R.; Coskun, F. SD-OCT assessment of macular and optic nerve alterations in patients recovered from COVID-19. *Can. J. Ophthalmol.* **2022**, *57*, 75–81. [CrossRef]
86. Ugurlu, A.; Agcayazi, S.B.; Icel, E.; Budakoglu, O.; Unver, E.; Barkay, O.; Karakeçili, F.; Bayrakceken, K. Assessment of the optic nerve, macular, and retinal vascular effects of COVID-19. *Can. J. Ophthalmol.* 2022. Available online: https://www.sciencedirect.com/science/article/pii/S0008418222001892 (accessed on 24 April 2023).
87. Taskiran-Sag, A.; Eroglu, E.; Ozulken, K.; Canlar, S.; Poyraz, B.M.; Sekerlisoy, M.B.; Mumcuoglu, T. Headache and cognitive disturbance correlate with ganglion cell layer thickness in patients who recovered from COVID-19. *Clin. Neurol. Neurosurg.* **2022**, *217*, 107263. [CrossRef] [PubMed]
88. Kanra, A.Y.; Altınel, M.G.; Alparslan, F. Evaluation of retinal and choroidal parameters as neurodegeneration biomarkers in patients with post-COVID-19 syndrome. *Photodiagnos. Photodyn. Ther.* **2022**, *40*, 103108. [CrossRef]
89. Schlick, S.; Lucio, M.; Wallukat, G.; Bartsch, A.; Skornia, A.; Hoffmanns, J.; Szewczykowski, C.; Schröder, T.; Raith, F.; Rogge, L.; et al. Post-COVID-19 Syndrome: Retinal Microcirculation as a Potential Marker for Chronic Fatigue. *Int. J. Mol. Sci.* **2022**, *23*, 13683. [CrossRef] [PubMed]
90. Jevnikar, K.; Meglič, A.; Lapajne, L.; Logar, M.; Valentinčič, N.V.; Petrovič, M.G.; Mekjavić, P.J. The Comparison of Retinal Microvascular Findings in Acute COVID-19 and 1-Year after Hospital Discharge Assessed with Multimodal Imaging—A Prospective Longitudinal Cohort Study. *Int. J. Mol. Sci.* **2023**, *24*, 4032. [CrossRef] [PubMed]
91. Jevnikar, K.; Mekjavic, P.J.; Valentincic, N.V.; Petrovski, G.; Petrovic, M.G. An Update on COVID-19 Related Ophthalmic Manifestations. *Ocul. Immunol. Inflamm.* **2021**, *29*, 684–689. [CrossRef] [PubMed]
92. de Almeida-Pititto, B.; Dualib, P.M.; Zajdenverg, L.; Dantas, J.R.; de Souza, F.D.; Rodacki, M.; Bertoluci, M.C.; Brazilian Diabetes Society Study Group (SBD). Severity and mortality of COVID 19 in patients with diabetes, hypertension and cardiovascular disease: A meta-analysis. *Diabetol. Metab. Syndr.* **2020**, *12*, 75. [CrossRef]
93. Savastano, M.C.; Gambini, G.; Cozzupoli, G.M.; Crincoli, E.; Savastano, A.; De Vico, U.; Culiersi, C.; Falsini, B.; Martelli, F.; Minnella, A.M.; et al. Retinal capillary involvement in early post-COVID-19 patients: A healthy controlled study. *Graefe's Arch. Clin. Exp. Ophthalmol.* **2021**, *259*, 2157–2165. [CrossRef]
94. Szkodny, D.; Wylęgała, E.; Sujka-Franczak, P.; Chlasta-Twardzik, E.; Fiolka, R.; Tomczyk, T.; Wylęgała, A. Retinal OCT Findings in Patients after COVID Infection. *J. Clin. Med.* **2021**, *10*, 3233. [CrossRef]
95. Kal, M.; Winiarczyk, M.; Cieśla, E.; Płatkowska-Adamska, B.; Walczyk, A.; Biskup, M.; Pabjan, P.; Głuszek, S.; Odrobina, D.; Mackiewicz, J.; et al. Retinal Microvascular Changes in COVID-19 Bilateral Pneumonia Based on Optical Coherence Tomography Angiography. *J. Clin. Med.* **2022**, *11*, 3621. [CrossRef]
96. Chiosi, F.; Campagna, G.; Rinaldi, M.; Manzi, G.; Dell'Omo, R.; Fiorentino, G.; Toro, M.; Tranfa, F.; D'Andrea, L.; Rejdak, M.; et al. Optical Coherence Tomography Angiography Analysis of Vessel Density Indices in Early Post-COVID-19 Patients. *Front. Med.* **2022**, *9*, 927121. Available online: https://www.frontiersin.org/articles/10.3389/fmed.2022.927121 (accessed on 24 April 2023). [CrossRef] [PubMed]

Journal of
Imaging

MDPI

Article

Evaluating Retinal Disease Diagnosis with an Interpretable Lightweight CNN Model Resistant to Adversarial Attacks

Mohan Bhandari [1], Tej Bahadur Shahi [2,3,*] and Arjun Neupane [2]

[1] Department of Science and Technology, Samriddhi College, Bhaktapur 44800, Nepal; mail2mohanbhandari@gmail.com
[2] School of Engineering and Technology, Central Queensland University, Norman Gardens, Rockhampton, QLD 4701, Australia; a.neupane@cqu.edu.au
[3] Central Department of Computer Science and IT, Tribhuvan University, Kathmandu 44600, Nepal
* Correspondence: t.shahi@cqu.edu.au or tejshahi@cdcsit.edu.np

Abstract: Optical Coherence Tomography (OCT) is an imperative symptomatic tool empowering the diagnosis of retinal diseases and anomalies. The manual decision towards those anomalies by specialists is the norm, but its labor-intensive nature calls for more proficient strategies. Consequently, the study recommends employing a Convolutional Neural Network (CNN) for the classification of OCT images derived from the OCT dataset into distinct categories, including Choroidal Neo-Vascularization (CNV), Diabetic Macular Edema (DME), Drusen, and Normal. The average k-fold (k = 10) training accuracy, test accuracy, validation accuracy, training loss, test loss, and validation loss values of the proposed model are 96.33%, 94.29%, 94.12%, 0.1073, 0.2002, and 0.1927, respectively. Fast Gradient Sign Method (FGSM) is employed to introduce non-random noise aligned with the cost function's data gradient, with varying epsilon values scaling the noise, and the model correctly handles all noise levels below 0.1 epsilon. Explainable AI algorithms: Local Interpretable Model-Agnostic Explanations (LIME) and SHapley Additive exPlanations (SHAP) are utilized to provide human interpretable explanations approximating the behaviour of the model within the region of a particular retinal image. Additionally, two supplementary datasets, namely, COVID-19 and Kidney Stone, are assimilated to enhance the model's robustness and versatility, resulting in a level of precision comparable to state-of-the-art methodologies. Incorporating a lightweight CNN model with 983,716 parameters, 2.37×10^8 floating point operations per second (FLOPs) and leveraging explainable AI strategies, this study contributes to efficient OCT-based diagnosis, underscores its potential in advancing medical diagnostics, and offers assistance in the Internet-of-Medical-Things.

Keywords: adversarial attacks; deep learning; health informatics; lightweight CNN; retinal image classification

1. Introduction

The retina, situated at the posterior aspect of the ocular globe, comprises photoreceptor cells adept at transducing luminous stimuli into intricate electrical signals, subsequently dispatched to the cerebral cortex via the optic nerve. This intricate process serves as the foundation for human visual perception, wherein the brain deciphers these electrical transmissions as coherent visual representations. Retinal diseases can seriously affect vision and in some cases, can lead to permanent blindness [1] which is a big problem for the general health of the public. Getting a prompt and accurate diagnosis with the help of automated tools is a great assist for medical specialists in making wise medical decisions. The advancement of digital medical imaging has brought about a significant change in ophthalmology as it has introduced effective technologies that help in the detection of such diseases. By improving early detection through image analysis and identifying minuscule anomalies, Artificial Intelligence (AI) has considerably coped with retinal diseases. AI has also enhanced treatment planning by analyzing patient data and enabling tailored care. Additionally, AI-driven systems help track the development of diseases, resulting in therapies that are more successful [2].

Different Machine Learning (ML) and Convolutional Neural Networks (CNNs) are efficient at analyzing images and are particularly incredible at recognizing complex patterns in medical images [3]. Their ability to diagnose complicated retinal diseases is efficient without a doubt, but in medical practice, using CNNs depends not only on how well they can diagnose the issues but also on how useful they are in places with limited computational resources. Not only CNN, but different variants of CNN like ResNet [4], VGG [5] and more have produced good accuracies statistically. These CNNs and their variants have a very high number of training parameters, and many layers which make it time-consuming in real-time predictions [6] and integration with the Internet-of-Medical-Things (IoMT) [7].

As AI technology advances, it has become essential to not only achieve better diagnostic abilities but also to understand how these AI systems make predictions and decisions [8–10]. As these models can be hard to understand because of their statistical nature making them black boxes [11], the addition of Explainable Artificial Intelligence (XAI) into these models can solve the problem. The combination of small and efficient CNN models in IoMT devices with XAI, as a bio-marker, helps retinal disease diagnosis to be more accurate and more accessible for medical experts, practitioners, and even ordinary people.

To resolve all of these issues, this study aims to achieve the following three key objectives:

1. To develop an efficient CNN model with minimal parameters for detecting retinal abnormalities such as CNV, DME, and Drusen using OCT datasets.
2. To incorporate Explainable Artificial Intelligence (XAI) methodologies such as Local Interpretable Model-agnostic Explanations (LIME) and Shapley additive explanations (SHAP) into the realm of clinical interpretation, with the aim of comprehending the prediction by the Convolutional Neural Network (CNN) model.
3. To generalize the model's reliability and applicability; two new additional datasets were trained and evaluated for the model.

2. Related Works

Researchers in health informatics are leveraging the predictive power of Deep Learning (DL) to address the automated diagnosis of various diseases such as COVID-19 [12], monkeypox [13], kidney stone [14] and so on. Here, we summarise the recent DL methods that have been employed for retinal disease diagnosis using various image modalities. These methods can be categorized into two broad classes: pre-trained DL models (that leverage the transfer learning strategies) and custom-designed CNN (which needs training from scratch).

Subramanian et al. [15] utilised four CNN models such as VGG16, DenseNet-201, Inception-V3, and Xception, to classify seven different retinal diseases. Moreover, Bayesian optimization was employed to fine-tune the hyperparameters of these CNN models, coupled with image augmentation techniques to enhance their ability to generalize. The use of DenseNet-201 in classifying retinal diseases on the Retinal OCT Image dataset resulted in an accuracy exceeding 99%, demonstrating superior performance compared to alternative methods. Puneet et al. [16] implemented the combination of Attention and Transfer Learning approaches into a DCNN for categorizing retinal diseases such as CNV, DME, and Drusen using OCT images. Their proposal achieved notable results, attaining accuracies of 97.79% during training and 95.6% during testing. Kayadibi et al. [17] implemented a hybrid fine-tuned CNN for retinal disease classification from OCT images. They utilized PCA to reduce the feature size and enhance the performance. The benchmarking outcomes for two OCT datasets demonstrated a high level of promise in terms of accuracy. Specifically, the UCSD dataset yielded an impressive accuracy of 99.70% according to Kermany et al.'s study [18], while the Duke dataset achieved a perfect 100% accuracy as reported by Srinivasan et al. [19]. In their research, Kim and colleagues [20] harnessed a variety of Convolutional Neural Networks (CNNs) like VGG16, ResNet50, DenseNet121, and Inception-v3 as feature extractors. Subsequently, they employed these features to develop binary OCT image classification models. A binary classifier model is developed for each category (CNV, DME, Drusen and Normal) and the VGG-16-based model for CNV vs. other classes achieved 98.6% accuracy. They achieve high accuracy using the pre-trained DL models. However, their proposal needs the training of individual models for each class which incurs high computational complexity. A pre-trained VGG-16 network was implemented by Li et al. [21] for retinal image classification on OCT images. They validated the model's performance on 1000 independent OCT images. Their work revealed that the transfer learning with the VGG-16 model has a promising accuracy of 98.6%, sensitivity of 97.8%, and specificity of 00.4%. With such commendable performance of the model, deep learning can automate the diagnosis of retinal disease. Li et al. [22] adopted the ensemble models for retinal disease classification using OCT images. They trained four DL- models based on improved ResNet50 to build the ensemble and achieved the highest accuracy of 96.3%, sensitivity of 96.6%, and specificity of 98.7%. However, the ResNet50 model is itself the heavyweight model.

In addition to employing pre-trained deep learning models, only a limited number of researchers have created custom CNNs for the classification of retinal images. For example, a deep CNN with six convolution blocks (including the Relu, batch normalization, and pooling operation) was implemented by Sujina et al. [23]. Their proposal achieved a promising accuracy of 99.69% with a low misclassification rate. However, the generalisability of the CNN on additional datasets is not reported. Altan et al. [24] implemented a deep learning architecture to detect the macular edema on OCT images and reported an accuracy of 99.20%.

Hybrid deep learning models for retinal image classification have also been proposed recently. For instance, a hybrid deep learning model for OCT image classification was implemented by Khan et al. [25]. They extracted retinal features from OCT images using three pre-trained deep learning models (DenseNet121, InceptionV3, and ResNet50), and ant colony optimization was used for best feature selection. Finally, the SVM and KNN were employed for classification. Their proposal achieved high performance on OCT image classification. However, the approach is not applicable to end-to-end training of the model.

3. Materials and Methods

The entire material and methods adopted in the study are depicted in Figure 1, which includes the stages ranging from data preparation to model evaluation.

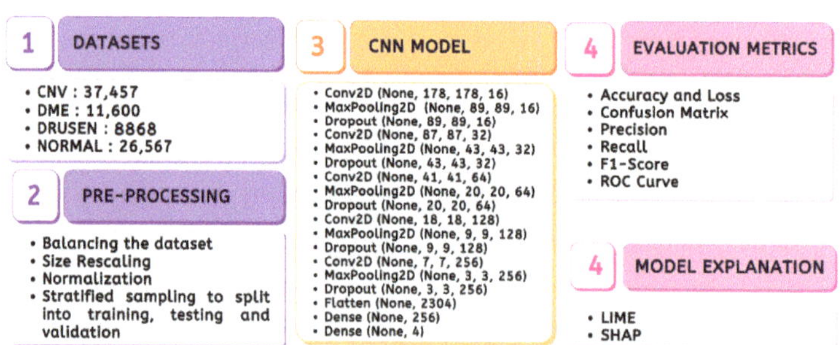

Figure 1. Methodology.

3.1. Dataset

3.1.1. Disease Description

Figure 2 [26] shows the different diseases considered in this study. CNV, depicted in Figure 2a, arises from the emergence of fresh blood vessels in proximity to the choroid. CNV is caused by flaws in the innermost section of the choroid known as Bruch's membrane, along with conditions like severe nearsightedness and heightened vascular endothelial growth. DME, Figure 2b, primarily affects individuals with diabetes. It leads to vision distortion as fluid accumulates in the macula. This accumulation impairs cone cells' light-sensing abilities, causing blurred vision. DME arises from the expansion of blood vessels at the posterior region. In Figure 2c, we can observe Drusen, a condition primarily linked to the aging process. It involves the accumulation of yellow extracellular particles between the Bruch's membrane and the retinal pigment in the eye. Drusen has the potential to hinder the transport system, which could lead to a deprivation of oxygen for the cone cells responsible for colour vision, ultimately resulting in their deterioration.

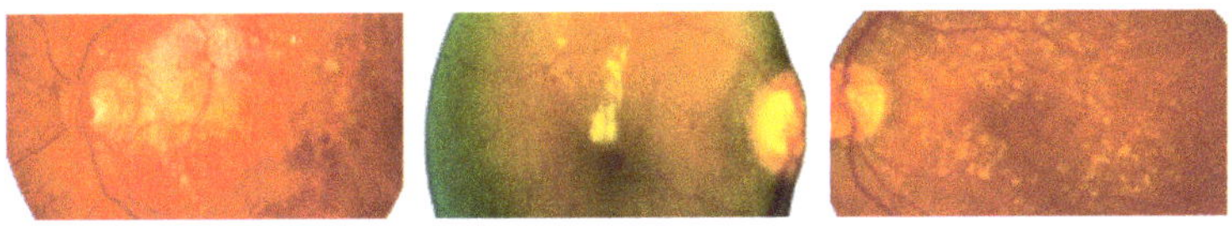

(**a**) CNV (**b**) DME (**c**) Drusen

Figure 2. Representative images for diseases.

3.1.2. Dataset Description

The publicly accessible dataset [27] encompasses detailed cross-sectional images of living patients' retinas, which have been classified into four distinct categories: Normal, CNV, Drusen, and DME. These categories are visually represented in Figure 3. The dataset comprises a grand total of 84,492 images, distributed as follows: CNV contains 37,457 images, Normal contains 26,567 images, DME includes 11,600 images, and Drusen encompasses 8868 images.

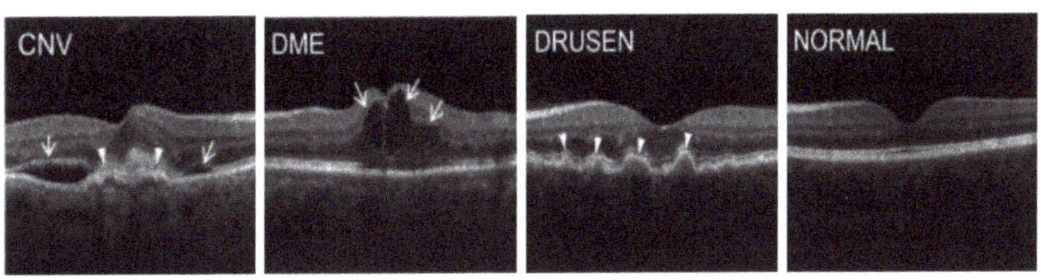

(**a**) (**b**) (**c**) (**d**)

Figure 3. Illustrative examples from the retinal image dataset are presented. Figure (**a**) showcasing Choroidal NeoVascularization, characterized by the presence of neovascular membranes (indicated by white arrowheads) along with associated sub-retinal fluid (marked by arrows). Figure (**b**) illustrates Diabetic Macular Edema, which manifests as intra-retinal fluid associated with retinal thickening (denoted by arrows). Figure (**c**) displays multiple instances of drusen (highlighted by arrowheads), while Figure (**d**) illustrates a normal, pristine retina with an undisturbed foveal structure and no signs of retinal fluid or edema.

3.2. Dataset Pre-Processing

The originally imbalanced dataset was transformed into a balanced one, where each category contained exactly 8868 images. In this balanced dataset, all images were resized to dimensions of 180×180 and then normalized to fall within the range [0, 1]. To ensure representative sampling, a stratified approach was employed, allocating 80% of the data for training, and the remaining 20% for testing, with half of the testing data reserved for validation.

3.3. CNN Model

The proposed lightweight model reported in Table 1 provides a concise overview of the CNN architecture and Figure 4 shows the graphical result. The model holds only five convolution layers to perform convolution operations on the input image, with increasing filter depths (16, 32, 64, 128, 256) to capture hierarchical features. Each convolution layer is followed by max-pooling layers to down-sample the feature maps, aiding in information compression. Dropout layers help mitigate over-fitting by randomly deactivating neurons during training with values of 0.2 on each. The final dense layers (256, 4) process flattened features for classification, culminating in four output classes. The model contains around 983,716 total trainable parameters, contributing to its complexity and predictive capacity.

Table 1. The proposed lightweight CNN Model Architecture. Note the "Param #." represents the parameters involved in the given CNN.

Layer (Type)	Output Shape	Param #
Conv2D	(None, 178, 178, 16)	448
MaxPooling2D	(None, 89, 89, 16)	0
Dropout	(None, 89, 89, 16)	0
Conv2D	(None, 87, 87, 32)	4640
MaxPooling2D	(None, 43, 43, 32)	0
Dropout	(None, 43, 43, 32)	0
Conv2D	(None, 41, 41, 64)	18,496
MaxPooling2D	(None, 20, 20, 64)	0
Dropout	(None, 20, 20, 64)	0
Conv2D	(None, 18, 18, 128)	73,856
MaxPooling2D	(None, 9, 9, 128)	0
Dropout	(None, 9, 9, 128)	0
Conv2D	(None, 7, 7, 256)	295,168
MaxPooling2D	(None, 3, 3, 256)	0
Dropout	(None, 3, 3, 256)	0
Flatten	(None, 2304)	0
Dense	(None, 256)	590,080
Dense	(None, 4)	1028
Total params.: 983,716		
Trainable params.: 983,716		
Non-trainable params.: 0		

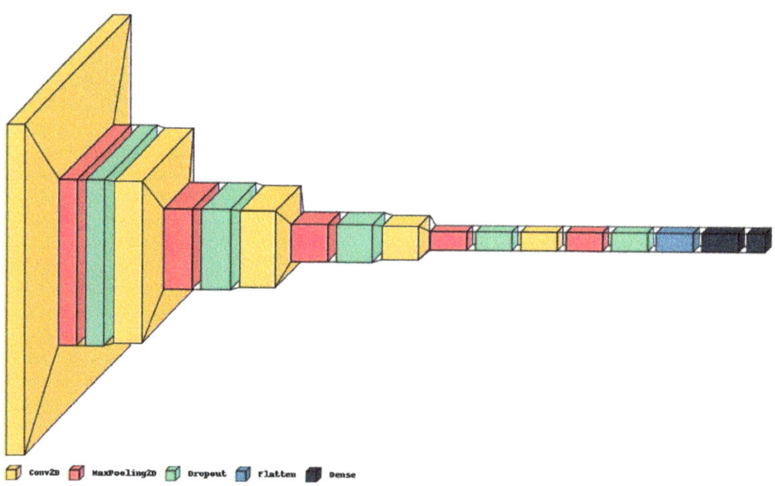

Figure 4. Graphical visualization of the proposed model.

3.4. Implementation Setup

The CNN model and XAI algorithms in this study were implemented using Python 3.10.12, along with Keras version 2.13.1 and TensorFlow version 2.13.0. The computational resources employed for this research included the runtime environment provided by Google Colab, supported by a robust NVIDIA K80 GPU with an impressive 12 GB of RAM. To assess the performance of the CT dataset, a cross-validation technique with a designated value of K = 10 [28] was employed, which entailed distinct and random allocations for both training and testing subsets. For the purpose of regularization, an early stopping strategy was employed, which relied on monitoring the validation loss for 10 consecutive epochs.

3.5. Performance Evaluation Metrics

3.5.1. Accuracy

The performance of classification models is typically evaluated using the metric of accuracy. Out of all the examples in a dataset, it calculates the percentage of accurately predicted instances. Mathematically, accuracy (Acc) is calculated as Equation (1).

$$Acc = \frac{\text{Number of Correct Predictions}}{\text{Total Number of Predictions}} \tag{1}$$

3.5.2. Precision

Precision is a performance measure that assesses the correctness of a model's positive predictions. It is determined by the ratio of true positive predictions (TP) to the sum of true positive and false positive predictions ($TP + FP$).

3.5.3. Recall

Recall evaluates a capacity to accurately identify every positive occurrence present in the dataset. It is described as the proportion of genuine positives (positives that were correctly detected) to all real positives.

3.5.4. F1-Score

The F1 Score is a classification task evaluation metric that balances precision and recall. It is calculated as the harmonic mean of precision and recall, offering a single measure of model performance that takes both false positives and false negatives into account.

3.5.5. ROC Curve

The Receiver Operating Characteristic (ROC) curve is a graph that shows how well a model works. It shows the trade-off between two things: how often the model correctly says "yes" when it should (Sensitivity), and how often it incorrectly says "yes" when it shouldn't (1-Specificity). This graph helps us see how good the model is at telling things apart in different settings. To make the ROC curve, we draw a graph of the True Positive Rate (TPR) against the False Positive Rate (FPR) for different settings. The area under the ROC curve (AUC-ROC) gives us a single number that tells us how well the model is doing.

3.5.6. FLOPS

Algorithm 1 calculates the Floating Point Operations (FLOPs) for the CNN model [29]. It defines two functions, CalculateCNNLayerFLOPs, and CalculateDenseLayerFLOPs, to compute FLOPs for Conv2D and Dense layers, respectively, based on their parameters. The CalculateTotalFLOPs function iterates through the model's layers, identifying Conv2D and Dense layers, and accumulates their respective FLOPs. This provides an estimate of the total computational complexity of the CNN model. The algorithm is valuable for assessing the computational efficiency of the CNN in terms of the number of operations needed for inference.

Algorithm 1 Calculate FLOPs for CNN Model [29]

```
function CALCULATECNNLAYERFLOPS(layer)
    Input: CNN layer
    Output: FLOPs for the layer
    return 2 × layer.filters × layer.kernel_size[0] × layer.kernel_size[1] × layer.input_shape[−1] ×
  layer.output_shape[1] × layer.output_shape[2]
end function
function CALCULATEDENSELAYERFLOPS(layer)
    Input: Dense layer
    Output: FLOPs for the layer
    return 2 × layer.input_shape[−1] × layer.output_shape[−1]
end function
function CALCULATETOTALFLOPS(model)
    Input: CNN model
    Output: Total FLOPs for the model
    total_flops ← 0
    for layer in model.layers do
        if layer is Conv2D then
            total_flops += CalculateCNNLayerFLOPs(layer)
        else if layer is Dense then
            total_flops += CalculateDenseLayerFLOPs(layer)
        end if
    end for
    return total_flops
end function
```

3.5.7. Explainable AI

Although there are certain challenges associated with XAI, such as its sensitivity to individual cases, the trade-off involving complexity, and the assumption of highly interdependent features, XAI delves into the visual computational approach of Deep Learning models. Consequently, the study incorporates the use of LIME and SHAP.

1. LIME

   In the pursuit of enhancing the transparency and interpretability of modern machine learning models, LIME has emerged as a powerful technique. LIME addresses the challenge of understanding complex black-box models' predictions by approximating their behaviour through locally interpretable models. This approach allows us to shed light on how specific features influence predictions, especially in contexts involving intricate data types such as images.

   LIME operates by selecting a target instance x, model f and generating perturbed instances x_i' in its vicinity. The model's predictions $f(x)$ and $f(x_i')$ are obtained, and interpretable features z_i are extracted from the perturbed instances. An interpretable CNN model $g(z)$, was trained using pairs $(z_i, f(x_i'))$ to approximate the complex behavior of f in the local neighborhood of x. To analyse $g(z)$, coefficients β_i in $g(z)$, the importance of the corresponding features z_i in influencing the predictions were reflected. Larger absolute values of β_i indicate stronger influences [30].

2. SHAP

   For the retinal input retinal image x with N number of pixels and f prediction, SHAP values for each pixel in the image were calculated. The values of SHAP show the contribution of the model to define how much each pixel i in the retinal image x contributes and is calculated using Equation (2).

$$\phi_i(x) = \sum_{S \subseteq \{1,2,\ldots,N\} \setminus \{i\}} \frac{|S|!(N - |S| - 1)!}{N!} [f(x_S \cup \{i\}) - f(x_S)] \quad (2)$$

   S represents a subset of pixels excluding pixel i, x_S is the image with the pixels in subset S unchanged, $f(x_S \cup \{i\})$ is the model's prediction when pixel i is included in the subset S and $f(x_S)$ is the model's prediction when pixel i is excluded from the subset S [31].

These SHAP values provide insights into the contribution of each pixel to the model's prediction. Positive SHAP values indicate that a pixel's presence positively influenced the forecast, while negative values suggest the opposite.

4. Results

4.1. Statistical Results

We analyzed classic statistical validation measures, which included the model's performance in terms of error and correctness throughout training, as well as across the validation and test datasets. Furthermore, we incorporated precision, F1-score, recall, confusion matrix, and k-fold validation into our evaluation.

Among ten different folds, 10th fold, stopped early in 15th epoch, holding the lowest accuracy, and the same is considered to plot the evaluation metrics. The training process spanned 20 epochs, with each fold configured to terminate early if the validation loss persisted for five consecutive epochs.

Figure 5, shows the training and validation accuracy of 10th fold where training phase yielded 95.64% and a loss of 0.1201, and the validation accuracy stood at 94.12% with a corresponding loss of 0.2185.

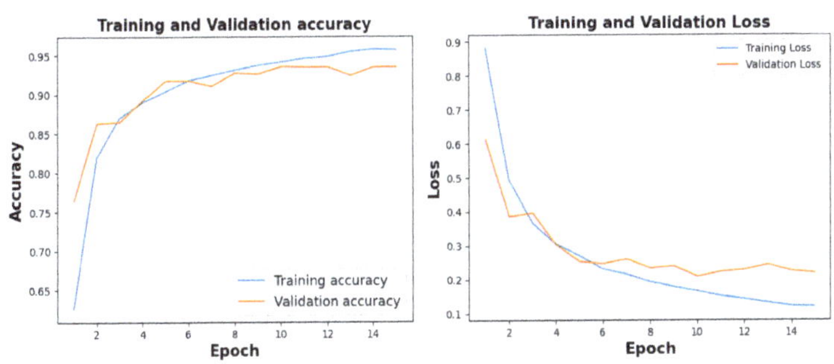

(a) (b)

Figure 5. Training and Validation Result. (**a**) represents training and validation accuracy (**b**) shows training and validation loss

Figure 6 shows the performance matrix of 10th fold. Among 892 Drusen samples, 843 were accurately predicted. Only 52 normal samples were misclassified, out of which 18 were predicted as Drusen, 4 were predicted as CNV, and 30 were predicted as DME. Out of 900 samples, a total of 807 samples were predicted correctly for the CNV class. Considering 876 samples, 843 samples were predicted correctly for DME.

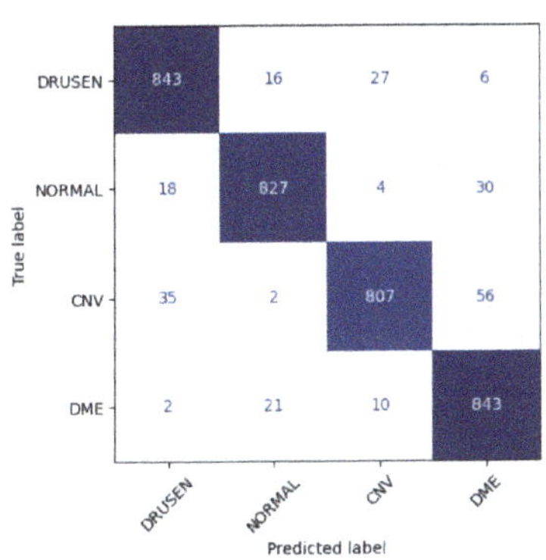

Figure 6. Class-wise performance of CNN model.

Following 10 folds, the model demonstrated an average training accuracy of 96.33%, a validation accuracy of 94.12%, a training loss of 0.1073, and a validation loss of 0.1927. The average testing accuracy and testing loss stood at 94.29% and 0.2002 respectively (Table 2).

Table 2. Performance Metrics for Different Folds. Symbols: TA, TL, TP, TR, VA, VL, VP, VR, TeA, TeL, TeP, and TeR represent training accuracy, training loss, training precision, training recall, validation accuracy, validation loss, validation precision, validation recall, test accuracy, test loss, test precision, and test recall, respectively in percentages.

K	TA	TL	TP	TR	VA	VL	VP	VR	TeA	TeL	TeP	TeR
1	96.31	0.1011	96.49	96.13	94.11	0.1820	94.37	93.99	94.28	0.1974	94.43	94.17
2	96.22	0.1010	96.39	96.04	94.01	0.1819	94.27	93.90	94.18	0.1972	94.33	94.07
3	96.31	0.1011	96.49	96.13	94.11	0.1820	94.37	93.99	94.28	0.1974	94.43	94.17
4	96.41	0.1012	96.59	96.23	94.20	0.1822	94.46	94.09	94.37	0.1976	94.52	94.26
5	96.51	0.1013	96.69	96.33	94.30	0.1824	94.56	94.19	94.47	0.1978	94.62	94.36
6	96.31	0.1135	96.49	96.13	94.10	0.1908	94.36	93.99	94.27	0.2029	94.42	94.16
7	96.81	0.1012	96.99	96.63	94.60	0.1781	94.86	94.49	94.77	0.2015	94.92	94.66
8	96.14	0.1176	96.32	95.96	93.93	0.2160	94.19	93.82	94.10	0.2101	94.25	93.99
9	96.64	0.1151	96.82	96.46	94.43	0.2135	94.69	94.32	94.60	0.1975	94.75	94.49
10	95.64	0.1201	95.82	95.46	93.43	0.2185	93.69	93.32	93.60	0.2025	93.75	93.49
Average	96.33	0.1073	96.51	96.15	94.12	0.1927	94.38	94.01	94.29	0.2002	94.44	94.18

Table 3 shows the classification report where "Drusen" and "Normal" show high Precision (0.94–0.95) indicating accurate positive predictions and DME shows high Recall (0.96), capturing most positives. F1-Score ranges from 0.94 to 0.95 showing the harmony between Precision and Recall.

Table 3. Classification Report of 10th fold.

Class	Precision	Recall	F1-Score
Drusen	0.94	0.95	0.94
Normal	0.95	0.94	0.95
CNV	0.95	0.90	0.92
DME	0.90	0.96	0.93
Accuracy			0.94
Macro Avg	0.94	0.94	0.94
Weighted Avg	0.94	0.94	0.94

As the imbalanced datasets were made balanced, the ROC curve (for 10th fold) was plotted as shown in Figure 7 to calculate the area under the curve and evaluate the model performance.

The model showed outstanding performance in distinguishing between positive and negative categories, as proven by its impressive AUC score of 0.99 in all areas. The ROC curve, which depends on TPR and FPR, displayed the model's predictions on the test dataset having an exceptionally high TPR, covering the entire range of AUC values. This demonstrates the model's exceptional effectiveness.

Table 4 presents sensitivity and specificity values of four classes in 10th fold. DME and Normal classes exhibit high sensitivity (>0.98), indicating accurate detection of relevant cases. Drusen has slightly lower sensitivity, while CNV has the highest specificity (>0.97), suggesting strong performance in distinguishing its class.

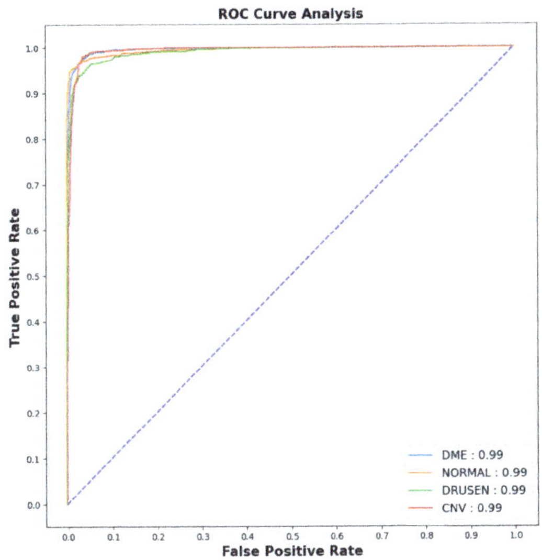

Figure 7. ROC-AUC curve.

Table 4. Sensitivity and Specificity of 10th fold.

Class	Sensitivity	Specificity
DME	0.989963	0.916084
NORMAL	0.99096	0.948488
DRUSEN	0.974254	0.925115
CNV	0.966781	0.974166

4.2. Explainable Results

4.2.1. SHAP

Because it is hard to understand how the CNN model predicted the output, XAI techniques are used to explain it [32]. The testing images are on the left, and each explanation has a transparent grey background (see Figures 8–11).

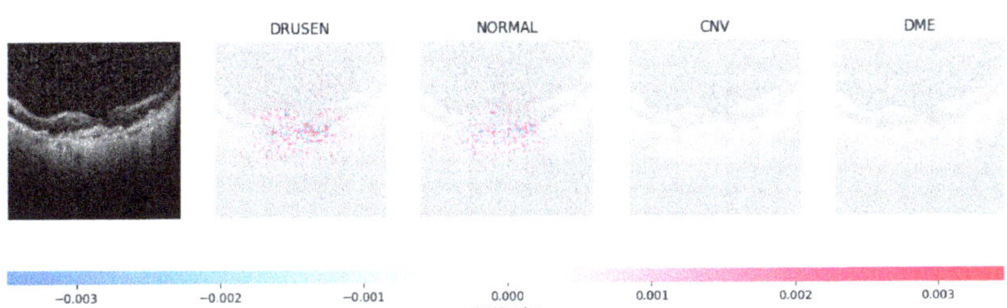

Figure 8. The model determined the presence of Drusen in the image by analyzing the OCT image and noting a significant concentration of red pixels (scattered in central regions) in the explanatory image (second in a row), which is located in the second column.

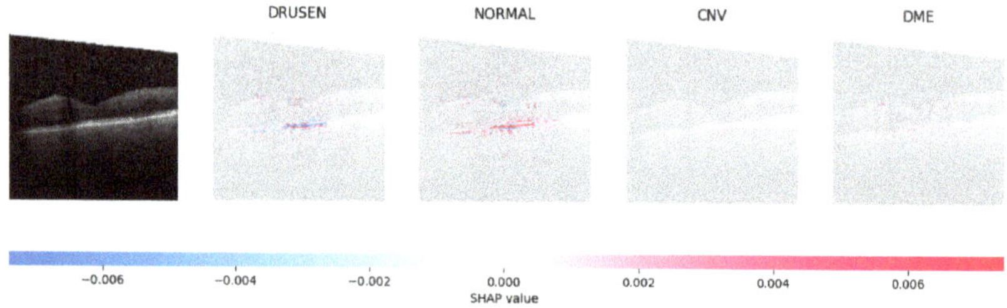

Figure 9. The model's examination of the OCT image revealed a significant number of red pixels in the explanatory image (third in a row) to suggest the presence of a healthy eye.

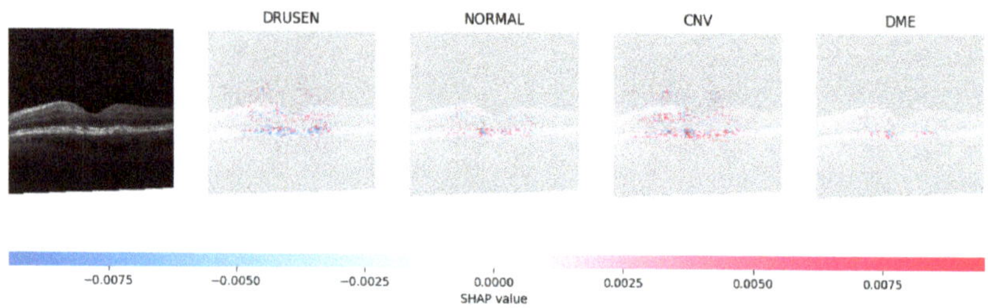

Figure 10. The model determined that the OCT image exhibited indications of CNV, which is a retinal disorder, due to the significant abundance of red pixels in the third explanatory image located in the fourth column.

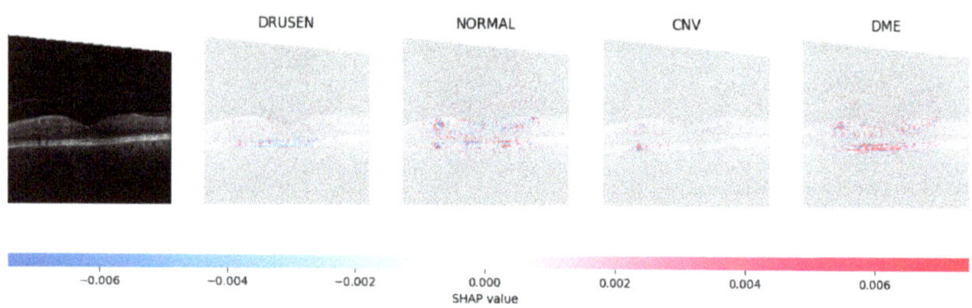

Figure 11. The model determined that the OCT image indicated the presence of DME due to the significant number of red pixels observed in the fifth column of the fourth explanatory image.

The red pixels in the first explanation image (refer to Figure 8) increase the probability of predicting a Drusen. In the second explanation, the model somehow attempts to indicate that the image is normal, but the red pixels' concentration in the first explanation is higher. Third, and fourth explanation images do not contain any red or blue pixels, so the probability of classifying the input image as CNV and DME is low.

In Figure 9, the presence of concentrated red pixels suggests that the image is normal. Conversely, in Figures 10 and 11, the absence of concentrated red pixels indicates that they correspond to CNV and DME, respectively.

4.2.2. LIME

LIME is employed to extract the features and perturbations from the training dataset. These perturbations are randomly generated from a standardized image, and subsequent operations involving mean-centering and scaling are conducted. The simple linear iterative clustering (SLIC) [33] is used to compute the initial three characteristics which delineate the most influential boundaries and incorporate them into the image. The second column of Table 5 displays the original test images corresponding to each category. The segmented image segment in the third column of Table 5 represents the segmentation obtained through LIME. As illustrated in Table 5, LIME furnishes visual explanations of the model's decision-making process, spotlighting the image regions that make a significant contribution to a specific class prediction.

Table 5. Interpretation of LIME results alongside input image and segmented image.

Category	OCT Image	LIME Interpretation
DRUSEN	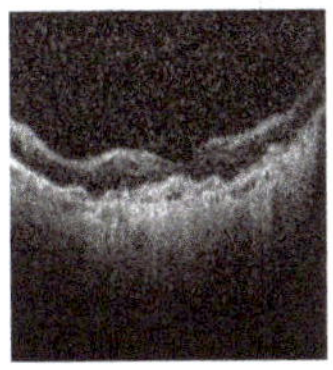	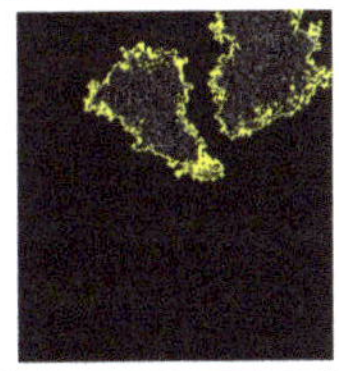
Normal	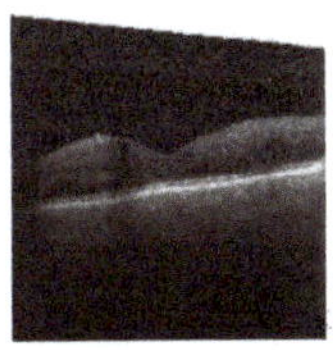	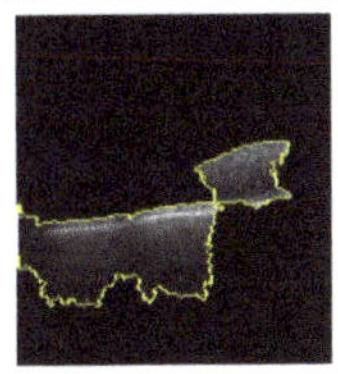
CNV	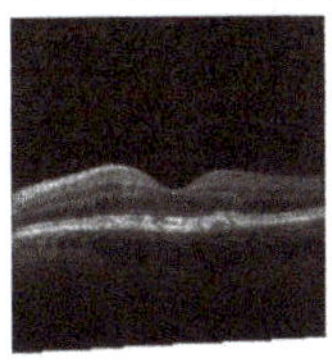	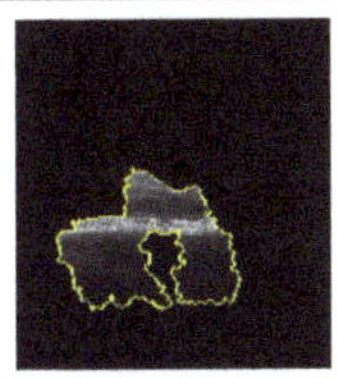
DME	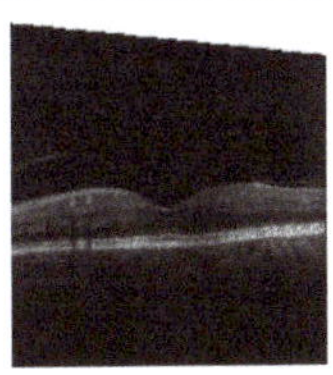	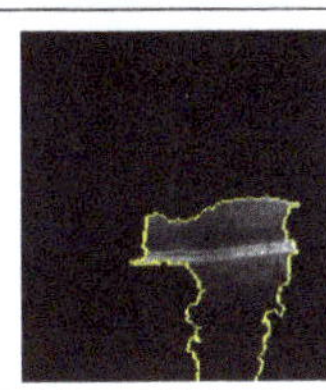

5. FLOPS Calculation

The proposed model's Floating-point Operations (FLOPs) were determined by considering all arithmetic operations involving floating-point values, such as addition, subtraction, division, multiplication, and any other relevant operations. The model executed a total of 2.37×10^8 operations, and this calculation was accomplished using Algorithm 1.

6. Generation of Fast Gradient Sign Method for Adversarial Examples

To evaluate the model's robustness, we conducted tests using adversarial examples. We computed the gradient of the loss function with respect to the input images to capture subtle variations. We introduced epsilon as a hyperparameter to quantify the perturbation's intensity, which was generated by taking the sign of the gradient and adjusting its magnitude. Subsequently, we incorporated this perturbation into the image and forwarded it for prediction. Our findings indicate that the model exhibited resilience up to an epsilon value of 0.1, as illustrated in Table 6 for DRUSEN and CNV, whereas all categories remained stable below the threshold of 0.1.

Table 6. Adversarial Examples.

Class	Original Image	Epsilon = 0.1	Adversarial Prediction
Drusen	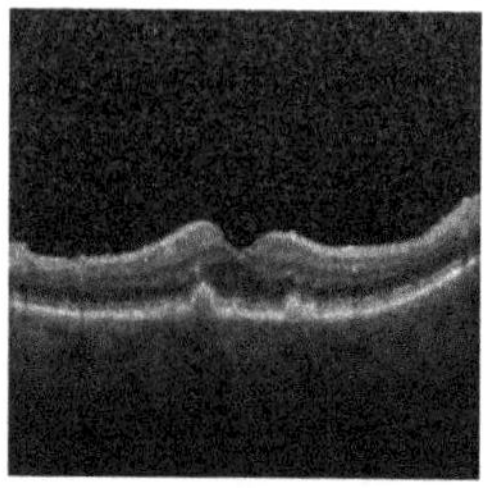	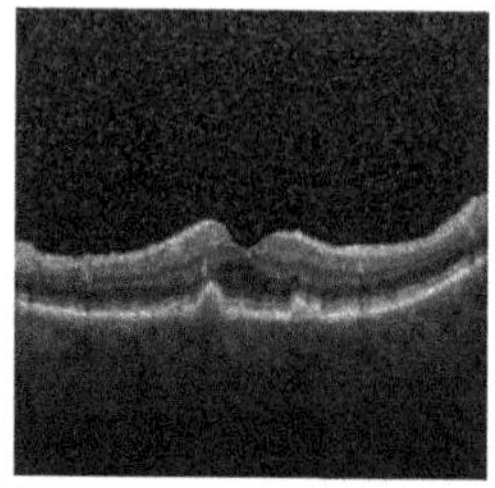	Drusen
Normal	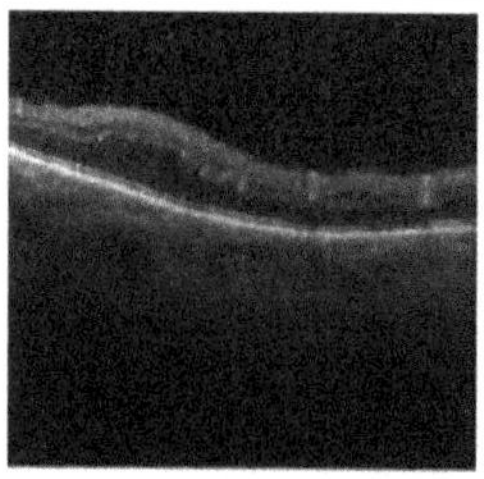	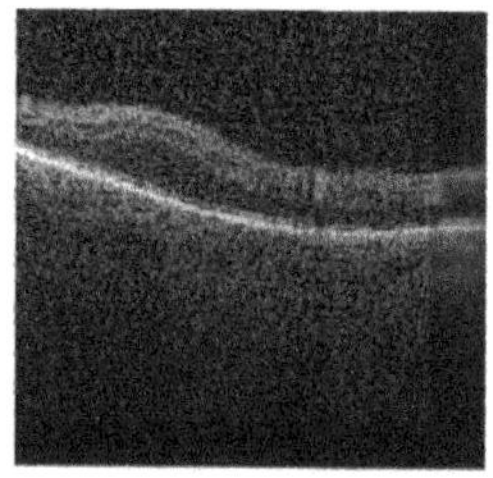	Drusen
CNV	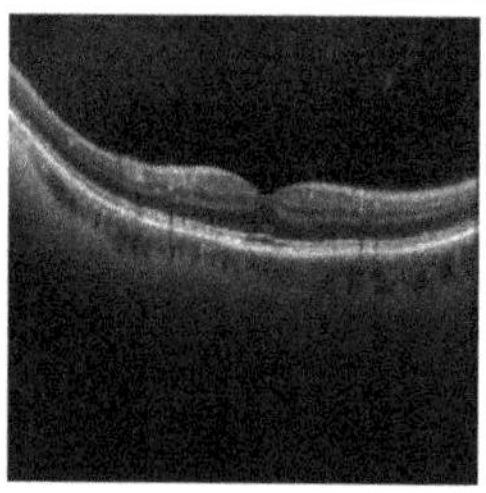	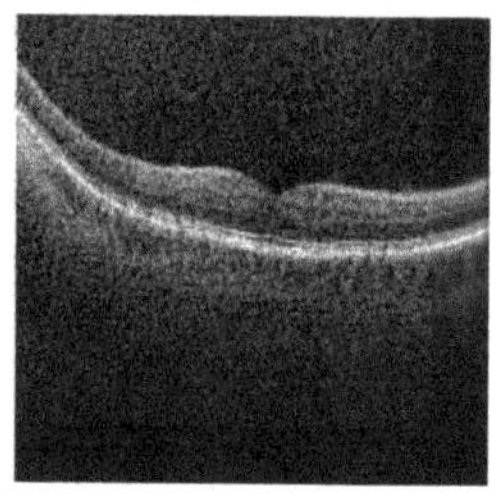	CNV
DME	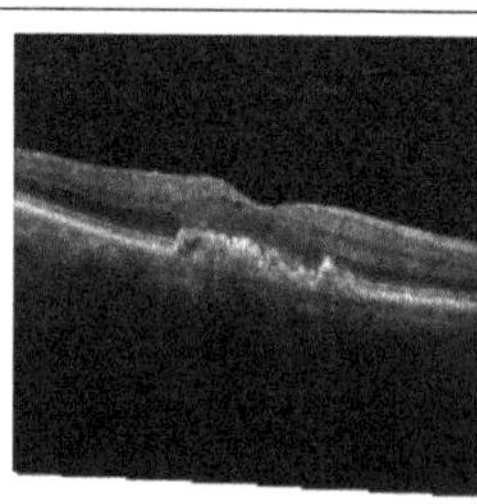	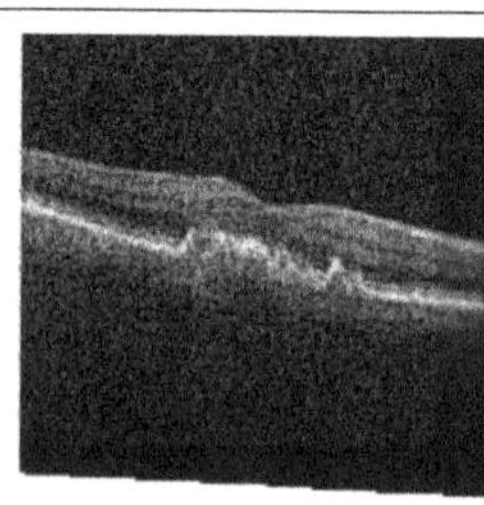	CNV

7. Generalizability Investigation

To see if the proposed model could be used to diagnose other common datasets with the same number of categories, the model was trained under the same constraints as before and the results were analysed for two additional datasets, COVID-19 and Kidney Stone.

7.1. COVID-19

Publicly available COVID-19 dataset [34], with 4273 pneumonia samples, 1583 normal samples, 703 tuberculosis samples, and 576 COVID-19 samples was balanced with an equal number of 576 images for each category. With a training accuracy of 97.18%, a test accuracy of 92.54%, and a validation accuracy of 94.78% as shown in Figure 12a, the model achieved a training loss of 0.0804, test loss of 0.2180 and validation loss of 0.1960 as shown in Figure 12b.

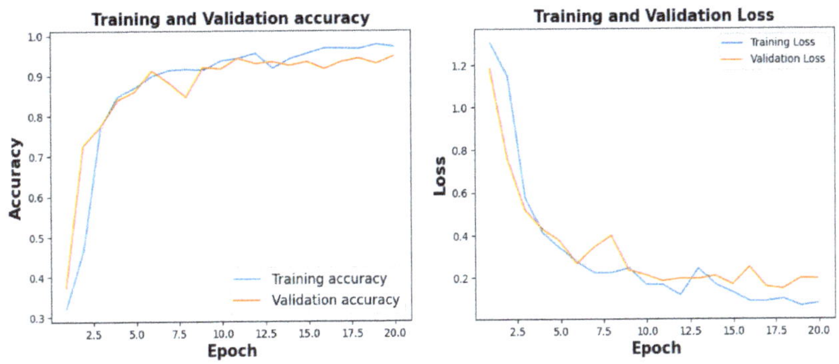

(a) Training and Validation Accuracy (b) Training and Validation Loss

Figure 12. Training and Validation Results for COVID-19.

Figure 13a represents the confusion matrix. Here, six instances of Tuberculosis samples were inaccurately predicted out of a comprehensive pool of 56 samples. Seven mispredictions were observed among pneumonia samples out of 55. Similarly, for COVID-19 cases, seven errors emerged out of 66 samples, whereas all 54 samples categorized as normal were accurately predicted. To offer a more detailed insight into the findings, it is worth noting that the AUC-ROC values for tuberculosis, pneumonia, COVID-19, and normal cases stand at 0.99, 0.98, 0.97, and 1.00, respectively, as shown in Figure 13b.

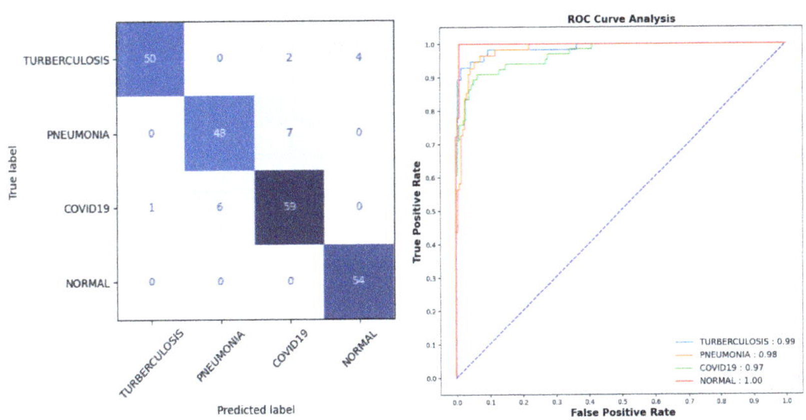

(a) Confusion Matrix (b) ROC Curve

Figure 13. Confusion Matrix and ROC for COVID-19.

Figures 14–17 show the SHAP explanation of tuberculosis, pneumonia, COVID-19, and Normal samples respectively.

Figure 18 shows the LIME segmented results for individual categories of the COVID-19 dataset. The segmented results highlight the infected regions in respective images.

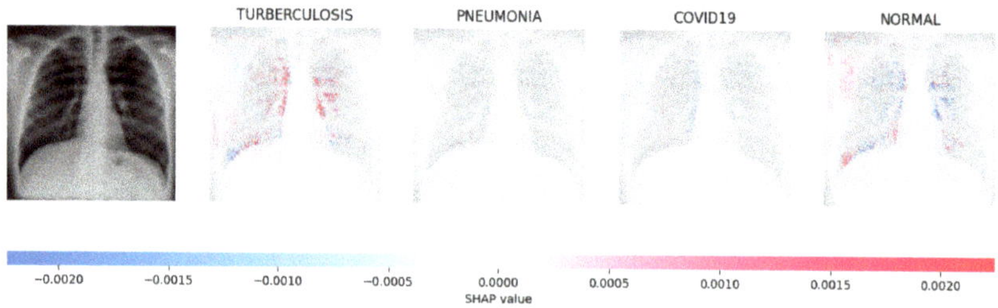

Figure 14. The analysis conducted by the model on the X-ray image indicated elevated concentrations of red pixels in the initial explanatory image (located in the second column). These red pixels are likely to represent regions of the image that are suggestive of tuberculosis.

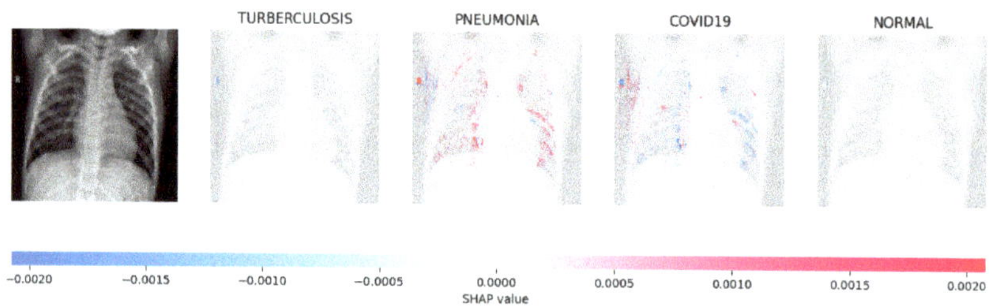

Figure 15. The model's prediction is based on the presence of a substantial number of red pixels in the second explanatory image, situated in the third column, suggesting that the X-ray image depicts pneumonia.

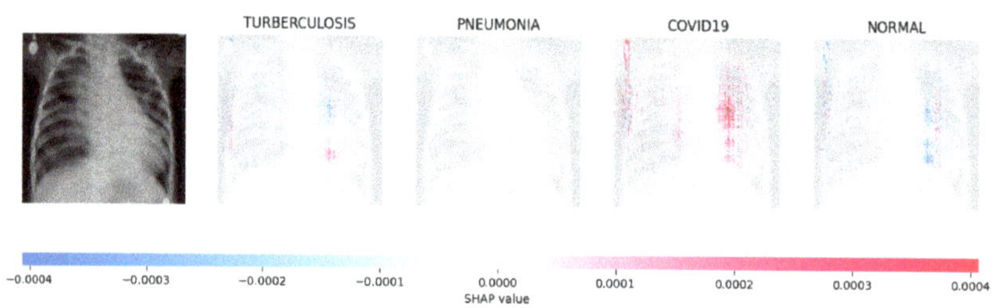

Figure 16. The model's forecast of pneumonia was substantiated by the elevated density of red pixels in the third explanatory image.

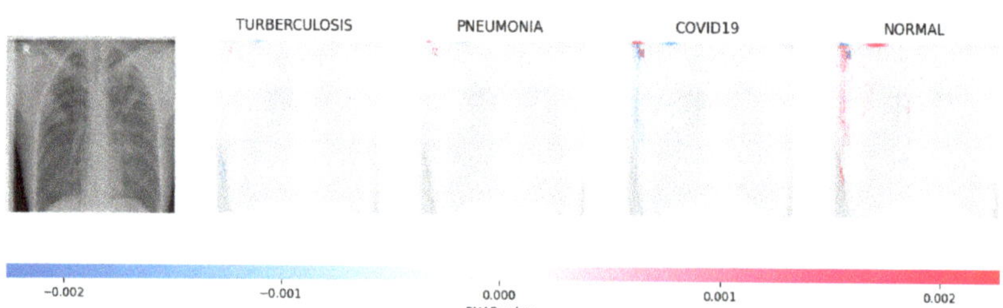

Figure 17. The model determined that the X-ray image was classified as "Normal" because there was a notable concentration of red pixels in the fourth explanatory image.

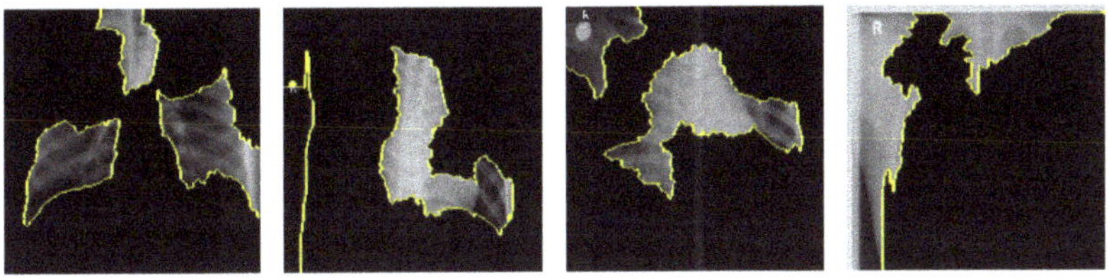

(a) Tuberculosis (b) Pneumonia (c) COVID-19 (d) Normal

Figure 18. Results for LIME segmentation for Figures 14, 15, 16, and 17 respectively.

The proposed model shows better results in comparison with other SOTA approaches and the results are tabulated in Table 7 showing training accuracy of 97.18% and number of parameters (0.983 million).

Table 7. Comparative Analysis of COVID-19 Result to other SOTA methods.

Ref	Algorithm	Accuracy (%)	Parameters (Millions)
[35]	CNN-based CoroNet	89.6	33.97
[36]	Custom CNN	94.53	34.73
[37]	Attention based VGG	85.43	VGG-16 = 18 VGG-19 = 21.2
[12]	Custom CNN	95.94	3.7
Proposed	Custom CNN	97.18	0.983

7.2. Kidney Stone

The publicly available Kidney Stone dataset [14], with 5077 normal samples, 3709 cyst samples, 2283 tumor samples and 1377 stone samples was balanced with an equal number of 1377 images for each category. With a training accuracy of 99.70%, a test accuracy of 99.64%, and a validation accuracy of 99.82% as shown in Figure 19a, the model achieved a training loss of 0.0056, test loss of 0.0345 and validation loss of 0.0078 as shown in Figure 19b.

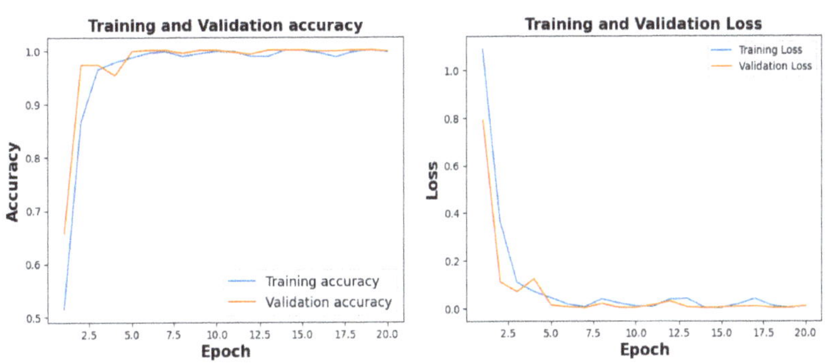

(a) Training and Validation Accuracy (b) Training and Validation Loss

Figure 19. Training and Validation Result for Kidney Stone.

Figure 20a represents the confusion matrix for the kidney stone dataset. All cyst and stone samples were correctly classified. One tumor sample was misclassified as normal, and one normal sample was misclassified as stone. The AUCROC values for all categories are 1.00, which indicates that the model has perfect accuracy as shown in Figure 20b.

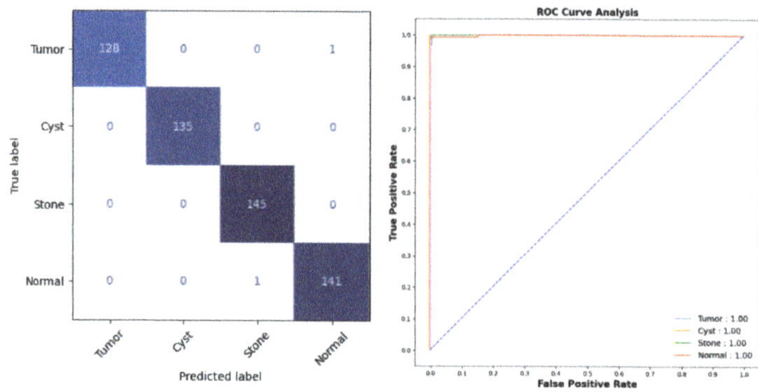

(a) Confusion Matrix (b) ROC Curve

Figure 20. Confusion Matrix and ROC for COVID-19.

Figures 21, 22, 23, and 24 show the SHAP explanation of tumor, cyst, and Normal samples respectively.

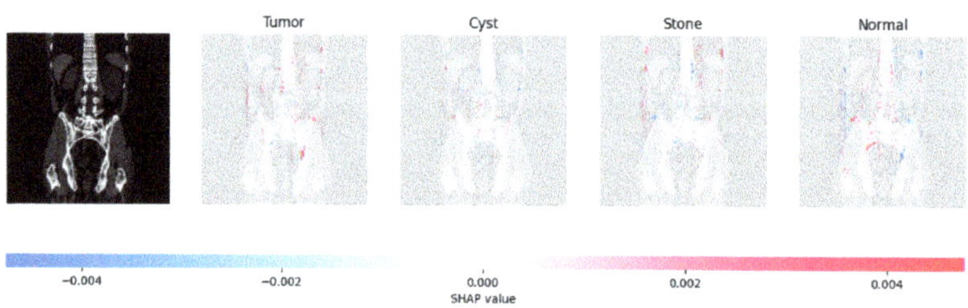

Figure 21. The model found that there were a lot of red pixels in the first explanation image (second column) of the CT scan. These red pixels are likely to be areas of the image that are indicative of a tumor.

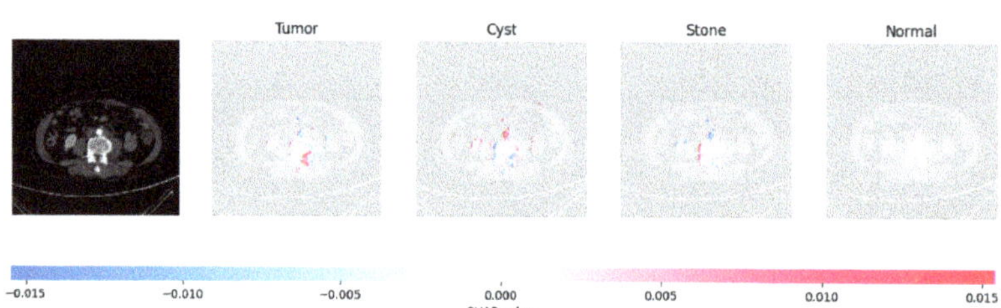

Figure 22. The model's prediction that the CT image is a cyst is supported by the high concentration of red pixels in the explanation image in the third column.

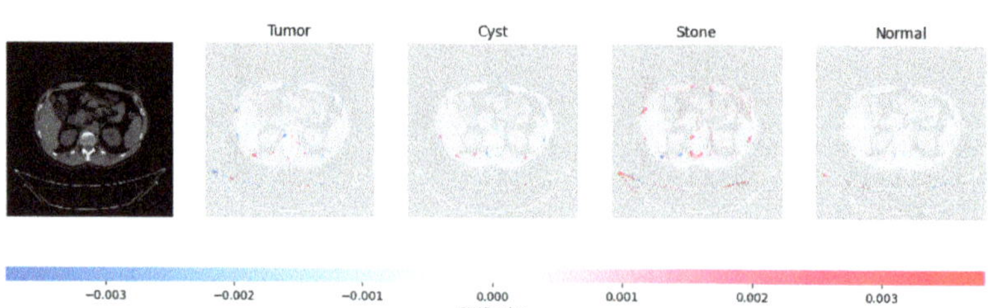

Figure 23. The model's prediction of Stone was supported by the high concentration of red pixels in the third explanation image.

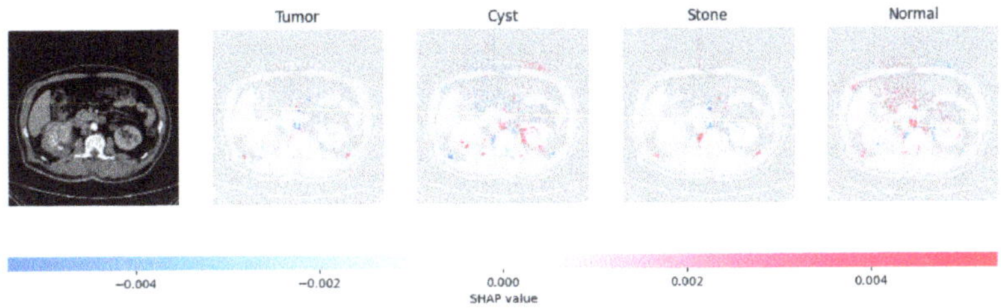

Figure 24. The model found that the CT image is predicted as Normal as the high concentration of red pixels is located in the fourth explanation image.

Figure 25 shows the LIME segmented results for individual categories of the kidney stone dataset. The segmented results highlight the infected regions in respective images.

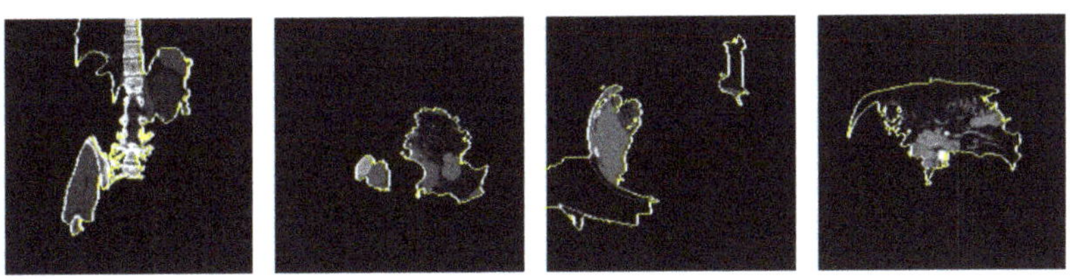

(**a**) Tumor (**b**) Cyst (**c**) Stone (**d**) Normal

Figure 25. Results for LIME segmentation for Figures 21, 22, 23, and 24, respectively.

The proposed model shows the competitive results in comparison with other state-of-the-art methods as tabulated in Table 8 in terms of training accuracy (99.70%) and number of parameters (0.983 million).

Table 8. Comparative Analysis of the Kidney Stone dataset.

Ref	Algorithm	Accuracy (%)	Parameters (Millions)
[14]	Inception V3	61.6	22.32
	VGG16	98.2	14.74
	Resnet	73.8	23.71
	EANet	77.02	6
	Swin Transformers	99.3	4.12
	CCT	96.54	4.07
[38]	DenseNet201-Random Forest	99.44	20
[39]	VGG16NB	96.26	14.74
	DenseNet121-KNN	96.64	20
	VGG-DN-KNN	100	14.74
Proposed	Custom CNN	99.70	0.983

8. Conclusions

The study presents a significant advancement in OCT-based diagnostic methodologies to address the labor-intensive nature of manual anomaly classification in retinal images. Achieving remarkable average accuracies and minimal losses across training, validation and test sets, the proposed model demonstrates its efficacy in classifying CNV, DME, Drusen, and Normal retinal conditions. The integration of XAI techniques provides interpretable insights into the model's decision-making process. Moreover, the model's robustness and generalizability are substantiated by its consistent performance on additional datasets related to COVID-19 and Kidney Stone conditions. With a focus on efficiency and lightweight, the model can play a significant role in IoMT devices.

With more real-time datasets, augmentation, and generative adversarial networks, other lightweight transfer learning models like MobileNet can be tested further as real-time sensations.

Author Contributions: M.B. came up with the research topic, designing the methodology, and writing the code. M.B. and T.B.S. wrote the first version of the document. M.B., T.B.S. and A.N. were responsible for gathering resources, examining the outcome, and making final changes/corrections. All authors have read and agreed to the published version of the manuscript.

Funding: This research received no external funding.

Institutional Review Board Statement: Not applicable.

Informed Consent Statement: The research investigation employed a readily accessible dataset for its analysis, devoid of any direct participation from actual patients in the study.

Data Availability Statement: The data used in the study is publicly accessed on 29 June 2023 https://data.mendeley.com/datasets/rscbjbr9sj/2. The COVID-19 dataset and Kidney Stone dataset are publicly available and cited with URLs in respective sections. Codes will be made available upon request.

Conflicts of Interest: The authors declare no conflict of interest.

References

1. Varela, M.D.; Conti, G.M.; Malka, S.; Vaclavik, V.; Mahroo, O.A.; Webster, A.R.; Tran, V.; Michaelides, M. Coats-like vasculopathy in inherited retinal disease: Prevalence, characteristics, genetics and management. *Ophthalmology* **2023**. [CrossRef]
2. Tariq, A.; Gill, A.Y.; Hussain, H.K. Evaluating the Potential of Artificial Intelligence in Orthopedic Surgery for Value-based Healthcare. *Int. J. Multidiscip. Sci. Arts* **2023**, *2*, 27–35. [CrossRef]
3. Mall, P.K.; Singh, P.K.; Srivastav, S.; Narayan, V.; Paprzycki, M.; Jaworska, T.; Ganzha, M. A comprehensive review of deep neural networks for medical image processing: Recent developments and future opportunities. *Healthc. Anal.* **2023**, *4*, 100216. [CrossRef]
4. Kumari, A.A.; Bhagat, A.; Henge, S.K.; Mandal, S.K. Automated Decision Making ResNet Feed-Forward Neural Network based Methodology for Diabetic Retinopathy Detection. *Int. J. Adv. Comput. Sci. Appl.* **2023**, *14*, 303–314. [CrossRef]
5. Udayaraju, P.; Jeyanthi, P.; Sekhar, B. A hybrid multilayered classification model with VGG-19 net for retinal diseases using optical coherence tomography images. *Soft Comput.* **2023**, *27*, 12559–12570. [CrossRef]
6. Narayan, V.; Mall, P.K.; Alkhayyat, A.; Abhishek, K.; Kumar, S.; Pandey, P. Enhance-Net: An Approach to Boost the Performance of Deep Learning Model Based on Real-Time Medical Images. *J. Sensors* **2023**, *2023*, 8276738. [CrossRef]
7. Sinha, A.; Garcia, D.W.; Kumar, B.; Banerjee, P. Application of Big Data Analytics and Internet of Medical Things (IoMT) in Healthcare with View of Explainable Artificial Intelligence: A Survey. In *Interpretable Cognitive Internet of Things for Healthcare*; Springer: Berlin/Heidelberg, Germany, 2023; pp. 129–163.
8. Carter, S.M.; Rogers, W.; Win, K.T.; Frazer, H.; Richards, B.; Houssami, N. The ethical, legal and social implications of using artificial intelligence systems in breast cancer care. *Breast* **2020**, *49*, 25–32. [CrossRef] [PubMed]
9. Nahavandi, S. Industry 5.0—A human-centric solution. *Sustainability* **2019**, *11*, 4371. [CrossRef]
10. Abou El Houda, Z.; Brik, B.; Khoukhi, L. "why should i trust your ids?": An explainable deep learning framework for intrusion detection systems in internet of things networks. *IEEE Open J. Commun. Soc.* **2022**, *3*, 1164–1176. [CrossRef]
11. Adadi, A.; Berrada, M. Peeking inside the black-box: A survey on explainable artificial intelligence (XAI). *IEEE Access* **2018**, *6*, 52138–52160. [CrossRef]
12. Bhandari, M.; Shahi, T.B.; Siku, B.; Neupane, A. Explanatory classification of CXR images into COVID-19, Pneumonia and Tuberculosis using deep learning and XAI. *Comput. Biol. Med.* **2022**, *150*, 106156. . [CrossRef]
13. Sitaula, C.; Shahi, T.B. Monkeypox virus detection using pre-trained deep learning-based approaches. *J. Med. Syst.* **2022**, *46*, 78. [CrossRef]
14. Islam, M.N.; Hasan, M.; Hossain, M.; Alam, M.; Rabiul, G.; Uddin, M.Z.; Soylu, A. Vision transformer and explainable transfer learning models for auto detection of kidney cyst, stone and tumor from CT-radiography. *Sci. Rep.* **2022**, *12*, 11440. [CrossRef] [PubMed]
15. Subramanian, M.; Kumar, M.S.; Sathishkumar, V.; Prabhu, J.; Karthick, A.; Ganesh, S.S.; Meem, M.A. Diagnosis of retinal diseases based on Bayesian optimization deep learning network using optical coherence tomography images. *Comput. Intell. Neurosci.* **2022**, *2022*, 8014979. [CrossRef] [PubMed]
16. Puneet; Kumar, R.; Gupta, M. Optical coherence tomography image based eye disease detection using deep convolutional neural network. *Health Inf. Sci. Syst.* **2022**, *10*, 13. [CrossRef] [PubMed]
17. Kayadibi, İ.; Güraksın, G.E.; Köse, U. A Hybrid R-FTCNN based on principal component analysis for retinal disease detection from OCT images. *Expert Syst. Appl.* **2023**, *230*, 120617. [CrossRef]

18. Kermany, D.S.; Goldbaum, M.; Cai, W.; Valentim, C.C.; Liang, H.; Baxter, S.L.; McKeown, A.; Yang, G.; Wu, X.; Yan, F.; et al. Identifying medical diagnoses and treatable diseases by image-based deep learning. *Cell* **2018**, *172*, 1122–1131. [CrossRef] [PubMed]
19. Srinivasan, P.P.; Kim, L.A.; Mettu, P.S.; Cousins, S.W.; Comer, G.M.; Izatt, J.A.; Farsiu, S. Fully automated detection of diabetic macular edema and dry age-related macular degeneration from optical coherence tomography images. *Biomed. Opt. Express* **2014**, *5*, 3568–3577. [CrossRef] [PubMed]
20. Kim, J.; Tran, L. Retinal disease classification from oct images using deep learning algorithms. In Proceedings of the 2021 IEEE Conference on Computational Intelligence in Bioinformatics and Computational Biology (CIBCB), Melbourne, Australia, 13–15 October 2021; pp. 1–6.
21. Li, F.; Chen, H.; Liu, Z.; Zhang, X.; Wu, Z. Fully automated detection of retinal disorders by image-based deep learning. *Graefe's Arch. Clin. Exp. Ophthalmol.* **2019**, *257*, 495–505. [CrossRef]
22. Li, F.; Chen, H.; Liu, Z.; Zhang, X.D.; Jiang, M.S.; Wu, Z.Z.; Zhou, K.Q. Deep learning-based automated detection of retinal diseases using optical coherence tomography images. *Biomed. Opt. Express* **2019**, *10*, 6204–6226. [CrossRef]
23. Sunija, A.; Kar, S.; Gayathri, S.; Gopi, V.P.; Palanisamy, P. Octnet: A lightweight cnn for retinal disease classification from optical coherence tomography images. *Comput. Methods Programs Biomed.* **2021**, *200*, 105877.
24. Altan, G. DeepOCT: An explainable deep learning architecture to analyze macular edema on OCT images. *Eng. Sci. Technol. Int. J.* **2022**, *34*, 101091. [CrossRef]
25. Khan, A.; Pin, K.; Aziz, A.; Han, J.W.; Nam, Y. Optical coherence tomography image classification using hybrid deep learning and ant colony optimization. *Sensors* **2023**, *23*, 6706. [CrossRef] [PubMed]
26. Bhowmik, A.; Kumar, S.; Bhat, N. Eye disease prediction from optical coherence tomography images with transfer learning. In Proceedings of the Engineering Applications of Neural Networks: 20th International Conference, EANN 2019, Xersonisos, Crete, Greece, 24–26 May 2019; Proceedings 20; Springer: Berlin/Heidelberg, Germany, 2019; pp. 104–114.
27. Kermany, D.; Zhang, K.; Goldbaum, M. Labeled Optical Coherence Tomography (OCT) and Chest X-ray Images for Classification. *Mendeley Data* **2018**, *2*, 651. [CrossRef]
28. Wong, T.T.; Yeh, P.Y. Reliable accuracy estimates from k-fold cross validation. *IEEE Trans. Knowl. Data Eng.* **2019**, *32*, 1586–1594. [CrossRef]
29. Patel, C.; Bhatt, D.; Sharma, U.; Patel, K.; Patel, R.; Patel, A.; Bhatt, U.; Pandya, S.; Modi, K.; Cholli, N.; et al. DBGC: Dimension Based Generic Convolution Block for Object Recognition. *Sensors* **2022**, *22*, 1780. [CrossRef] [PubMed]
30. Gramegna, A.; Giudici, P. SHAP and LIME: An evaluation of discriminative power in credit risk. *Front. Artif. Intell.* **2021**, *4*, 752558. [CrossRef] [PubMed]
31. Zhang, Y.; Xu, F.; Zou, J.; Petrosian, O.L.; Krinkin, K.V. XAI Evaluation: Evaluating Black-Box Model Explanations for Prediction. In Proceedings of the 2021 II International Conference on Neural Networks and Neurotechnologies (NeuroNT), Saint Petersburg, Russia, 16 June 2021; pp. 13–16.
32. Banerjee, P.; Barnwal, R.P. Methods and Metrics for Explaining Artificial Intelligence Models: A Review. *Explain. AI Found. Methodol. Appl.* **2023**, *232*, 61–88.
33. Sharma, V.; Mir, R.N.; Rout, R.K. Towards secured image steganography based on content-adaptive adversarial perturbation. *Comput. Electr. Eng.* **2023**, *105*, 108484. [CrossRef]
34. JTIPTJ. Chest X-ray (Pneumonia, COVID-19, Tuberculosis). Available online: https://www.kaggle.com/datasets/jtiptj/chest-xray-pneumoniacovid19tuberculosis (accessed on 29 June 2023).
35. Khan, A.I.; Shah, J.L.; Bhat, M.M. CoroNet: A deep neural network for detection and diagnosis of COVID-19 from chest X-ray images. *Comput. Methods Programs Biomed.* **2020**, *196*, 105581. [CrossRef]
36. Qaqos, N.N.; Kareem, O.S. COVID-19 diagnosis from chest X-ray images using deep learning approach. In Proceedings of the 2020 International Conference on Advanced Science and Engineering (ICOASE), Duhok, Iraq, 23–24 December 2020; pp. 110–116.
37. Sitaula, C.; Hossain, M.B. Attention-based VGG-16 model for COVID-19 chest X-ray image classification. *Appl. Intell.* **2021**, *51*, 2850–2863. [CrossRef]
38. Qadir, A.M.; Abd, D.F. Kidney Diseases Classification using Hybrid Transfer-Learning DenseNet201-Based and Random Forest Classifier. *Kurd. J. Appl. Res.* **2023**, *7*, 131–144. [CrossRef]
39. Rajinikanth, V.; Vincent PM, D.R.; Srinivasan, K.; Ananth Prabhu, G.; Chang, C.Y. Framework to Distinguish Healthy/Cancer Renal CT Images using Fused Deep Features. *Front. Public Health* **2023**, *11*, 1109236. [CrossRef] [PubMed]

Journal of
Imaging

MDPI

Article

Multi-Fundus Diseases Classification Using Retinal Optical Coherence Tomography Images with Swin Transformer V2

Zhenwei Li *, Yanqi Han and Xiaoli Yang

College of Medical Technology and Engineering, Henan University of Science and Technology, Luoyang 471023, China; 210321221641@stu.haust.edu.cn (Y.H.); yxl@haust.edu.cn (X.Y.)
* Correspondence: lizhenwei@haust.edu.cn

Abstract: Fundus diseases cause damage to any part of the retina. Untreated fundus diseases can lead to severe vision loss and even blindness. Analyzing optical coherence tomography (OCT) images using deep learning methods can provide early screening and diagnosis of fundus diseases. In this paper, a deep learning model based on Swin Transformer V2 was proposed to diagnose fundus diseases rapidly and accurately. In this method, calculating self-attention within local windows was used to reduce computational complexity and improve its classification efficiency. Meanwhile, the PolyLoss function was introduced to further improve the model's accuracy, and heat maps were generated to visualize the predictions of the model. Two independent public datasets, OCT 2017 and OCT-C8, were applied to train the model and evaluate its performance, respectively. The results showed that the proposed model achieved an average accuracy of 99.9% on OCT 2017 and 99.5% on OCT-C8, performing well in the automatic classification of multi-fundus diseases using retinal OCT images.

Keywords: multi-fundus diseases classification; optical coherence tomography; Swin Transformer V2; PolyLoss function; OCT2017 and OCT-C8

1. Introduction

Fundus diseases include conditions such as diabetic macular edema (DME), choroidal neovascularization (CNV), and drusen, which significantly impact the quality of life [1]. With the continuous development of ophthalmic medicine, OCT technology has become an important diagnostic tool, especially in the diagnosis of fundus diseases. OCT is a non-invasive imaging technique that provides high-resolution retinal images to help diagnose eye diseases, evaluate treatment outcomes, and monitor disease progression [2]. However, due to the large amount of data and complex structural and morphological features of retinal OCT images, manual diagnosis requires a significant amount of time and effort. Therefore, computer-aided diagnosis (CAD) techniques have significant value in the automatic classification of retinal OCT images.

CAD refers to the use of computer technology to analyze and process medical images to provide diagnostic assistance [3]. CAD is now widely used in the automatic analysis and diagnosis of medical images, such as breast cancer, lung cancer, and colorectal cancer. CAD systems can help doctors diagnose diseases quickly and accurately, improving diagnostic accuracy and efficiency. Deep learning is a machine learning technique that has been widely applied in the field of computer-aided diagnosis [4]. Convolutional neural networks (CNNs) are a type of deep learning technique that has been continuously developed since the 1980s. CNNs have achieved great success in the field of computer vision and are widely used in tasks such as image classification, object detection, and semantic segmentation. Some early CNN models include LeNet [5] and AlexNet [6]. As deep learning technology has continued to develop, many new CNN models have emerged, including VGGNet [7], GoogLeNet [8], ResNet [9], DenseNet [10], MobileNet [11], and EfficientNet [12]. Although the existing models have achieved great success, there is still room for improvement in the classification of fundus diseases using OCT images.

Unlike RNNs, which require recursive processing to obtain global information, or CNNs, which can only obtain local information, Transformer is a new neural network architecture that can directly obtain global information. Transformer is essentially an Attention structure that can perform parallel computations, and is therefore much faster than RNNs. Transformer network architecture was proposed by Ashish Vaswani et al. in their paper "Attention Is All You Need" and has been used for machine translation tasks. Unlike previous network architectures, the encoder and decoder in this architecture do not use RNN or CNN network architectures, but instead rely on an architecture that is completely dependent on the attention mechanism [13].

The Swin Transformer is a novel Transformer model that has achieved excellent performance in many computer vision tasks [14]. Compared to the traditional Vision Transformer (ViT) [15], the Swin Transformer utilizes a multi-scale design and integrates the multi-scale design into the Transformer. One of the main features of the Swin Transformer is its pyramidal structure, i.e., the deeper the network is, the smaller the size of the feature map is, and the more channels are available. This is different from the columnar structure of ViT, where the feature map size remains constant. In addition, the Swin Transformer borrows many techniques from CNNs, such as hierarchical feature extraction (FPN), Sliding Window + Attention Mask + Cyclic Shift. These techniques help the Swin Transformer to better capture local information in the image and extract multi-scale features. In conclusion, by adopting a multi-scale design and borrowing techniques from CNNs, the Swin Transformer achieves better performance than traditional ViT models in several computer vision tasks.

1.1. The Proposed Model

In this paper, we propose a multi-foveal disease classification model based on Swin Transformer V2 [16]. The dataset is first subjected to preprocessing operations such as data enhancement, and then the network is trained. Based on the results of training, the network parameters such as learning rate and batch size are fine-tuned to determine the appropriate training parameters. By comparing different loss functions, we finally adopted PolyLoss [17] as the loss function to obtain better performance in retinal OCT image classification. In order to improve the interpretability of the model and understand its decision-making process, visualization methods such as the confusion matrix and Grad-CAM heatmap [18] were used in the testing phase. Finally, after continuous optimization of network parameters and loss functions, the results were compared after multiple training sessions to obtain the optimal network model for multiple fundus disease classification.

The contributions of this paper are as follows:

1. The proposed method will first use the Swin Transformer V2 model to classify multiple diseases in retinal OCT images.
2. Based on the Swin Transformer V2 model, its loss function is improved by introducing PolyLoss, which improves the model's performance.
3. Experimental validation was performed with two datasets, OCT2017 and OCT-C8, and using Grad-CAM visualization to help understand decision-making mechanisms in network models.

1.2. Related Work

The use of deep learning algorithms for identifying OCT images has been extensively studied by many researchers. For example, Lee et al. used a deep neural network to classify OCT images as normal or AMD, achieving an accuracy of 87.63% [19]. Lu et al. and Bhadra et al. used a deep multi-layer CNN to categorize OCT images into healthy, dry AMD, wet AMD, and DME [20]. Kermany et al. applied deep transfer learning to automatically diagnose diabetic retinopathy in OCT images [21]. Rong et al. suggested a different auxiliary classification method, based on CNNs, for the automatic categorization of retinal OCT images [22]. Fang et al. proposed a novel lesion-aware convolutional neural network (LACNN) method for retinal OCT image classification, where retinal lesions in OCT images were used to guide the CNN to achieve more accurate classification [23]. Singh et al. studied attribute-explained deep learning: application to ophthalmic diagnosis and proposed a framework for explaining the classification decisions of a deep learning network on retinal OCT images [24]. Wang et al. proposed classifying volumetric OCT images via a recurrent neural network (VOCT-RNN), which can fully exploit temporal information among B-scans. This choice may introduce unnecessary model complexity, limiting the interpretation of such model results in clinical practice [25]. To investigate this hypothesis, Arefin et al. developed a configurable deep convolutional neural network (CNN) that classifies four types of macular diseases using retinal optical coherence tomography (OCT) images [26]. V et al. proposed a method to improve the automatic classification and detection of macular diseases using retinal optical coherence tomography (OCT) images by fusing two pre-trained deep learning networks [27]. Identifying macular diseases and segmenting lesion areas to assist ophthalmologists in clinical diagnosis is necessary. Liu et al. studied joint disease classification and lesion segmentation in OCT images via a one-stage attention-based convolutional neural network [28]. Deep-learning-based methods have been proposed to address this problem. To evaluate the proposed method, Esfahani et al. used publicly available data including 45 OCT volumes, 15 age-related macular degeneration, 15 diabetic macular edema, and 15 normal volumes captured by Heidelberg OCT imaging equipment [29]. He et al. proposed a method for classifying retinal OCT images using an interpretable Swin-Poly Transformer network [30]. This is a significant contribution to the field of retinal OCT image classification. At the same time, our work has been inspired by this study, and we have improved upon it. Other influential works include those by Lbrahim, Ai, Z, etc. [31,32]. However, to achieve fast and accurate detection results, it is necessary to break out of the existing CNN framework, which is challenging.

The Transformer is a type of model architecture in the field of natural language processing (NLP). Its relatively mature theoretical support and technological development in the field of natural language processing have brought it to the attention of researchers, and it has been shown that Transformer methods can be applied to computer vision tasks, outperforming existing CNN methods in some tasks [33]. The Vision Transformer (ViT) is a model proposed by the Google team in 2020 that applies the Transformer to image classification. Its model is "simple" and effective, with strong scalability (the larger the model, the better the performance), and performs well in the field of computer vision. The Swin Transformer is a new type of visual Transformer that can serve as a general backbone network for computer vision. It adopts a hierarchical structure and shifted windows to effectively extract multi-scale features. In addition, some researchers have attempted to combine Transformers and CNNs to improve prediction performance. For example, when performing object detection in drone images, a Transformer-based model can be fused with a CNN-based model [34]. Swin Transformer V2 is a large model for computer vision that addresses three main issues in training and applying large visual models, including training instability, the resolution gap between pre-training and fine-tuning, and the need for labeled data. Swin Transformer V2 can better handle complex image data and achieve excellent performance in the automatic classification of retinal OCT images.

2. Materials and Methods

The overall framework of the proposed method is illustrated in Figure 1. The PolyLoss loss function is employed during the experiment to enhance the training efficiency of the model. Data augmentation methods are applied during the training phase to increase the diversity of the training data and enhance the network's ability to generalize. After training, Grad-CAM is utilized to visualize and explain the results.

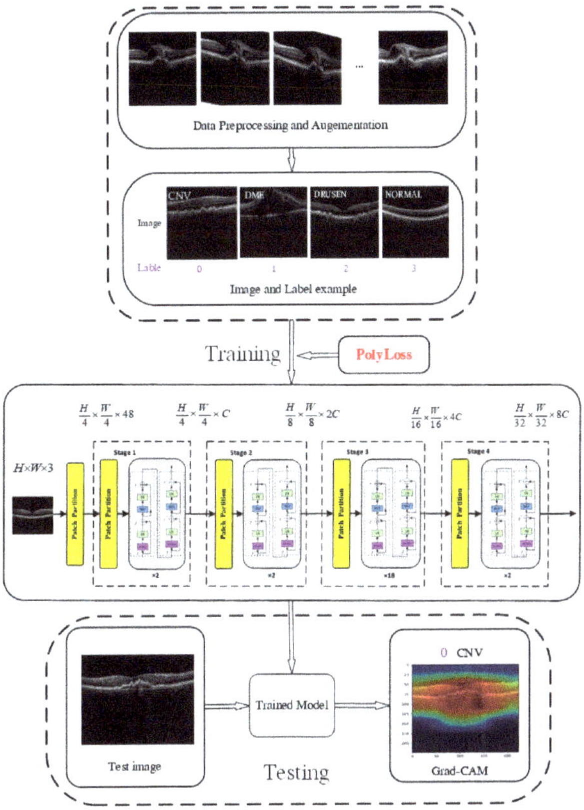

Figure 1. The overall framework of the proposed method.

2.1. Architecture of Swin Transformer V2

Swin Transformer V2 is an upgraded version of Swin Transformer. It improves upon version 1.0 by making the model larger and able to adapt to different image resolutions and window sizes. The Swin Transformer V2 block incorporates two Swin Transformer modules, the window multi-head self-attention (W-MSA) module and the shifted window multi-head self-attention (SW-MSA) module, in place of the standard multi-head self-attention (MSA) module found in ViT. In addition, when calculating Attention in the Transformer block in ViT, the dot(Q,K) operation is used, which is replaced by cosine(Q,K)/τ in Swin V2, where τ is a learnable parameter that is not shared between blocks. The cosine operation inherently includes normalization, which further stabilizes the attention output values.

Figure 2 illustrates the overall structure of the Swin Transformer V2 model [14]. The input image, with a size of 256 $\times$ 256, is first divided into non-overlapping 4 $\times$ 4 patches by the patch partitioning module. These patches are then treated as 'tokens' and projected into C dimensions using a linear embedding layer. Two consecutive Swin Transformer V2 blocks with self-attention computation are applied to these patch tokens, controlling their number as shown in Figure 2b. A 'stage' consists of a linear embedding layer and Swin Transformer V2 blocks. The design of Swin Transformer V2 resembles the layer structure of CNNs, where the resolution is halved, and the number of channels is doubled at each stage. To produce hierarchical representations, the Swin Transformer reduces the number of tokens by merging patch layers, making the network deeper. Figure 3a shows an example of a hierarchical representation. Differing from the 224 $\times$ 224 input resolution used by He et al. [30], we employ Swin Transformer V2, which uses a higher resolution of

256 × 256. The advantage of this is that the network has access to more features, and increasing the feature extraction capability of the network improves the performance of the model.

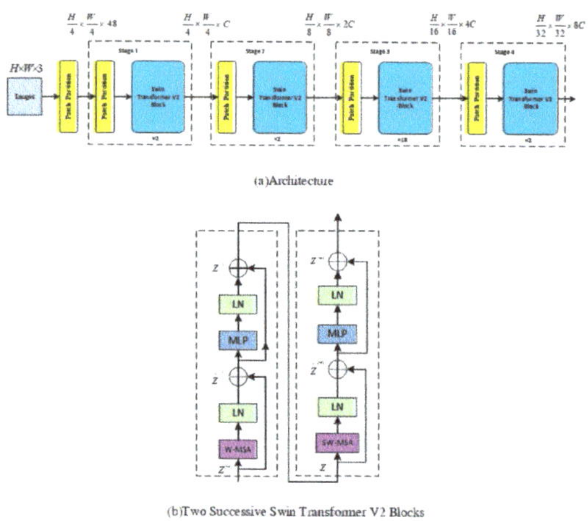

Figure 2. (**a**) he overall architecture of Swin Transformer V2. (**b**) Two successive Swin Transformer V2 blocks.

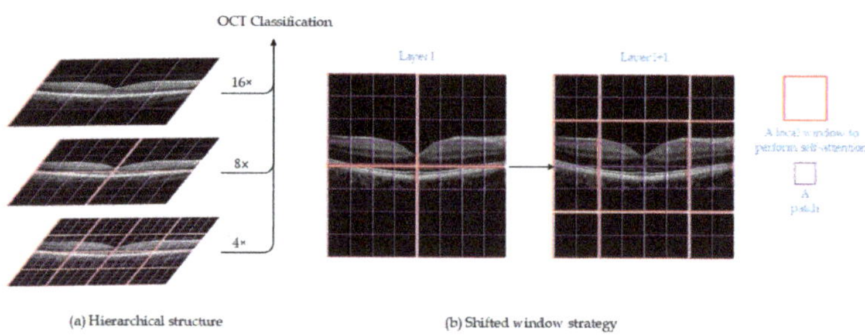

Figure 3. (**a**) The hierarchical structure of Swin Transformer V2 for extracting multi-scale feature representation. (**b**) An illustration of the shifted window strategy for computing self-attention in the Swin Transformer V2 architecture.

Each Swin Transformer V2 block comprises two units, with each unit containing two normalization layers (LayerNorm), a self-attention module, and a multi-layer perceptron (MLP) layer. The standard multi-head self-attention (MSA) module from ViT is replaced by two consecutive Swin Transformer V2 modules in the Swin Transformer V2 block: the window multi-head self-attention (W-MSA) module and the shifted window multi-head self-attention (SW-MSA) module, as shown in Figure 2b. The first unit utilizes the window MSA (W-MSA) module, while the second unit employs the shifted window MSA (SW-MSA) module. In contrast to the Swin Transformer, Swin Transformer V2 incorporates a LayerNorm layer after each MSA module and MLP layer and implements residual connections after each module.

2.2. Shifted-Window-Based Self-Attention

A method of calculating self-attention within local windows is used to reduce computational complexity and improve modeling efficiency. The moving window strategy used to calculate self-attention in this experiment is shown in Figure 3a. In the ViT architecture, the standard MSA module is used for global attention, resulting in an unbearable amount of computation and quadratic computational complexity. In W-MSA, this relationship is linear, and the amount of computation is acceptable. Assuming that each window includes M × M patches, windows are organized in a non-overlapping manner to split the image in an equal amount. On an image with hardware patches, the global MSA module's computational complexity and the window-based MSA module's computational complexity are, respectively:

$$\Omega(\text{MSA}) = 4hwC^2 + 2(hw)^2C \tag{1}$$

$$\Omega(\text{W-MSA}) = 4hwC^2 + 2M^2hwC \tag{2}$$

where h × w is the total number of patches in the picture, and C denotes the patch channel's channel. When M is constant (the default value is 7), the complexity of Equation (2) is linear as opposed to Equation (1), where the difficulty is quadratic with respect to the number of patches h × w.

The window-based self-attention module lacks cross-window connections, ignoring the relationships between different windows and limiting modeling capabilities. This approach switches between two partition configurations in succeeding Swin Transformer V2 blocks to set up cross-window connections while retaining the computational efficiency of non-overlapping windows. As identified in Figure 4 [14], the first module equally divides the 8 × 8 feature map into 2 × 2 windows of size 4 × 4 (M = 4) using a standard window partitioning approach starting from the top-left pixel. Then, the next module adopts a window configuration that is offset from the previous layer's window configuration by shifting the window from the regular partitioned window by $(\left[\frac{M}{2}\right], \left[\frac{M}{2}\right])$pixels. In the new window, the self-attention calculation also takes into account the boundary of the previous window, thus considering the connection information between different windows. Using the shifted window partitioning method, consecutive Swin Transformer V2 blocks are calculated as:

$$\hat{Z}^l = \text{W-MSA}(\text{LN}(Z^{l-1})) + Z^{l-1} \tag{3}$$

$$Z^l = \text{MLP}(\text{LN}(\hat{Z}^l)) + \hat{Z}^l \tag{4}$$

$$\hat{Z}^{l+1} = \text{SW-MSA}(\text{LN}(\hat{Z}^l)) + \hat{Z}^l \tag{5}$$

$$Z^{l+1} = \text{MLP}(\text{LN}(\hat{Z}^{l+1})) + \hat{Z}^{l+1} \tag{6}$$

where W-MSA and SW-MSA indicate window-based multi-head self-attention utilizing normal and shifted window partitioning configurations, respectively; and $\hat{Z}^l$ and Z^l denote the output characteristics of the (S)W-MSA module and MLP in the l layer, respectively.

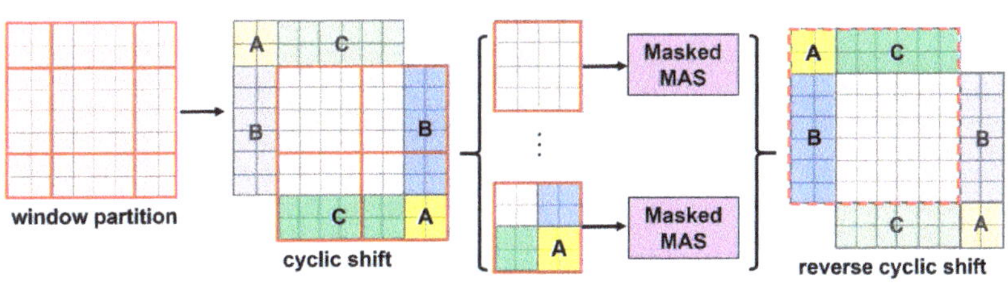

Figure 4. Illustration of an efficient batch computation approach for self-attention in shifted window partitioning.

A number of new windows are produced by the window partitioning technique, some of which are smaller than M × M. One typical method for calculating self-attention is to flatten all windows to M × M. This method, however, results in more windows. For instance, in Figure 3b, the window transformation technique results in a large rise in the computational cost of the model when the number of windows goes from 2 × 2 to 3 × 3. As demonstrated in Figure 4, we apply an effective batch computation technique that cyclically shifts to the top left to address this problem. The batch-calculated windows may include a number of non-adjacent windows in the feature map after shifting. Therefore, to confuse the self-attention calculation for each sub-window, we use a masking method. The computational efficiency is increased for cyclic shifting since the number of batch windows and regular window divisions stays constant.

2.3. PolyLoss

The PolyLoss function has been demonstrated to outperform cross-entropy loss and focal loss in tasks such as 3D detection, 2D picture classification, instance segmentation, and object identification. As a result, in this experiment, we adopted PolyLoss as the loss function for our model to improve the OCT classification model's classification accuracy. The coefficients of the polynomial are represented by, and the PolyLoss formula is expressed as follows:

$$L_{Poly} = \alpha_1(1-P_t) + \alpha_2(1-P_t)^2 + \cdots + \alpha_N(1-P_t)^N + \cdots = \sum_{j=1}^{\infty} \alpha_j(1-P_t)^j \quad (7)$$

There are an endless number of polynomial coefficients that need to be changed in this formula. Tuning multiple polynomial coefficients would still result in a dauntingly large search space, which is not feasible. Additionally, cross-entropy loss does not perform better than many coefficients being tuned simultaneously. This problem is solved by perturbing the leading polynomial coefficient in the cross-entropy loss while leaving the other coefficients constant. The loss formula is written as Poly-N, where N is the quantity of leading coefficients that need to be changed.

$$\begin{aligned} L_{Poly-N} &= \underbrace{(\varepsilon_1+1)(1-P_t) + \cdots + (\varepsilon_N + 1/N)(1-P_t)^N}_{perturbed\ by\ \varepsilon_j} + \underbrace{1/(N+1)(1-P_t)^{N+1} + \cdots}_{same\ as\ CrossEntropy} \\ &= -\log(P_t) + \sum_{j=1}^{N} \varepsilon_j (1-P_t)^j \end{aligned} \quad (8)$$

In particular, we update the cross-entropy loss's j polynomial coefficient from $1/j$ to $1/j + \varepsilon_j$, where $\varepsilon_j \in [-1/j, \infty)$ is the perturbation term. Equation (8) demonstrates how the first N polynomials may be precisely computed without having to worry about an endless number of higher-order (j > N + 1) coefficients. The largest increase is possible for the first polynomial term. The final PolyLoss formula is as follows with further simplification of the Poly-N formula and concentration on Poly-1 evaluation, where only the first polynomial coefficient in the cross-entropy loss is changed:

$$L_{Poly-1} = (1+\varepsilon_1)(1-P_t) + 1/2(1-P_t)^2 + \cdots = -\log(P_t) + \varepsilon_1(1-P_t) \quad (9)$$

In this experiment, we accomplish OCT image classification using the value of $\varepsilon_1 = 2$.

2.4. Datasets

In this paper, two public datasets, OCT2017 [35] and OCT-C8 [36], were used to train and test the network model. Dataset 1, as shown in Figure 5, depicts examples of three fundus diseases and normal retina, while Dataset 2, as shown in Figure 6 [37], depicts OCT images of seven diseases and one normal category of retinal OCT images. The OCT2017 dataset contains images of three diseases: choroidal neovascularization (CNV), diabetic macular edema (DME), Drusen, and a class of normal fundus. The OCT2017 dataset contains 84,452 retinal OCT images of 4 classes (as shown in Figure 5): 83,484 training images and 968 test images. The training set includes 36,205 CNV images, 10,348 DME images, 7616 DRUSEN images, and 25,315 NORMAL images for training and four classes of 1000 images each for validation. Details of the two datasets have been shown in Table 1.

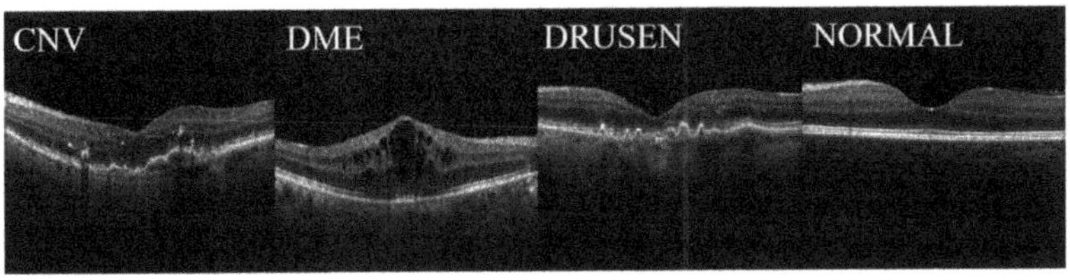

Figure 5. Optical coherence tomography images from the OCT2017 dataset. The panels display images of choroidal neovascularization (CNV) on the far left, diabetic macular edema (DME) on the middle left, drusen on the middle right, and a normal image on the far right.

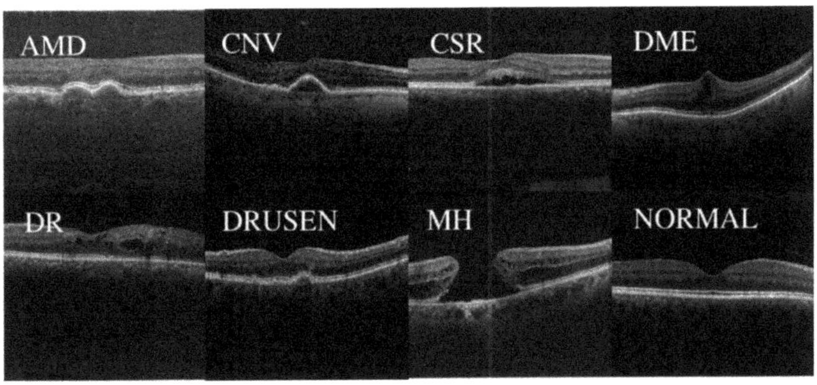

Figure 6. Displays examples of the eight classes in the OCT-C8 dataset, including AMD, CNV, CSR, DME, DR, DRUSEN, MH, and NORMAL.

Table 1. Classification and dataset setup for datasets OCT2017 and OCT-C8.

Dataset	Class	Number	Train	Validation	Test
OCT2017	CNV	37,447	36,205	1000	242
	DME	11,590	10,348	1000	242
	DRUSEN	8858	7616	1000	242
	NORMAL	26,557	25,315	1000	242
OCT-C8	AMD	3000	2300	350	350
	CNV	3000	2300	350	350
	CSR	3000	2300	350	350
	DME	3000	2300	350	350
	DR	3000	2300	350	350
	DRUSEN	3000	2300	350	350
	MH	3000	2300	350	350
	NORMAL	3000	2300	350	350

The OCT-C8 dataset contains 24,000 images of eight categories (as shown in Figure 6), including AMD, choroidal neovascularization (CNV), central serous retinopathy (CSR), DME, diabetic retinopathy (DR), drusen, macular hole (MH), and one for healthy classes. The training set consists of 2300 images per category for a total of 18,400 images for training and 2800 images each for testing and validation containing 350 images per category for the network model. Before training the model, we preprocessed and augmented the data. Obtaining a large number of labeled medical images is challenging due to the time-consuming nature of the labeling process and the need for professional medical expertise, which can be costly. To increase the diversity of the training data, data augmentation methods such as random rotation, cropping, and mirroring were used. Additionally, the images were resized to 256 × 256 and normalized to match the model's input requirements. In the final step, the data were converted into tensors and fed into the model for training. This process helps to enhance the model's ability to generalize and improve its stability.

2.5. Evaluation Metrics

To evaluate the performance of the model in classification, we use Accuracy, Precision, and Recall as evaluation metrics. The formulas for these evaluation metrics are shown below.

$$\text{Accuracy} = \frac{\text{TP} + \text{TN}}{\text{TP} + \text{TN} + \text{FP} + \text{FN}} \tag{10}$$

$$\text{Precision} = \frac{\text{TP}}{\text{TP} + \text{FP}} \tag{11}$$

$$\text{Recall} = \frac{\text{TP}}{\text{TP} + \text{FN}} \tag{12}$$

$$\text{F1-score} = \frac{2\text{TP}}{2\text{TP} + \text{FP} + \text{FN}} \tag{13}$$

The numbers TP, TN, FP, and FN stand for the corresponding amounts of true positives, true negatives, false positives, and false negatives. For OCT classification, TP is defined as the proportion of cases that the model correctly classified as positive, TN as the proportion of cases that the model correctly classified as negative, FP as the proportion of negative samples that the model incorrectly classified as positive, and FN as the proportion of positive cases that the model incorrectly classified as negative.

3. Results

In this research, the network was trained and evaluated on a Windows 10 operating system with 64 GB of memory, an NVIDIA 4090 24 GB GPU, a 2 TB solid-state drive, Python 3.7, and PyTorch 1.10.1 + cu102. At the start of each experiment, we imported ImageNet-22K pre-trained models through transfer learning. The input resolution for the EfficientNetV2 is set to 384 × 384, the VIT and Swin Transformer models are set to 224 × 224, and the V2 model supports higher resolution image input than the Swin Transformer, set to 256 × 256. The batch size was set to 32 and each model was trained for 200 epochs. During training, we saved the models with the highest accuracy and lowest loss function and selected the model with the highest test accuracy as the optimal model through comparison.

The performance of each category in the OCT2017 dataset was tested using pre-trained EfficientNetV2 [38], Vision Transformer (VIT), Swin Transformer, and our improved Swin Transformer V2 network. Table 2 shows the experimental results for the three retinal disease and normal category diagnoses when the CrossEntropy loss function is used for the four network models on the dataset OCT2017. Table 3 shows the experimental results obtained for different network models on the same dataset when using the PolyLoss function.

To further validate our models, we also tested and analyzed the performance of the VIT, Swin Transformer, and Swin Transformer V2 network models on the OCT-C8 dataset using CrossEntropyLoss, with the results shown in Table 4, and the PolyLoss loss function, with the results shown in Table 5, to categorize the performance of the VIT, Swin Transformer, and Swin Transformer V2 network models.

In order to visualize the performance of each model more intuitively, we use the confusion matrix to visualize the matching results between the model predictions and the true categories. The results obtained by our models on the OCT2017 and OCT-C8 datasets using different loss functions, respectively, are shown in Figure 7a,c are the results when CrossEntropy is applied, and Figure 7b,d represent the results obtained by the PolyLoss function. The diagonal elements in the confusion matrix represent the correct classification, and the remaining elements represent the misclassification.

Table 2. Classification results using OCT2017 with a CrossEntropy loss function. Significant values are in [bold].

Dataset	Method	Class	Accuracy	Precision	Recall	Specificity	F1-Score
OCT2017	EfficientNetV2	CNV	0.975	0.913	0.996	0.968	0.953
		DME	0.986	0.996	0.946	0.968	0.970
		DRUSEN	0.977	**1.0**	0.909	0.999	0.952
		NORMAL	0.988	0.953	**1.0**	0.983	0.976
	VIT	CNV	0.950	0.839	0.992	0.937	0.909
		DME	0.975	0.987	0.913	0.996	0.949
		DRUSEN	0.951	0.990	0.814	0.997	0.893
		NORMAL	0.982	0.934	**1.0**	0.977	0.966
	Swin Transformer	CNV	0.995	0.980	**1.0**	0.993	0.990
		DME	**0.999**	**1.0**	0.996	**1.0**	**0.998**
		DRUSEN	0.996	**1.0**	0.983	**1.0**	0.991
		NORMAL	**1.0**	**1.0**	**1.0**	**1.0**	**1.0**
	Swin Transformer V2	CNV	**0.996**	**0.984**	**1.0**	**0.994**	**0.992**
		DME	0.997	**1.0**	**0.988**	**1.0**	0.994
		DRUSEN	**0.999**	**1.0**	**0.996**	**1.0**	**0.998**
		NORMAL	**1.0**	**1.0**	**1.0**	**1.0**	**1.0**

Table 3. Classification results using OCT2017 with a PolyLoss function. Significant values are in [bold].

Dataset	Method	Class	Accuracy	Precision	Recall	Specificity	F1-Score
OCT2017	EfficientNetV2	CNV	0.971	0.896	1.0	0.961	0.945
		DME	0.987	**1.0**	0.946	**1.0**	0.972
		DRUSEN	0.976	**1.0**	0.905	**1.0**	0.950
		NORMAL	0.992	0.968	1.0	0.980	0.984
	VIT	CNV	0.952	0.845	0.992	0.939	0.913
		DME	0.978	0.987	0.926	0.996	0.956
		DRUSEN	0.950	0.985	0.814	0.996	0.891
		NORMAL	0.985	0.942	**1.0**	0.979	0.970
	Swin Transformer	CNV	0.997	0.988	**1.0**	**0.996**	**0.994**
		DME	**0.999**	**1.0**	0.996	**1.0**	**0.998**
		DRUSEN	0.998	**1.0**	0.992	**1.0**	**0.996**
		NORMAL	**1.0**	**1.0**	**1.0**	**1.0**	**1.0**
	Ours	CNV	**0.999**	**0.996**	**1.0**	**0.996**	**0.994**
		DME	**0.999**	**1.0**	**1.0**	**1.0**	**0.998**
		DRUSEN	**1.0**	**1.0**	**1.0**	**1.0**	**0.996**
		NORMAL	**1.0**	**1.0**	**1.0**	**1.0**	**1.0**

Table 4. Classification results using OCT-C8 with a CrossEntropy loss function. Significant values are in [bold].

Dataset	Method	Class	Accuracy	Precision	Recall	Specificity	F1-Score
OCT-C8	VIT	AMD	**1.0**	**1.0**	**1.0**	**1.0**	**1.0**
		CNV	0.965	0.873	0.846	0.982	0.859
		CSR	0.993	0.958	0.986	0.994	0.972
		DME	0.962	0.901	0.783	0.988	0.838
		DR	0.989	0.954	0.954	0.993	0.954
		DRUSEN	0.943	0.775	0.769	0.968	0.772
		MH	0.991	0.977	0.951	0.997	0.964
		NORMAL	0.959	0.787	0.920	0.964	0.848
	Swin Transformer	AMD	**1.0**	**1.0**	**1.0**	**1.0**	**1.0**
		CNV	**0.988**	0.954	**0.951**	0.993	**0.952**
		CSR	**1.0**	**1.0**	**1.0**	**1.0**	**1.0**
		DME	0.990	0.968	0.957	0.996	0.962
		DR	**1.0**	**1.0**	**1.0**	**1.0**	**1.0**
		DRUSEN	**0.985**	**0.956**	0.937	**0.992**	**0.946**
		MH	**1.0**	**1.0**	**1.0**	**1.0**	**1.0**
		NORMAL	0.987	0.945	**0.977**	0.992	0.961
	Swin Transformer V2	AMD	**1.0**	**1.0**	**1.0**	**1.0**	**1.0**
		CNV	**0.988**	**0.959**	0.940	**0.994**	0.949
		CSR	**1.0**	**1.0**	**1.0**	**1.0**	**1.0**
		DME	**0.992**	**0.974**	**0.963**	**0.996**	**0.968**
		DR	**1.0**	**1.0**	**1.0**	**1.0**	**1.0**
		DRUSEN	**0.985**	0.938	**0.946**	0.991	0.942
		MH	**1.0**	**1.0**	**1.0**	**1.0**	**1.0**
		NORMAL	**0.991**	**0.955**	**0.977**	**0.993**	**0.966**

Table 5. Classification results using OCT-C8 with a PolyLoss loss function. Significant values are in [bold].

Dataset	Method	Class	Accuracy	Precision	Recall	Specificity	F1-Score
OCT-C8	VIT	AMD	**1.0**	**1.0**	**1.0**	**1.0**	**1.0**
		CNV	0.967	0.893	0.834	0.986	0.862
		CSR	0.994	0.961	0.991	0.994	0.976
		DME	0.962	0.894	0.794	0.987	0.841
		DR	0.989	0.957	0.957	0.994	0.957
		DRUSEN	0.943	0.772	0.774	0.967	0.773
		MH	0.992	0.985	0.954	0.998	0.969
		NORMAL	0.958	0.781	0.917	0.963	0.844
	Swin Transformer	AMD	**1.0**	**1.0**	**1.0**	**1.0**	**1.0**
		CNV	0.988	0.959	0.943	**0.995**	0.952
		CSR	**1.0**	**1.0**	**1.0**	**1.0**	**1.0**
		DME	0.991	**0.974**	0.957	**0.996**	0.965
		DR	**1.0**	**1.0**	**1.0**	**1.0**	**1.0**
		DRUSEN	**0.988**	0.954	**0.940**	0.993	0.947
		MH	**1.0**	**1.0**	**1.0**	**1.0**	**1.0**
		NORMAL	0.990	0.938	**0.986**	**0.993**	0.961
	Ours	AMD	**1.0**	**1.0**	**1.0**	**1.0**	**1.0**
		CNV	**0.989**	**0.965**	**0.949**	**0.995**	**0.957**
		CSR	**1.0**	**1.0**	**1.0**	**1.0**	**1.0**
		DME	**0.992**	0.963	**0.977**	0.995	**0.970**
		DR	**1.0**	**1.0**	**1.0**	**1.0**	**1.0**
		DRUSEN	**0.988**	**0.965**	0.934	**0.995**	**0.949**
		MH	**1.0**	**1.0**	**1.0**	**1.0**	**1.0**
		NORMAL	**0.991**	**0.948**	0.980	0.992	**0.964**

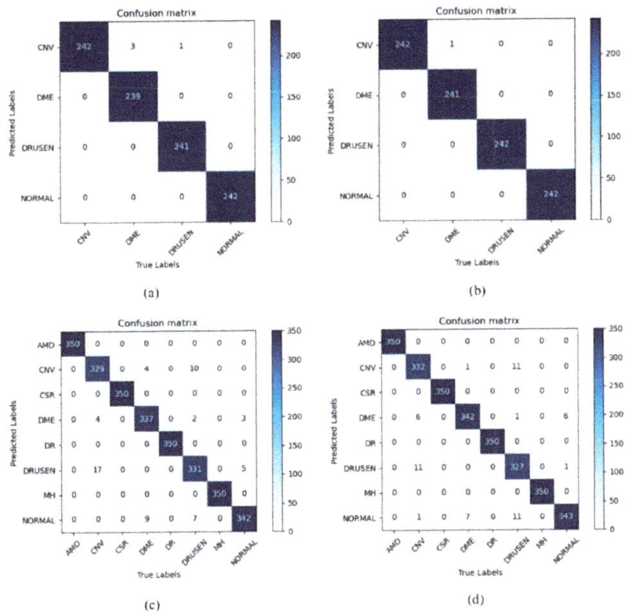

Figure 7. The Confusion matrix of our model on (**a**) OCT2017 (CrossEntropyLoss), (**b**) OCT2017 (PolyLoss), (**c**) OCT-C8 (CrossEntropyLoss), and (**d**) OCT-C8 (PolyLoss).

4. Discussion

As can be seen from Table 2, EfficientNetV2 achieved an accuracy of 0.975 in the CNV category, and the highest accuracy of 0.988 was obtained in the normal category, with an F1-Score of 0.953 and 0.976 in the CNV and normal, respectively. The category accuracies of 0.986 and 0.977 were achieved in the DME and DRUSEN, respectively, while the VIT model obtained an overall lower evaluation metric than EfficientNetV2 on all four categories. Both Swin Transformer and our model achieved more than 99% accuracy on a single category, and the evaluation metrics achieved a score of 1 on the normal category. Table 3 shows that when using the PolyLoss function, EfficientNetV2 shows a slight decrease in diagnostic performance on the CNV and DRUSEN categories and a slight increase on the DME and NORMAL categories. The evaluation metrics for the three retinal disease diagnoses improved on Swin Transformer and our model. Compared to the Swin Transformer, our model obtained a higher performance evaluation with a category diagnostic accuracy of 0.999 for both CNV and DME. An accuracy score of 1 was obtained on DEUSEN and normal fundus.

Table 6 is the average of the experimental results obtained using the CrossEntropy and PolyLoss functions on the OCT2017 and OCT-C8 datasets, respectively. We observed that the performance of the EfficientNetV2 network was better than that of VIT when using CrossEntropy loss, with average accuracies of 98.2% and 96.5%, respectively. However, the Swin Transformer model achieved a 3.3% average accuracy improvement over EfficientNetV2 and performed better. We achieved an average accuracy of 99.8% using Swin Transformer V2, which improved on Precision, Recall, Specificity, and F1-Score compared to the Swin Transformer. When the loss function was changed from CrossEntropyLoss to Polyloss, although the Swin Transformer network achieved the same accuracy, it improved in several other evaluation metrics. It can be seen that when using PolyLoss, compared with CrossEntropyLoss, Swin Transformer V2 showed an improvement in Performance, with a 0.3% increase in Precision, a 0.4% increase in Recall, and a 0.1% increase in F1-Score. Swin Transformer V2 achieved 100% Precision, Recall, and Sensitivity in the DME, DRUSEN, and NORMAL categories and achieved near 1.0 accuracy in the CNV, DME, DRUSEN, and

NORMAL categories. This proves the excellent classification ability of Swin Transformer V2 on the OCT dataset and that using the PolyLoss loss function can further improve the performance of the network.

Table 6. Average of experimental results using CrossEntropy and PolyLoss functions on datasets OCT2017 and OCT-C8, respectively. Significant values are in bold.

Dataset	Method	Loss	Accuracy	Precision	Recall	Specificity	F1-Score
OCT2017	EfficientNetV2	CrossEntropy	0.982	0.966	0.963	0.980	0.963
		PolyLoss	0.981	0.966	0.963	0.985	0.963
	VIT	CrossEntropy	0.965	0.938	0.930	0.977	0.917
		PolyLoss	0.966	0.940	0.933	0.978	0.933
	Swin Transformer Paper [30]	CrossEntropy	0.998	0.995	0.995	0.998	0.995
		PolyLoss	0.998	0.997	0.997	**0.999**	**0.997**
	Swin Transformer V2 Ours	CrossEntropy	0.998	0.996	0.996	0.999	0.996
		PolyLoss	**0.999**	**0.999**	**1.0**	**0.999**	**0.997**
OCT-C8	VIT	CrossEntropy	0.975	0.903	0.901	0.986	0.901
		PolyLoss	0.976	0.905	0.903	0.986	0.903
	Swin Transformer Paper [30]	CrossEntropy	0.994	0.978	0.978	0.997	0.978
		PolyLoss	0.994	0.978	0.978	**0.997**	0.978
	Swin Transformer V2 Ours	CrossEntropy	0.995	0.978	0.978	0.997	0.978
		PolyLoss	**0.995**	**0.980**	**0.980**	**0.997**	**0.980**

On the OCT-C8 dataset, this method outperformed VIT and Swin Transformer, and using the PolyLoss loss function further improved performance, resulting in the best average performance. After using the PolyLoss loss function, Swin Transformer and our Swin Transformer V2 achieved 100% accuracy in the ADM, CSR, DR, and MH categories. In summary, in our experiments, Swin Transformer V2 demonstrated excellent classification ability on the OCT dataset. In addition, we found that using the PolyLoss loss function can further improve the performance of the network.

In addition, we compared our results with other studies. Table 7 shows the results of our comparison. Through comparison, we found that our Swin Transformer V2 improved with PolyLoss, achieving better accuracy and sensitivity performance. This demonstrates the reliability of our method in OCT image classification. These results indicate that our method has high reliability and accuracy in OCT image classification. Our Swin Transformer V2 improved with PolyLoss not only performs well in terms of accuracy, but also achieves good results in terms of sensitivity. These achievements provide strong support for our research in the field of OCT image classification and lay a solid foundation for future research.

Figure 7a,b are the confusion matrices of Swin Transformer V2 using CrossEntropyLoss and PolyLoss when tested with 968 images in the OCT2017 dataset, respectively. Figure 8b represents that the model judged a DME image as CNV disease, while it made zero errors in other categories, thus proving the excellent classification ability of the network. Figure 7c,d are the confusion matrices using two loss functions on 2800 test images in OCT-C8, respectively. As can be seen, the network has successfully classified AMD, CSR, DR, and MH data.

Table 7. Experimental results using different models on the OCT2017 and OCT-C8 datasets, respectively. Significant values are indicated in bold.

Dataset	Model	Accuracy	Sensitivity
OCT2017	InceptionV3 [39]	0.934	0.978
	MobileNet-v2 [40]	0.985	0.994
	ResNet50-v1 [9]	0.993	0.993
	Joint-Attention-Network ResNet-v1 [41]	0.924	
	Xception [42]	0.997	0.997
	OpticNet-71 [43]	0.998	0.998
	Swin Transformer V1 [30]	0.998	0.998
	Ours	**0.999**	**0.999**
OCT-C8	VIT	0.975	0.986
	GAN [44]	0.939	
	Swin Transformer	0.994	0.997
	Deep CNN [45]	0.938	
	CenterNet [46]	0.981	
	Ours	**0.995**	**0.997**

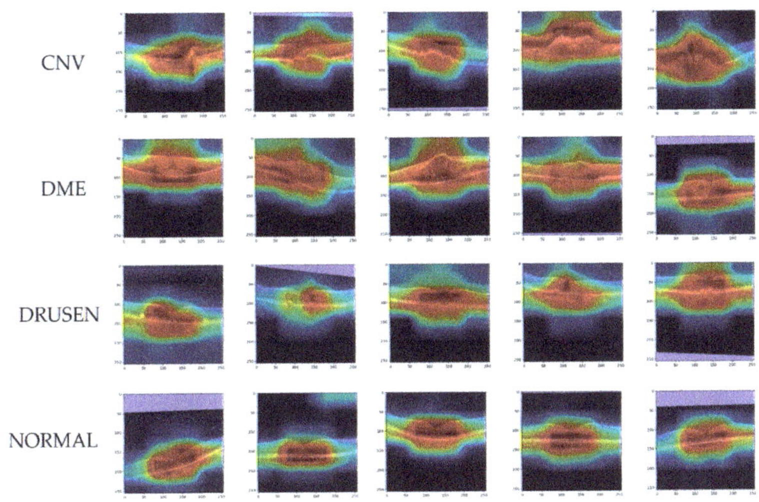

Figure 8. Gradient-weight class activation mapping on OCT2017 of our proposed networks.

For the trained OCT model, we use Grad-CAM to visualize the decision-making mechanism of the prediction. Grad-CAM is a gradient-based deep network visualization method that explains the classification basis of deep neural network models in the form of heat maps, making category judgments through the pixels of the image. Figures 8 and 9 show heatmaps of the prediction results for the OCT2017 and OCT-C8 datasets, respectively. The colors of the heatmap represent regions of interest, with red indicating high correlation with the target category and blue indicating less attention to the region. The purple area is the result of filling the blank area after data enhancement of the image. Meanwhile, lesion regions show up as a darker red color in disease OCT images. As shown in Figure 8, the second row of images shows the Grad-CAM of the DME image, and from the third image, it can be observed that the region of susceptibility contains the macular edema lesion. The image in the third row and fourth column of Figure 8 shows the region of interest for Drusen and also the region where the lesion occurred. Figure 9 is a partial image of the heat maps of the eight disease categories on the OCT-C8 dataset, showing the prediction of the heat maps of the lesion regions of each disease by our trained model. Grad-CAM helps us to see the regions of interest that the model focuses on when making a prediction, and thus to understand the decision-making process of the prediction. It is worth noting that this focus on the region of interest is also consistent with the ophthalmologist's observation and diagnostic process.

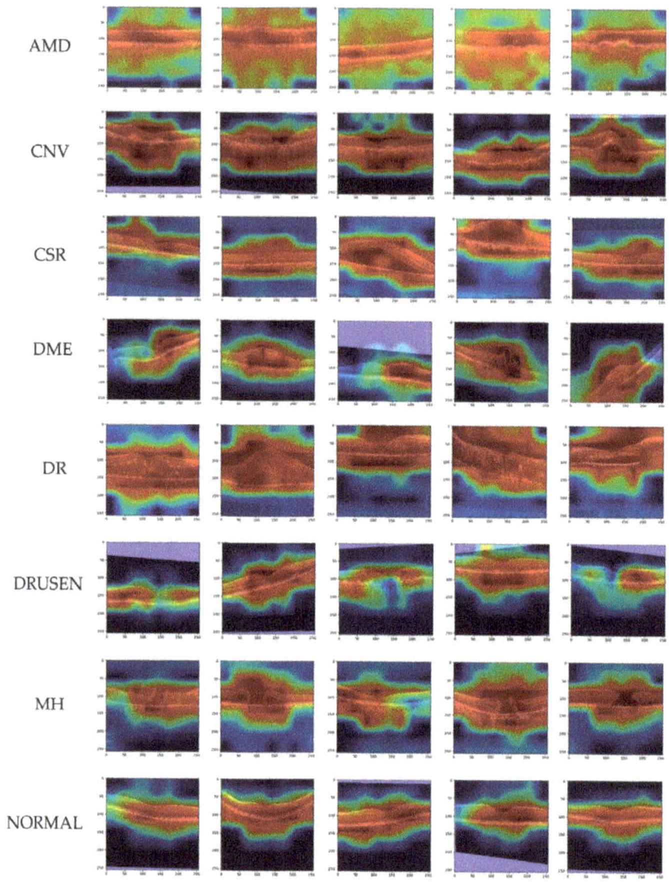

Figure 9. Gradient-weight class activation mapping on OCT-C8 of our proposed networks.

5. Conclusions

In this paper, a multi-fundus disease classification model based on Swin Transformer V2 and the PolyLoss loss function was proposed. By comparing two different loss functions, it has been demonstrated that the PolyLoss function can enhance the model's functionality. In the final experiment, an evaluation index close to 1 was achieved on the OCT2017 dataset, proving the good performance of the model in classifying OCT images. To validate the generalization ability of the network, it was trained and evaluated on OCT-C8, attaining a score of 1 for accuracy and other assessment metrics in half of the OCT illness categories and an average accuracy of 99.5% on the OCT-C8 dataset, proving the effectiveness of our designed model in classifying fundus diseases on OCT images.

The basic Swin Transformer V2 demonstrated strong performance on the publicly available OCT2017 dataset, making further improvements challenging. In clinical practice, misdiagnosis and missed diagnosis can lead to serious medical accidents and cause great pain to patients. The aim of our work is to improve the accuracy of model automatic diagnosis as much as possible to reduce the occurrence of misdiagnosis and missed diagnosis. However, by using polynomial loss and optimizing the network parameters, we were able to achieve a comprehensive improvement in performance metrics at a high-performance level of 99.7%, achieving a score close to 1. This indicates that our modified network model exhibits superior diagnostic capabilities. Although the magnitude of improvement is relatively small, it has positive implications for reducing misdiagnosis and improving diagnosis.

However, despite the good progress made by deep learning models in identifying abnormalities on retinal OCT images, due to the limited dataset, it is not possible to verify how well they perform on other retinal OCT data. In the future, more retinal OCT data will be sought to validate and improve the network. In addition, turning a network model into a powerful tool in the hands of clinical ophthalmologists in real life is also a major challenge, requiring more professionals to work together to turn theoretical methods into products that improve ophthalmic diagnosis.

Author Contributions: Conceptualization, Z.L., Y.H. and X.Y.; Methodology, Z.L., Y.H. and X.Y.; Data curation, Y.H.; Formal analysis, Y.H.; Writing—original draft preparation, Y.H.; Writing—review and editing, Z.L. and X.Y. All authors have read and agreed to the published version of the manuscript.

Funding: This research received no external funding.

Institutional Review Board Statement: Not applicable.

Informed Consent Statement: Informed consent was obtained from all subjects involved in the study.

Data Availability Statement: Data is contained within the article.

Conflicts of Interest: The authors declare no conflict of interest.

References

1. Prem Senthil, M.; Khadka, J.; Gilhotra, J.S.; Simon, S.; Pesudovs, K. Exploring the quality of life issues in people with retinal diseases: A qualitative study. *J. Patient-Rep. Outcomes* **2017**, *1*, 15. [CrossRef] [PubMed]
2. Huang, D.; Swanson, E.A.; Lin, C.P.; Schuman, J.S.; Stinson, W.G.; Chang, W.; Hee, M.R.; Flotte, T.; Gregory, K.; Puliafito, C.A. Optical coherence tomography. *Science* **1991**, *254*, 1178–1181. [CrossRef] [PubMed]
3. Doi, K. Computer-aided diagnosis in medical imaging: Historical review, current status and future potential. *Comput. Med. Imaging Graph. Off. J. Comput. Med. Imaging Soc.* **2007**, *31*, 198–211. [CrossRef] [PubMed]
4. LeCun, Y.; Bengio, Y.; Hinton, G. Deep learning. *Nature* **2015**, *521*, 436–444. [CrossRef]
5. Lecun, Y.; Bottou, L.; Bengio, Y.; Haffner, P. Gradient-based learning applied to document recognition. *Proc. IEEE* **1998**, *86*, 2278–2324. [CrossRef]
6. Krizhevsky, A.; Sutskever, I.; Hinton, G.E. ImageNet Classification with Deep Convolutional Neural Networks. *Commun. ACM* **2017**, *60*, 84–90. [CrossRef]
7. Simonyan, K.; Zisserman, A. Very Deep Convolutional Networks for Large-Scale Image Recognition. *arXiv* **2015**, arXiv:1409.1556.
8. Szegedy, C.; Liu, W.; Jia, Y.Q.; Sermanet, P.; Reed, S.; Anguelov, D.; Erhan, D.; Vanhoucke, V.; Rabinovich, A. Going Deeper with Convolutions. In Proceedings of the IEEE Conference on Computer Vision and Pattern Recognition (CVPR), Boston, MA, USA, 7–12 June 2015; pp. 1–9.
9. He, K.M.; Zhang, X.Y.; Ren, S.Q.; Sun, J. Deep Residual Learning for Image Recognition. In Proceedings of the 2016 IEEE Conference on Computer Vision and Pattern Recognition (CVPR), Seattle, WA, USA, 27–30 June 2016; pp. 770–778.
10. Huang, G.; Liu, Z.; van der Maaten, L.; Weinberger, K.Q. Densely Connected Convolutional Networks. In Proceedings of the 30th IEEE/CVF Conference on Computer Vision and Pattern Recognition (CVPR), Honolulu, HI, USA, 21–26 July 2017; pp. 2261–2269.
11. Howard, A.G.; Zhu, M.; Chen, B.; Kalenichenko, D.; Wang, W.; Weyand, T.; Andreetto, M.; Adam, H. MobileNets: Efficient Convolutional Neural Networks for Mobile Vision Applications. *arXiv* **2017**, arXiv:1704.04861.
12. Tan, M.X.; Le, Q.V. EfficientNet: Rethinking Model Scaling for Convolutional Neural Networks. In Proceedings of the 36th International Conference on Machine Learning (ICML), Long Beach, CA, USA, 9–15 June 2019.
13. Vaswani, A.; Shazeer, N.; Parmar, N.; Uszkoreit, J.; Jones, L.; Gomez, A.N.; Kaiser, L.; Polosukhin, I. Attention Is All You Need. In Proceedings of the 31st Annual Conference on Neural Information Processing Systems (NIPS), Long Beach, CA, USA, 4–9 December 2017.
14. Liu, Z.; Lin, Y.T.; Cao, Y.; Hu, H.; Wei, Y.X.; Zhang, Z.; Lin, S.; Guo, B.N. Swin Transformer: Hierarchical Vision Transformer using Shifted Windows. In Proceedings of the 18th IEEE/CVF International Conference on Computer Vision (ICCV), Electr Network, Montreal, BC, Canada, 11–17 October 2021; pp. 9992–10002.
15. Dosovitskiy, A.; Beyer, L.; Kolesnikov, A.; Weissenborn, D.; Zhai, X.; Unterthiner, T.; Dehghani, M.; Minderer, M.; Heigold, G.; Gelly, S.; et al. An Image is Worth 16x16 Words: Transformers for Image Recognition at Scale. *arXiv* **2021**, arXiv:2010.11929.

16. Liu, Z.; Hu, H.; Lin, Y.; Yao, Z.; Xie, Z.; Wei, Y.; Ning, J.; Cao, Y.; Zhang, Z.; Dong, L.; et al. Swin Transformer V2: Scaling Up Capacity and Resolution. *arXiv* **2022**, arXiv:2111.09883.
17. Leng, Z.; Tan, M.; Liu, C.; Cubuk, E.D.; Shi, X.; Cheng, S.; Anguelov, D. PolyLoss: A Polynomial Expansion Perspective of Classification Loss Functions. *arXiv* **2022**, arXiv:2204.12511.
18. Selvaraju, R.R.; Cogswell, M.; Das, A.; Vedantam, R.; Parikh, D.; Batra, D. Grad-CAM: Visual Explanations from Deep Networks via Gradient-Based Localization. *Int. J. Comput. Vis.* **2020**, *128*, 336–359. [CrossRef]
19. Lee, C.S.; Baughman, D.M.; Lee, A.Y. Deep learning is effective for the classification of OCT images of normal versus Age-related Macular Degeneration. Ophthalmology. *Retina* **2017**, *1*, 322–327. [CrossRef] [PubMed]
20. Wang, D.; Wang, L. On OCT Image Classification via Deep Learning. *IEEE Photonics J.* **2019**, *11*, 1–14. [CrossRef]
21. Islam, K.T.; Wijewickrema, S.; Leary, S.O. Identifying Diabetic Retinopathy from OCT Images using Deep Transfer Learning with Artificial Neural Networks. In Proceedings of the 2019 IEEE 32nd International Symposium on Computer-Based Medical Systems (CBMS), Cordoba, Spain, 5–7 June 2019; pp. 281–286.
22. Rong, Y.B.; Xiang, D.H.; Zhu, W.F.; Yu, K.; Shi, F.; Fan, Z.; Chen, X.J. Surrogate-Assisted Retinal OCT Image Classification Based on Convolutional Neural Networks. *IEEE J. Biomed. Health Inform.* **2019**, *23*, 253–263. [CrossRef]
23. Fang, L.Y.; Wang, C.; Li, S.T.; Rabbani, H.; Chen, X.D.; Liu, Z.M. Attention to Lesion: Lesion-Aware Convolutional Neural Network for Retinal Optical Coherence Tomography Image Classification. *IEEE Trans. Med. Imaging* **2019**, *38*, 1959–1970. [CrossRef]
24. Singh, A.; Rasheed, M.A.; Zelek, J.; Lakshminarayanan, V. Interpretation of deep learning using attributions: Application to ophthalmic diagnosis. In Proceedings of the Conference on Applications of Machine Learning, Electr Network, Online, 24 August–4 September 2020.
25. Wang, C.; Jin, Y.; Chen, X.; Liu, Z. Automatic Classification of Volumetric Optical Coherence Tomography Images via Recurrent Neural Network. *Sens. Imaging* **2020**, *21*, 32. [CrossRef]
26. Arefin, R.; Samad, M.D.; Akyelken, F.A.; Davanian, A.; Soc, I.C. Non-transfer Deep Learning of Optical Coherence Tomography for Post-hoc Explanation of Macular Disease Classification. In Proceedings of the 9th IEEE International Conference on Healthcare Informatics (IEEE ICHI), Electr Network, Victoria, BC, Canada, 9–12 August 2021; pp. 48–52.
27. Latha, V.; Ashok, L.R.; Sreeni, K.G.; IEEE. Automated Macular Disease Detection using Retinal Optical Coherence Tomography images by Fusion of Deep Learning Networks. In Proceedings of the 27th National Conference on Communications (NCC), Electr Network, Kanpur, India, 27–30 July 2021; pp. 333–338.
28. Liu, X.M.; Bai, Y.J.; Cao, J.; Yao, J.P.; Zhang, Y.; Wang, M. Joint disease classification and lesion segmentation via one-stage attention-based convolutional neural network in OCT images. *Biomed. Signal Process. Control* **2022**, *71*, 103087. [CrossRef]
29. Esfahani, E.N.; Daneshmand, P.G.; Rabbani, H.; Plonka, G. Automatic Classification of Macular Diseases from OCT Images Using CNN Guided with Edge Convolutional Layer. In Proceedings of the Annual International Conference of the IEEE Engineering in Medicine and Biology Society, Glasgow, UK, 11–15 July 2022; pp. 3858–3861. [CrossRef]
30. He, J.Z.; Wang, J.X.; Han, Z.Y.; Ma, J.; Wang, C.J.; Qi, M. An interpretable transformer network for the retinal disease classification using optical coherence tomography. *Sci. Rep.* **2023**, *13*, 3637. [CrossRef]
31. Ibrahim, M.R.; Fathalla, K.M.; Youssef, S.M. HyCAD-OCT: A Hybrid Computer-Aided Diagnosis of Retinopathy by Optical Coherence Tomography Integrating Machine Learning and Feature Maps Localization. *Appl. Sci.* **2020**, *10*, 4716. [CrossRef]
32. Ai, Z.; Huang, X.; Feng, J.; Wang, H.; Tao, Y.; Zeng, F.X.; Lu, Y.P. FN-OCT: Disease Detection Algorithm for Retinal Optical Coherence Tomography Based on a Fusion Network. *Front. Neuroinform.* **2022**, *16*, 876927. [CrossRef] [PubMed]
33. Arkin, E.; Yadikar, N.; Xu, X.B.; Aysa, A.; Ubul, K. A survey: Object detection methods from CNN to transformer. *Multimed. Tools Appl.* **2023**, *82*, 21353–21383. [CrossRef]
34. Hendria, W.F.; Phan, Q.T.; Adzaka, F.; Jeong, C. Combining transformer and CNN for object detection in UAV imagery. *ICT Express* **2023**, *9*, 258–263. [CrossRef]
35. Kermany, D.S.; Goldbaum, M.; Cai, W.J.; Valentim, C.C.S.; Liang, H.Y.; Baxter, S.L.; McKeown, A.; Yang, G.; Wu, X.K.; Yan, F.B.; et al. Identifying Medical Diagnoses and Treatable Diseases by Image-Based Deep Learning. *Cell* **2018**, *172*, 1122–1131.e9. [CrossRef] [PubMed]
36. Subramanian, M.; Shanmugavadivel, K.; Naren, O.S.; Premkumar, K.; Rankish, K. Classification of Retinal OCT Images Using Deep Learning. In Proceedings of the 2022 International Conference on Computer Communication and Informatics (ICCCI), Coimbatore, India, 25–27 January 2022; pp. 1–7.
37. Subramanian, M.; Kumar, M.S.; Sathishkumar, V.E.; Prabhu, J.; Karthick, A.; Ganesh, S.S.; Meem, M.A. Diagnosis of Retinal Diseases Based on Bayesian Optimization Deep Learning Network Using Optical Coherence Tomography Images. *Comput. Intell. Neurosci.* **2022**, *2022*, 8014979. [CrossRef] [PubMed]
38. Tan, M.X.; Le, Q.V. EfficientNetV2: Smaller Models and Faster Training. In Proceedings of the International Conference on Machine Learning (ICML), Electr Network, Virtual Event, 18–24 July 2021; pp. 7102–7110.
39. Szegedy, C.; Vanhoucke, V.; Ioffe, S.; Shlens, J.; Wojna, Z.; IEEE. Rethinking the Inception Architecture for Computer Vision. In Proceedings of the 2016 IEEE Conference on Computer Vision and Pattern Recognition (CVPR), Seattle, WA, USA, 27–30 June 2016; pp. 2818–2826.
40. Sandler, M.; Howard, A.; Zhu, M.L.; Zhmoginov, A.; Chen, L.C. MobileNetV2: Inverted Residuals and Linear Bottlenecks. In Proceedings of the 31st IEEE/CVF Conference on Computer Vision and Pattern Recognition (CVPR), Salt Lake City, Salt Lake City, UT, USA, 18–23 June 2018; pp. 4510–4520.

41. Kamran, S.A.; Tavakkoli, A.; Zuckerbrod, S.L. Improving robustness using joint attention network for detecting retinal degeneration from optical coherence tomography images. In Proceedings of the IEEE International Conference on Image Processing (ICIP), Electr Network, Abu Dhabi, United Arab Emirates, 25–28 October 2020; pp. 2476–2480.
42. Chollet, F. Xception: Deep Learning with Depthwise Separable Convolutions. In Proceedings of the 30th IEEE/CVF Conference on Computer Vision and Pattern Recognition (CVPR), Honolulu, HI, USA, 21–26 July 2017; pp. 1800–1807.
43. Amit Kamran, S.; Saha, S.; Shihab Sabbir, A.; Tavakkoli, A. Optic-Net: A Novel Convolutional Neural Network for Diagnosis of Retinal Diseases from Optical Tomography Images. *arXiv* **2019**, arXiv:1910.05672.
44. Yoo, T.K.; Choi, J.Y.; Kim, H.K. Feasibility study to improve deep learning in OCT diagnosis of rare retinal diseases with few-shot classification. *Med. Biol. Eng. Comput.* **2021**, *59*, 401–415. [CrossRef]
45. Sathishkumar, V.E.; Park, J.; Cho, Y. Seoul bike trip duration prediction using data mining techniques. *IET Intell. Transp. Syst.* **2020**, *14*, 1465–1474. [CrossRef]
46. Nazir, T.; Nawaz, M.; Rashid, J.; Mahum, R.; Masood, M.; Mehmood, A.; Ali, F.; Kim, J.; Kwon, H.Y.; Hussain, A. Detection of Diabetic Eye Disease from Retinal Images Using a Deep Learning based CenterNet Model. *Sensors* **2021**, *21*, 5283. [CrossRef]

Journal of
Imaging

MDPI

Article

Advancements in Cataract Detection: The Systematic Development of LeNet-Convolutional Neural Network Models

Thittaporn Ganokratanaa [1], Mahasak Ketcham [2,*] and Patiyuth Pramkeaw [3]

[1] Applied Computer Science Programme, King Mongkut's University of Technology Thonburi, Bangkok 10140, Thailand; thittaporn.gan@kmutt.ac.th
[2] Department of Information Technology Management, King Mongkut's University of Technology North Bangkok, Bangkok 10800, Thailand
[3] Media Technology Programme, King Mongkut's University of Technology Thonburi, Bangkok 10150, Thailand; patiyuth.pra@kmutt.ac.th
* Correspondence: mahasak.k@itd.kmutnb.ac.th

Abstract: Regular screening and timely treatment play a crucial role in addressing the progression and visual impairment caused by cataracts, the leading cause of blindness in Thailand and many other countries. Despite the potential for prevention and successful treatment, patients often delay seeking medical attention due to the gradual and relatively asymptomatic nature of cataracts. To address this challenge, this research focuses on the identification of cataract abnormalities using image processing techniques and machine learning for preliminary assessment. The LeNet-convolutional neural network (LeNet-CNN) model is employed to train a dataset of digital camera images, and its performance is compared to the support vector machine (SVM) model in categorizing cataract abnormalities. The evaluation demonstrates that the LeNet-CNN model achieves impressive results in the testing phase. It attains an accuracy rate of 96%, exhibiting a sensitivity of 95% for detecting positive cases and a specificity of 96% for accurately identifying negative cases. These outcomes surpass the performance of previous studies in this field. This highlights the accuracy and effectiveness of the proposed approach, particularly the superior performance of LeNet-CNN. By utilizing image processing technology and convolutional neural networks, this research provides an effective tool for initial cataract screening. Patients can independently assess their eye health by capturing self-images, facilitating early intervention and medical consultations. The proposed method holds promise in enhancing the preliminary assessment of cataracts, enabling early detection and timely access to appropriate care.

Keywords: cataract detection; systematic development; LeNet–convolutional neural network

1. Introduction

Cataract represents a prevalent ocular disorder characterized by the opacification or clouding of the crystalline lens within the eye. The crystalline lens is an integral component responsible for refracting and focusing incoming light onto the retina, thereby enabling visual perception. Cataracts emerge when this normally transparent and refractive lens undergoes degenerative changes, leading to a loss of transparency and impairment of optical functionality. The etiology of cataracts is multifactorial, encompassing various causative factors. Age-related cataracts, the most common form, develop gradually as a result of intrinsic aging processes, such as oxidative stress and protein denaturation. Additionally, other factors, including genetic predisposition, metabolic disorders (e.g., diabetes mellitus), trauma, long-term exposure to ultraviolet radiation, and certain medications (e.g., corticosteroids) have been implicated in the pathogenesis of cataracts. Clinically, individuals afflicted with cataracts experience a range of visual disturbances that adversely impact their visual acuity and quality of life. The hallmark symptom is a progressive blurring of vision, which gradually worsens over time. As the cataract matures, the affected individuals often report a reduction in visual sharpness and clarity, leading to difficulties in discerning fine details and performing daily activities that require visual precision, such as reading, driving, or recognizing faces. In addition to blurred vision, cataract-related visual impairments encompass alterations in color perception. Colors may appear faded, washed out, or desaturated, hindering the accurate discrimination of hues. Moreover, the presence of cataracts can elicit an increased sensitivity to light, resulting in glare and halos around light sources, especially in low-light conditions or in the presence of bright illumination. Furthermore, individuals suffering from cataracts frequently describe a sensation of visual discomfort, including the perception of darkened or dimmed vision. This reduced contrast sensitivity impairs the ability to distinguish objects against backgrounds of similar tonal value, leading to difficulties in perceiving objects in low-contrast environments. Additionally, cataracts can give rise to visual aberrations, such as the presence of irregularities or lines within the visual field, contributing to visual distortion and further hindrance in accurate visual perception. It is worth noting that cataracts may affect one or both eyes, and the severity and progression of the condition can vary among individuals. Furthermore, the location and extent of lens opacification can vary, ranging from subtle localized changes to widespread clouding of the lens. Given the prevalence of cataracts and their impact on visual function, timely diagnosis and appropriate management are crucial. Cataract surgery, involving the removal of the cloudy lens and its replacement with an intraocular lens, represents the most effective treatment option, offering significant improvements in visual acuity and quality of life for affected individuals. Figure 1 shows the eye conditions of normal eyes and cataracts.

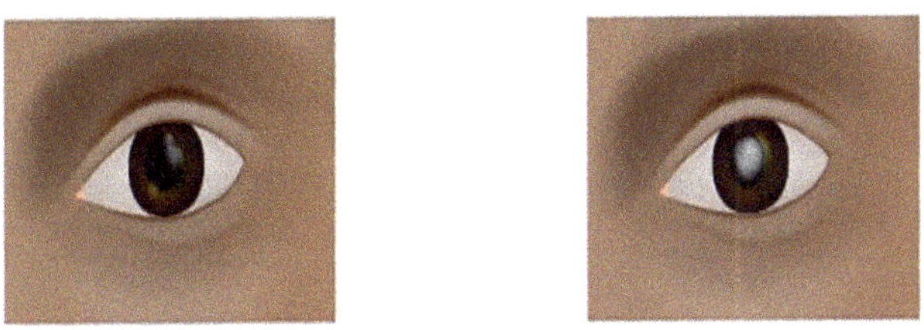

(a) (b)

Figure 1. Illustration of eye conditions: (**a**) normal eyes and (**b**) cataracts.

Investigating the unique occurrences of cataracts in the elderly population unveils a deeper understanding of the intricate interplay between aging, ocular health, and visual impairment. This exploration underscores the imperative of a comprehensive and differentiated assessment, enabling a more nuanced approach to diagnosing and treating these specific cataract variations. As we delve into the specific types of cataracts, we encounter nuclear cataracts as the starting point. This variety affects the nucleus, the lens's central part near the inner corner of the eye. The opacity in this region contributes to cloudiness and a yellow or brown tinge. This distortion not only distorts the passage of light but also hints at the underlying physiological changes occurring within the lens. This type of cataract carries implications beyond mere visual impairment, touching on the complex mechanisms that influence refractive errors and their impact on near and distant vision. It is intriguing how a localized issue can have ramifications that extend beyond the immediate affected area. Transitioning to cortical cataracts, we enter into a realm of lens anatomy that might not be as familiar to the general population. The cortex, a critical component of the lens, undergoes swelling, leading to an opaque appearance. The pattern of clouding, resembling radial grooves, invokes curiosity about the underlying biomechanical processes. It is evident that cataracts are not solely about blurred vision; they offer a glimpse into the intricate molecular changes that unfold within the lens over time. Posterior subcapsular cataracts take us even further, revealing a process occurring around the lens fibers at the back of the lens. The description of "small golden and white grains" opens up a visual representation of the microscopic world within our eyes. This imagery contrasts with our typical perception of the eye as a translucent organ, revealing the complex internal dynamics that contribute to visual acuity and clarity. Understanding these processes adds a layer of complexity to the simple notion of "clouded vision." What is particularly captivating is the interconnectedness between these cataract variations and other ocular conditions. The link between cataracts and glaucoma, for instance, demonstrates how a single condition can cascade into more severe consequences. The intricate web of cause-and-effect relationships calls for a more holistic approach to diagnosis and treatment, recognizing that eye health is not isolated but part of a broader systemic landscape. By meticulously examining the progression of these cataract variations, we gain insight into the challenges faced by untreated older adults. The cascading effects of abnormalities within the lens are akin to a domino effect, ultimately impacting visual function and overall ocular health. This insight emphasizes the significance of proactive measures, particularly for a demographic that is more susceptible to such disorders. In light of this, the emphasis on early management and treatment becomes not just a medical recommendation but a strategic imperative. It is a call to action to integrate knowledge, awareness, and accessible healthcare to address the specific needs of older adults and mitigate the potential for long-term visual impairment. This in-depth exploration of atypical cataracts and their implications invites us to see beyond the surface of visual impairment and delve into the intricacies of ocular health and aging. It underscores the value of multidisciplinary collaboration and continuous research in enhancing our understanding and management of these complex ocular disorders.

The manifestation of cataract encompasses a perceptible deterioration in visual acuity, characterized by progressive opacification of the eye's lens. While the initial stages of this ocular condition do not detrimentally impact visual function, its relentless progression eventually impairs visual perception. The incidence of cataracts predominantly afflicts individuals aged 40 years and above [1]. Early identification of cataract pathology holds the potential to avert the exigency for invasive and arduous surgical procedures while concurrently curbing the substantial financial burden. Moreover, timely detection serves as a prophylactic measure against escalating visual debilitation in consonance with the disease's severity spectrum [2]. The World Health Organization (WHO) [3] reports a staggering global prevalence of approximately 285 million individuals grappling with visual impairment. Among this population, 39 million individuals experience visual constriction, while the remaining fraction endures aberrant visual phenotypes. Notably, cataract pathology assumes responsibility for 33% of visual impairment cases, with an even more daunting statistic of 51% attributing to ocular blindness [4]. In 2020, Flaxman et al. [5] prognosticated the forthcoming figures of individuals afflicted with moderate to severe visual impairment and blindness, approximating 237.1 million and 38.5 million, respectively. Within these prognoses, cataract-induced visual morbidity would impact an estimated 57.1 million individuals (24%) and 13.4 million individuals (35%). The extrapolation warrants the alarming projection that the global population of visually impaired individuals will surpass the 40-million threshold by the year 2025 [6]. Previous research studies have substantiated the minor advancements made in the realm of ocular healthcare and visual impairment management over the past decade. Among the principal etiological factors leading to visual disability, cataract assumes a prominent position [7,8]. The prevailing ocular landscape encompasses three primary categories of cataracts: nuclear cataracts [9], cortical cataracts [10], and posterior subcapsular (PSC) cataracts [11]. These cataract types are commonly attributed to prevalent factors, including advancing age, diabetes mellitus, and tobacco consumption [5]. Early detection of cataracts assumes pivotal significance in mitigating the risk of visual impairment. However, the development of automated cataract detection systems presents formidable challenges, encompassing the intricate interplay of three pivotal aspects: (i) the wavelength regime of cataractous opacities and accompanying color variations, (ii) the morphological dimensions and spatial localization of cataract lesions, and (iii) intrinsic factors, such as the patient's age, gender, and ocular characteristics. Contemporary investigations have explored an array of imaging modalities as potential candidates for automated cataract detection and classification, including slit-lamp biomicroscopy, retro-illumination imaging, ultrasound imaging, and fundus imaging. In particular, fundus imaging has garnered significant attention in this domain, owing to its facile applicability, even enabling patient self-assessment [12]. In contrast, the utilization of slit-lamp biomicroscopy necessitates the presence of experienced and proficient ophthalmologists, thus impeding its widespread adoption, particularly in developing nations with inadequate healthcare infrastructure [4]. To streamline the intricate landscape of primary cataract screening and facilitate the advent of an automated cataract detection system, a paradigm integrating advanced image processing methodologies is imperative. A multitude of deep learning-based automated cataract detection systems have been documented, predominantly relying on central feature descriptors, such as the Discrete Cosine Transform (DCT) [13], specific local standards, such as local pattern deviation [14], and sophisticated deep convolutional neural networks (CNN) that capitalize on their inherent capacity for profound feature extraction, consequently affording heightened diagnostic accuracy. Nonetheless, despite the extensive proliferation of deep learning-based automated cataract detection systems in the literature, certain inherent limitations persist, encompassing suboptimal detection accuracy, an excessive proliferation of high-dimensional parameters, and computationally intensive operations that engender exorbitant computational costs. Cognizant of the paramount significance and ramifications of timely cataract management, researchers harbor a resolute aspiration to delve into diverse process models that can be harnessed for ocular image processing, thereby empowering patients to conduct preliminary self-assessments of their ocular condition. In this context, researchers have proffered an innovative approach characterized by an image processing workflow tailored for detecting and classifying cataract abnormalities, enabling patients to engage in rudimentary evaluations of their own cataractous manifestations. By enabling preliminary assessments that manifest congruence with visual impairments and somatic distress, this framework imparts patients with the necessary impetus to seek expeditious therapeutic interventions, thereby curbing the trajectory of disease progression and preempting the advent of latent sequelae [4].

The researchers have identified the inherent challenges and profound significance associated with the management of ocular disorders, particularly corneal diseases. Consequently, their primary objective entails a comprehensive investigation of diverse paradigms and methodologies that can be harnessed for the purpose of corneal image analysis. Such endeavors aim to empower patients to engage in preliminary self-assessment of their corneal status, thus facilitating an initial evaluation of their ocular health. To this end, the researchers have meticulously devised a discerning framework, leveraging cutting-edge image processing technologies to discern and classify aberrant corneal conditions. By virtue of this preliminary assessment, complemented by concurrent visual anomalies and corporeal discomfort, affected individuals are prompted to expeditiously seek specialized medical intervention, thereby mitigating the gravity and latent perils that may ensue from deferred therapeutic measures.

The contemporary automated corneal detection systems encompass a tripartite process comprising preprocessing, feature extraction, and classification [15–20]. These methodologies can be bifurcated into two cohorts contingent upon the algorithms harnessed for either feature extraction or classification stages: machine learning (ML) and deep learning (DL) paradigms. Recent scholarly investigations have extensively expounded upon these modalities [21–24]. We aim to succinctly summarize the foremost contributions emanating from both factions.

Machine Learning (ML)-Based Methods: ML-based algorithms have ubiquitously permeated corneal detection systems, serving as the bedrock for extracting pertinent features and effectuating corneal abnormality classification. Prominent studies have harnessed ML techniques, encompassing support vector machines (SVM), random forests (RF), and K-nearest neighbor (KNN) [21]. These endeavors have evinced compelling outcomes in terms of accuracy and efficiency, affording precise identification and taxonomizing of corneal manifestations.

Gao et al. [25] presented a computer-aided corneal detection system designed for large-scale screening and preparatory grading. The system integrates enhanced feature extraction with linear discriminant analysis (LDA) as its primary analytical framework. Experimental evaluations on clinical databases showcase its efficiency, boasting a remarkable accuracy rate of 84.8% in detecting corneal abnormalities.

Yang et al. [26] presented an automated corneal detection methodology utilizing a sequential three-step approach that incorporates the top-bottom hat transformation technique to optimize the contrast between the corneal region and its background. The method considers brightness values and distinctive features during the detection process and employs a backpropagation-based neural network (BBNN) for severity classification, differentiating corneal conditions into mild, moderate, or severe categories. They work on cataracts and highlight a neural network classifier designed for automatic cataract detection in retinal images. This classifier uses enhanced image transformation, feature extraction techniques, and a two-layer BP neural network to grade cataracts by severity. Together, these advancements signify strides in computer-assisted corneal detection and cataract grading, with promising outcomes in diagnostic efficiency and patient care, rooted in techniques such as LDA and BBNN.

Guo et al. [27] introduced a computer-aided corneal categorization system that employs image-based referencing, followed by wavelet-based and morphology-based feature extraction. Subsequently, a multi-level classification algorithm is utilized for corneal detection and grading, yielding correct classification rates (CCRs) of 90.9% and 77.1% for wavelet-based features and 86.1% and 74.0% for morphology-based features, respectively.

Furthermore, Fuadah et al. [28] employed the KNN algorithm for corneal classification and achieved feature fusion by considering image-based characteristics, such as sharpness difference and uniformity. This system was successfully implemented on smartphones, yielding an impressive accuracy rate of 97.5%.

Moreover, a highly competitive corneal detection system was presented by the researchers in [29], where statistical analysis of content-based features was utilized. The system incorporated the Gray-Level Co-occurrence Matrix (GLCM) for content feature extraction. The test dataset was categorized into normal or corneal conditions using KNN, resulting in an average accuracy of 94.5%. It should be noted that manual intervention by users/experts is required during the processing stage, specifically for outlining and segmenting the specific areas of interest. Consequently, the system cannot be considered fully automated. Nevertheless, this system successfully operated on standard eye images captured without employing a slit-lamp or confocal microscopy. These scholarly contributions illustrate significant progress in computer-assisted corneal detection and classification systems. By employing various techniques, such as wavelet transform, image morphology, KNN, and GLCM, these studies have demonstrated promising outcomes in achieving accurate corneal analysis and grading. Consequently, these advancements hold great potential in enhancing diagnostic capabilities and facilitating clinical decision-making processes.

Yang et al. [4] proposed a novel approach for corneal detection and grading using ensemble learning methods. The study involved the extraction of features from distinct feature sets and the construction of dual learning models for each group. By leveraging ensemble methods, such as combining multiple learning models, image categorization was achieved with high accuracy, as evidenced by the correct classification rates (CCRs) of 93.2% for detection and 84.5% for grading.

Similarly, Caixinha et al. [30] developed an automated in vivo system for the detection and categorization of nuclear cataracts. This system incorporated machine learning techniques and ultrasound-based algorithms. By extracting 27 frequency and time-domain features, the researchers explored classification methods, including SVM, Bayes, multilayer perceptron, and random forest, to effectively categorize corneal conditions. Notably, while the ultrasound-based method demonstrated commendable accuracy in corneal screening, it incurred substantial costs and involved a complex image-processing workflow.

Moreover, in [31], a corneal image classification framework utilizing SVM and subsequent grading by an RBF network was presented. This framework achieved a noteworthy specificity rate of 93.33% in assessing the severity of corneal abnormalities. Additionally, Rana and Galib [1] introduced a smartphone application compatible with Android, iOS, and Windows platforms, empowering users to self-diagnose corneal conditions. Through the consideration of texture characteristics and the analysis of relevant data, the application exhibited a commendable accuracy rate of 85%. These investigations underscore the significance of machine learning techniques, ensemble learning, ultrasound-based algorithms, and smartphone applications in the realm of corneal analysis. The integration of these advanced methodologies holds promise for accurate corneal detection, reliable grading, and improved accessibility to corneal diagnostics.

Jagadale et al. [32] proposed the utilization of Hough circle detection transformation as a means to detect the centers and radii of lenses. Subsequently, statistical feature extraction techniques were applied and integrated into an SVM classifier for precise corneal detection, achieving a commendable accuracy rate of 90.25%. Similarly, Sigit et al. [33] presented an innovative approach for corneal detection on Android smartphones. The categorization process was facilitated by a streamlined single perceptron layer methodology, yielding an impressive accuracy of 85%. In a more recent study, the work introduced in [6] introduced a hierarchical feature extraction strategy for corneal grading. The challenge of categorizing corneal severity into four distinct levels was transformed into a more manageable two-level grouping task. This was effectively addressed through the individual training of neural networks, followed by their collective integration. Remarkably, the proposed system achieved detection and grading accuracies of 94.83% and 85.98%, respectively. These research endeavors exemplify the successful application of Hough circle detection, statistical feature extraction techniques, SVM classifiers, perceptron layers, and hierarchical neural networks in the domain of corneal analysis. The adoption of these methodologies has yielded significant advancements in corneal detection, grading, and categorization. Consequently, these findings contribute substantively to the enhancement of corneal diagnostic accuracy and overall efficacy.

Deep Learning (DL)-Based Methods: DL methodologies have attained paramount significance owing to their intrinsic capacity to discern intricate patterns and glean salient representations from intricate datasets. CNNs, an influential subset of DL models, have exhibited extraordinary performance vis-à-vis corneal image analysis [21]. Astute researchers have judiciously employed CNN architectures, such as VGGNet, ResNet, and InceptionNet, to discern and classify corneal aberrations with unprecedented accuracy and resilience. The utilization of deep learning methodologies has demonstrated their efficacy in acquiring essential features and integrating feature learning procedures into the model creation process, thereby mitigating the inherent deficiencies of manual feature engineering and applying them across divergent medical imaging modalities [34,35]. Investigative endeavors by Gao et al. [36] have explored the utilization of deep learning for the categorization of nuclear cataract severity from slit-lamp images. A localized image patch grid was established, whereby image patches were assigned to a CNN to extract regional features. Subsequently, recurrent neural networks (RNNs) were employed to extract increasingly intricate features. The clustering of cataracts was executed utilizing support vector regression (SVR) techniques.

In a similar vein, Zhang et al. [37] proposed a deep convolutional neural network (DCNN) for the detection and grading of cataracts, capitalizing on feature maps derived from hierarchical data fusion structures. This approach proved to be temporally efficient, yielding commendable accuracies in cataract detection and grading, amounting to 93.52% and 86.69%, respectively. These investigations exemplify the successful application of deep learning methodologies in tackling the challenges pertaining to cataract detection and grading. By automating the feature learning process and leveraging advanced neural network architectures, such as CNNs and RNNs, these methodologies offer substantial enhancements in accuracy and efficiency, transcending conventional manual feature engineering techniques. Moreover, the incorporation of SVR and DCNN techniques further elevates the performance and robustness of the models, resulting in promising outcomes in cataract diagnosis and severity assessment. These contributions underscore the potential of deep learning approaches in advancing the realm of medical image analysis and fostering the development of automated systems for ophthalmic healthcare.

Ran et al. [38] developed a novel methodology aimed at accurately grading the severity of cataracts across six distinct levels. The approach employed a synergistic combination of DCNNs and RF to effectively extract and analyze features at various levels from cataract images. Specifically, the methodology consisted of three modules, each contributing to the feature extraction process. The DCNNs were responsible for generating a dataset comprising extracted features, which were subsequently utilized by RF to perform the intricate task of grading cataracts across the complex six-level scale. Notably, the proposed methodology achieved an average accuracy of 90.69%. The multi-level grading system offered by this approach has significant implications for improving the accuracy and precision with which ophthalmology experts comprehend the severity of cataracts in patients. By integrating deep learning techniques with ensemble learning methodologies, Ran et al. [38] have not only achieved high accuracy but have also provided a comprehensive understanding of cataract severity across multiple levels. Such advancements hold considerable potential for enhancing clinical decision-making and ultimately improving the quality of patient care in the field of ophthalmology. In conclusion, the amalgamation of DCNNs and RF in the proposed methodology by Ran et al. [38] represents a notable contribution to the field of cataract severity grading. The rigorous utilization of deep learning and ensemble learning techniques offers a robust framework for effectively analyzing and categorizing cataract images. This research underscores the importance of leveraging advanced computational approaches to address complex medical challenges, thereby paving the way for improved diagnostic accuracy and treatment outcomes.

Pratap and Kokil [39] propose an innovative methodology aimed at facilitating the detection and assessment of cataract severity, ranging from prototype images to advanced stages. Their approach leverages the concept of transfer learning to accomplish automatic cataract classification. The final classification stage is carried out through the utilization of feature extraction techniques in conjunction with SVM. Impressively, the methodology achieves a noteworthy accuracy rate of 92.91% throughout the four-step process. The utilization of transfer learning in this study showcases the potential of leveraging pre-existing knowledge and pre-trained models from prototype images to effectively classify cataracts across varying degrees of severity. By integrating advanced feature extraction methods with SVM-based classification, the proposed approach enables accurate categorization and augments the overall diagnostic capabilities of the system. The reported high accuracy rate exemplifies the efficacy and promise of the proposed methodology in automated cataract classification. Pratap and Kokil's work significantly contributes to the field of cataract diagnosis and classification by presenting a robust methodology that synergistically combines transfer learning and machine learning techniques. By achieving exceptional accuracy in automated categorization, their approach presents a valuable tool for clinicians and ophthalmologists, empowering them to accurately assess the severity of cataracts. The research demonstrates the potential of leveraging transfer learning to develop automated systems for cataract diagnosis, which ultimately translates into timely and precise treatment decisions for patients. In conclusion, the methodology proposed by Pratap and Kokil [39] underscores the efficacy of transfer learning in automating the classification of cataracts. The amalgamation of sophisticated feature extraction techniques with SVM-based classification underscores the potential of machine learning methodologies in enhancing the diagnostic process for cataract severity assessment. This research opens up new avenues for further advancements in computer-assisted diagnosis and contributes to the collective understanding and management of cataracts within the field.

Jun et al. [40] have presented a pioneering approach in the domain of cataract severity assessment, wherein they introduce a Tournament-based Ranking CNN system. This novel system addresses the critical task of accurately classifying the severity levels of cataracts through the fusion of tournament-based architecture and CNN modeling, thus facilitating robust evaluation of binary labeled cataract data. The Tournament-based Ranking CNN framework leverages the inherent advantages of a tournament structure, enabling comprehensive comparative evaluations of cataract images within the system. By assigning a ranking to each image based on its severity level, the framework establishes a reliable hierarchy that captures the varying degrees of cataract severity. Concurrently, the CNN model integrated within the framework plays a pivotal role in extracting salient features from the cataract images, contributing to precise classification outcomes. The adoption of a binary labeled dataset in this study empowers the system to discriminate between normal and cataract-affected images with remarkable efficacy. The incorporation of the Tournament-based Ranking CNN methodology marks a significant milestone in the field of cataract severity assessment, revolutionizing the traditional approaches prevalent in clinical settings. The system's ability to rank cataract images based on their severity not only offers quantitative metrics but also augments the decision-making process for clinicians, facilitating tailored treatment strategies. Furthermore, the integration of CNN-based feature extraction mechanisms enhances the system's discriminative capabilities by capturing high-level characteristics indicative of cataract severity. The research conducted by Jun et al. establishes a notable contribution to the expanding body of knowledge in the field of cataract diagnosis and severity assessment. The Tournament-based Ranking CNN system presents a robust and efficient methodology for objectively evaluating cataract severity levels. The utilization of binary labeled data, coupled with the incorporation of CNN-based feature extraction, accentuates the system's precision, thereby advancing the accuracy and reliability of cataract severity classification. To summarize, Jun et al. [40] introduce an innovative system in their research, encompassing the Tournament-based Ranking CNN framework for cataract severity classification. This cutting-edge methodology amalgamates tournament-based evaluations and CNN modeling, enabling accurate ranking and classification of cataract images according to their severity levels. The findings demonstrate promising outcomes, promising to automate and streamline cataract severity assessment, empowering clinicians with vital information for informed decision-making and optimal patient care.

Hossain et al. [41] have made notable contributions to the field of cataract detection by proposing a sophisticated system that leverages DCNNs and Residual Networks (ResNets). Their work focuses on achieving high accuracy in categorizing cataract severity levels. In a similar vein, Zhang et al. [42] have presented a cataract detection method that employs an ensemble approach, incorporating ultrasound images to enhance accuracy, which surpasses alternative deep learning methodologies with a remarkable 97.5% precision. The proposed system comprises three key components: an object detection network, multiple categorization networks, and a model fusion module. Performance evaluation of the system demonstrates satisfactory outcomes, particularly in scenarios with limited training data. However, it is important to acknowledge a primary limitation of this methodology, namely its reliance on image clarity as a reference for assessing cataract severity, as this criterion may be influenced by other ocular conditions such as corneal opacities and diabetic retinopathy. Consequently, the method may not be adept at distinguishing between different types of ocular diseases. The research conducted by Hossain et al. [41] and Zhang et al. [42] represents a significant advancement in cataract detection and severity assessment. The utilization of DCNNs, ResNets, and the ensemble methodology underscores their effectiveness in accurately categorizing cataracts. These findings hold great promise for enhancing the diagnostic process and assisting healthcare professionals in making informed decisions regarding patient care. Nonetheless, future investigations should address the challenges associated with classifying various ocular conditions and explore novel techniques to further enhance the overall performance and reliability of the system. In summary, the research conducted by Hossain et al. [41] and Zhang et al. [42] constitutes a notable contribution to the field of cataract detection. Their utilization of advanced deep learning architectures, such as DCNNs, ResNets, and the ensemble methodology, demonstrates the potential to achieve accurate categorization of cataract severity. However, it is essential to acknowledge the limitations imposed by the reliance on image clarity as a criterion for severity assessment, as this may hinder the differentiation of different ocular diseases. Further research efforts in this domain have the potential to refine cataract diagnosis and significantly impact patient outcomes.

Khan et al. [43] conducted a comprehensive investigation into the measurement of cataract severity from panoramic dental images. Their study employed the VGG-19 model with transfer learning techniques on the latest publicly available dataset from the KAGGLE platform [44]. Remarkably, their approach achieved a near-equivalent accuracy rate of 97.47%, establishing a new benchmark in cataract detection. In a related work, Pratap and Kokil [45] published a novel method for aiding cataract diagnosis in environments plagued by disruptive noise. By leveraging a dedicated CNN trained on both internal and external datasets, they demonstrated the resilience of their approach even under adverse conditions. This pioneering research represents a pioneering exploration into the robustness of cataract detection systems, particularly in the presence of acoustic interferences. The contributions made by Khan et al. [43] and Pratap and Kokil [45] constitute notable advancements in the field of cataract detection. The utilization of the VGG-19 model and dedicated CNN architectures in these studies attests to their efficacy in achieving highly accurate cataract diagnosis. Moreover, the investigation of system resilience to environmental noise holds substantial potential for enhancing the reliability and practicality of cataract detection methodologies. Further validation and refinement of these approaches within real-world clinical settings will yield valuable insights into their integration into the existing healthcare framework. In summary, the research conducted by Khan et al. [43] and Pratap and Kokil [45] significantly contributes to the domain of cataract detection. Their utilization of advanced models, such as VGG-19, along with tailored CNN architectures, highlights the potential for achieving exceptional accuracy in cataract diagnosis. Furthermore, the exploration of system robustness in the presence of environmental noise demonstrates a commendable effort to address practical challenges faced in clinical settings. Ongoing research endeavors in this area are expected to further enhance cataract detection methodologies, ultimately impacting clinical decision-making processes in a meaningful manner.

Observations reveal a conspicuous disparity [46–48] between the abundance of research endeavors following conventional machine learning paradigms and the paucity of investigations reporting on cataract detection and severity classification through deep learning methodologies. Consequently, a multitude of obstacles persist, necessitating diligent attention. Paramount among these challenges is the imperative to heighten model precision whilst concurrently mitigating complexity by curtailing the number of parameters during the training phase, the stratum count, the depth, the runtime, and the overall dimensions of the model. To augment the accuracy of deep learning models for cataract detection and severity classification, the employment of sophisticated regularization techniques emerges as a promising avenue. By incorporating regularization mechanisms such as dropout, L1 or L2 regularization, or batch normalization, models can effectively assuage overfitting and bolster generalization prowess. Moreover, the exploration of advanced optimization algorithms, including Adam, RMSprop, or learning rate scheduling, holds the potential to fortify model convergence and overall performance. Additionally, harnessing cutting-edge architectures like ResNet, DenseNet, or Inception offers prospects for refining feature representation and hierarchical learning, facilitating the discernment of nuanced patterns and intricate relationships embedded within cataract images. Fine-tuning pre-trained models on voluminous and heterogeneous datasets can further surmount limitations attributable to inadequate training samples, thus culminating in elevated performance and heightened generalizability [49–51]. Furthermore, grappling with the exigencies associated with interpretability and explainability in deep learning models tailored to cataract detection assumes paramount significance. The development of techniques enabling visualization and interpretation of learned features and decision-making processes in these models engenders profound insights into their inferential outputs, fostering trust and facilitating their adoption within clinical milieus. To recapitulate, despite the extensive examination of traditional machine learning approaches in cataract detection, an array of challenges intrinsic to deep learning methodologies remains extant. Enhanced model accuracy while minimizing complexity, the adept utilization of advanced regularization and optimization techniques, the exploration of avant-garde architectures, and the resolution of interpretability quandaries all constitute pivotal domains warranting sustained scrutiny [52–55]. By surmounting these impediments, the field can ascend towards the realm of more precise, efficient, and explicable deep learning models for cataract detection and severity classification.

Santra et al. [56] presented a novel research endeavor focused on the development of an automatic classification system for optic nerve diseases. Employing sophisticated computer vision algorithms, the study explored two distinctive approaches to tackle this problem. The first approach encompassed a traditional computer vision methodology, which was subsequently complemented by the integration of a machine learning (ML) model. Various ML algorithms, namely multinomial logistic regression, random forest, gradient boosting classifier, and support vector machine, were employed to train the data and extract valuable insights from the optic nerve images. The second approach, characterized by automatic feature engineering and classification, leveraged CNN. This technique facilitated the creation of multiple class-specific models, enabling the accurate identification of specific diseases as well as binary classification to determine disease presence or absence. To evaluate the performance of the developed models, a carefully labeled dataset comprising 600 retinal images was utilized for training purposes. Results indicated that the ML models exhibited remarkable accuracy, surpassing 90% in correctly identifying various optic nerve diseases. Conversely, the CNN models yielded less satisfactory outcomes during the testing phase, with an accuracy below 60%. Accuracy was calculated as the ratio of correctly classified images to the total number of images within the dataset. These findings underscore the efficacy of the ML approach in effectively discerning optic nerve diseases, thus highlighting its potential for clinical application. However, further refinement and optimization of the CNN models are warranted to enhance their accuracy and reliability. The work conducted by Santra et al. represents a significant contribution to the field of automatic classification of optic nerve diseases, opening avenues for future research and innovation in this domain. This study [57] addresses challenging hemorrhage segmentation in retinal images, particularly on mobile phones, with poor lighting. A novel KMMRC-INRG method enhances segmentation by addressing uneven illumination using KMMRC and improving boundary segmentation through INRG. The approach achieves high performance, with recall, precision, F1 measure, and IoU scores of 80.18%, 91.26%, 85.36%, and 80.08%, respectively. Notably, the F1 measure improves by 19.02% compared to DT-HSVE on the same dataset and up to 58.88% for images with large hemorrhages.

2. Materials and Methods

The current research endeavors reflect a considerable body of work focusing on the adoption of CNNs as a deep learning framework for the detection and severity assessment of glass fractures. CNNs are characterized by their utilization of convolutional and pooling operations across different layers, accompanied by non-linear activation functions, to establish a hierarchical feature representation. Notably, the integration of feature extraction and classification processes within the deep learning paradigm distinguishes this approach from conventional handcrafted feature extraction methods, which often treat these steps as distinct entities. Nevertheless, the investigation of deep learning-based approaches for glass fracture detection and severity assessment remains relatively limited compared to the extensive exploration of traditional machine learning techniques. To address this gap, the current study introduces a novel methodology that leverages convolutional neural networks, specifically a variant known as LeNet, to overcome the constraints associated with handcrafted feature extraction and computational complexity. LeNet, originally popularized as Lenet-5, is a specific instantiation of a feed-forward neural network architecture that excels in processing large-scale images, wherein individual neurons exhibit receptive fields and demonstrate remarkable performance within their local contexts. In the context of the present research, the LeNet-5 variant has been specifically tailored to accommodate the demands of glass fracture detection. Consequently, it encompasses a convolutional neural network comprising seven distinct layers, each contributing to the processing pipeline, and effectively operates on digital images with dimensions of 32×32 pixels. The adoption of LeNet within the research framework signifies its efficacy in overcoming the limitations associated with traditional handcrafted feature extraction techniques while simultaneously mitigating the computational complexity inherent in processing high-resolution images. By embracing the convolutional neural network paradigm, the study capitalizes on the innate strengths of deep learning models, ultimately facilitating enhanced performance and the generation of efficient feature representations. Methodologically, the research endeavor encompasses a systematic workflow with various stages. These stages involve comprehensive data preprocessing, the thoughtful design of network architecture, meticulous model training utilizing backpropagation, and rigorous model evaluation employing appropriate performance metrics. Importantly, the iterative nature of deep learning methodologies enables the refinement of models across multiple training epochs, allowing for the optimization of network weights and biases and the subsequent enhancement of predictive capabilities. To summarize, the present research contributes to the existing body of knowledge by integrating convolutional neural networks, specifically the LeNet architecture, within the deep learning paradigm to tackle the challenges associated with handcrafted feature extraction and computational complexity. Through the proposed methodology, the study highlights the potential of deep learning approaches in effectively processing high-resolution images and provides insights into the sequential stages involved in implementing the research framework as shown in Figure 2.

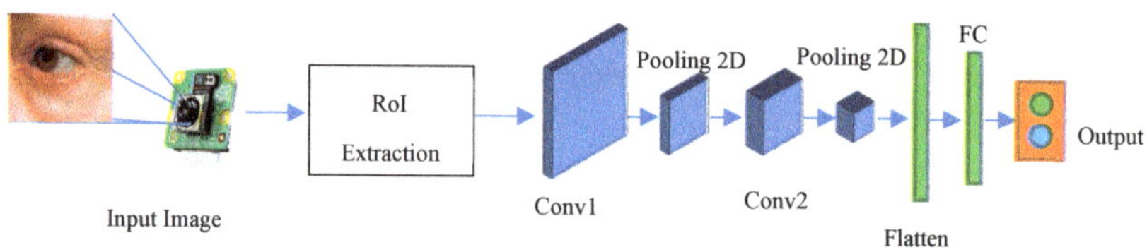

Figure 2. Overview framework of the proposed method.

The architecture of the implemented system involves the use of a CNN for classifying eye images into two categories: those with cataracts and those without. The architecture consists of several essential layers. Convolutional Layers: These layers are responsible for learning important features from the input images through convolution operations, which help in identifying patterns like edges, corners, and shapes. Activation Functions (ReLU): Rectified Linear Activation functions introduce non-linearity to the model, enhancing its capability to capture complex relationships in data. Max-pooling Layers: These layers reduce the dimensions of the feature maps, thus reducing computational load and focusing on the most relevant information. Flatten Layer: This layer converts the matrix-like feature maps into a single vector, preparing the data for fully connected layers. Fully Connected Layers: These layers process the extracted features and make class predictions based on the learned representations.

The training process involves several steps: (i) Data Loading: The dataset is loaded, resized, and preprocessed to be compatible with the CNN model. (ii) Data Splitting: The dataset is divided into training and testing subsets for training and evaluation purposes. (iii) Data Encoding: Labels are encoded into numerical format using techniques like LabelEncoder and OneHotEncoder. (iv) Model Compilation: The CNN model is built using Keras Sequential API, incorporating convolutional, activation, max pooling, and fully connected layers. (v) Optimizer and Loss Function: Stochastic Gradient Descent (SGD) is selected as the optimizer, and binary cross-entropy is chosen as the loss function due to the binary classification nature of the problem. (vi) Model Training: The model is trained on the training data using the defined optimizer and loss function, with a specified number of epochs and batch size. (vii) Evaluation: The trained model is evaluated on the testing data to assess its performance using metrics such as accuracy and classification reports.

Hyperparameter settings include the learning rate (lr), which is set to 0.1 in the SGD optimizer, controlling the step size during weight updates. We set momentum to 0.0, affecting the weights' update direction based on past updates. Decay is set to 0.0, controlling the learning rate decay during training. Batch size is set to 150, determining the number of samples processed before updating the model. Number of epochs is set to 50, representing the number of times the entire training dataset is passed through the model during training.

2.1. Preprocessing

Upon the initiation of data importation, the acquired dataset undergoes a fundamental stage known as data preprocessing. This pivotal process encompasses a series of methodical steps aimed at refining and organizing the dataset to ensure its suitability for subsequent analyses as shown in Figure 3. The following steps typically constitute.

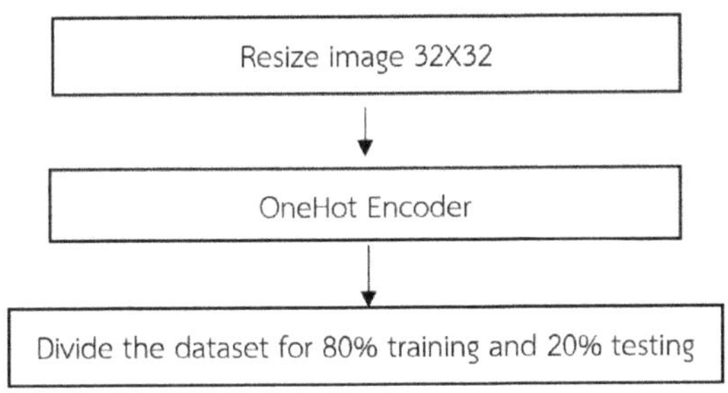

Figure 3. Preprocessing procedure of LeNet-CNN.

(1) In order to achieve uniformity and compatibility across the dataset, a crucial preprocessing step involves resizing the input images to a consistent dimension of 32 × 32. This resizing process is integral to the LeNet-CNN architecture and involves two primary steps: the convolutional layer and the pooling layer. For the convolutional layer, this layer employs convolutional filters to perform localized receptive field operations on the input images. By convolving the filters across the image, relevant features are extracted while simultaneously reducing the spatial dimensions of the image. This process plays a pivotal role in adjusting the image size to conform to the desired dimensions. Following the convolutional layer, the pooling layer further contributes to the resizing process. Utilizing techniques such as max pooling or average pooling, this layer reduces the dimensionality of the convolved feature maps. By aggregating information from neighboring regions, the pooling layer effectively downsamples the feature maps, resulting in images of the desired size. By sequentially applying these two fundamental steps within the LeNet-CNN architecture, the imported dataset undergoes a comprehensive resizing procedure, yielding images with a standardized dimension of 32 × 32. This standardized image size ensures consistency and facilitates subsequent stages of analysis, feature extraction, and classification.

The purpose of encoding 32 × 32 images is primarily to maintain uniformity and compatibility throughout a dataset, especially when using architectures such as the LeNet-CNN. This consistency ensures smooth data flow, optimal feature extraction, and efficient model training and performance. The LeNet-CNN architecture is designed to work optimally with 32 × 32 images, considered to be standard dimension for the model to perform efficiently. Specifically, the convolutional layer, the first step in adjusting the image size, uses filters to scan the image and capture localized features. As these filters move across the image, they reduce its spatial dimensions while preserving and emphasizing relevant features, bringing it closer to the desired 32 × 32 size. Following the convolutional process, the pooling layer further aids in achieving the desired image size. It simplifies the feature maps by condensing information. The goal is to obtain an abstracted yet informative representation of the original image. Once images are resized to a consistent 32 × 32 dimension, it becomes easier to apply further analysis, feature extraction, and classification operations. This standardization ensures that the entire dataset can be processed uniformly, reducing errors and complications that may arise from inconsistent image dimensions. In essence, encoding 32 × 32 images provides a consistent foundation for data processing, especially when using architectures like LeNet-CNN.

- The convolutional layer within the LeNet-CNN architecture utilizes small-sized convolutional filters to extract relevant features from input images. During the convolutional operation, the image undergoes pixel-wise multiplication with the weight values present in the convolutional filters. This process yields a new feature map that represents the extracted features. The filters are slid across the image, applying the convolution operation by using a predefined stride value. By convolving the filters, the convolutional layer effectively extracts important visual patterns and characteristics of the image, which are subsequently utilized in subsequent stages of the network. This process of convolution plays a vital role in feature extraction and enables the network to capture spatial dependencies and patterns within the image.
- The pooling layer is employed in the LeNet-CNN architecture to reduce the dimensionality of the feature maps obtained from the convolutional layer. In LeNet-CNN, the max-pooling technique is utilized, which selects the maximum value from a specified group of pixels within the feature map. The kernel size is defined to determine the size of the pixel group, and the stride value is employed to specify the displacement when moving to the next set of pixels. The pooling operation helps decrease the size of the feature maps while preserving important features. By reducing the dimensionality, the pooling layer aids in controlling the computational complexity of the network and achieving translation invariance, allowing the network to focus on the most salient features and discard redundant information.

(2) OneHot encoding is a data processing technique employed in image data to transform categorical labels into numerical representations for computational purposes. This method encodes each label as an independent binary vector, enabling the creation of machine learning models that exhibit enhanced accuracy and performance. OneHot encoding is used to convert categorical variables into a numerical format suitable for consumption by machine learning models. The underlying principle of OneHot encoding involves transforming each category value into a vector of equal length, representing all the categories as shown in Figure 4. Each position in the vector corresponds to a specific category, with a value of 1 indicating the presence of the variable in that category and a value of 0 indicating its absence. By representing categorical variables as binary vectors, OneHot encoding facilitates the utilization of these variables in machine learning algorithms, enabling effective model training and inference. We apply this encoding to the labels in both the training and the test datasets. This ensures that the model's output can be correctly matched and evaluated against the true labels.

Label
Cataract
Cataract
Normal
Normal

→

Cataract	Normal
1.0	0.0
1.0	0.0
0.0	1.0
0.0	1.0

Figure 4. Example of encoding data type OneHotEncoder.

(3) The dataset is divided into training and testing sets, following the conventional practice of an 80–20% split. This partitioning scheme ensures that 80% of the data are allocated for model training while the remaining 20% is reserved for evaluation purposes. By separating the dataset into these distinct subsets, we can effectively assess the performance and generalization capabilities of the trained model on unseen data. The training set serves as the foundation for model learning, enabling the optimization of model parameters through iterative processes such as gradient descent. On the other hand, the testing set serves as an unbiased evaluation metric, allowing us to gauge the model's performance on unseen data and estimate its ability to generalize to new instances. This division strategy helps mitigate issues related to overfitting, as the model is assessed on independent data that it has not been exposed to during the training phase. Consequently, the 80–20% split facilitates robust model development and evaluation in a rigorous and statistically sound manner.

2.2. Define the Layer of the CNN

In the CNN architecture, the composition of layers can be defined to incorporate hidden layers and the classification layer as shown in Figure 5. In the initial layer, a convolutional layer is created with a filter/kernel size of 3 × 3. The 3 × 3 filter/kernel is applied elementwise to the first 3 × 3 patch of the image, and the resulting values are summed. The sum is then placed in the first row, first column of the output matrix. Next, the 3 × 3 filter/kernel is shifted one position to the right in the first matrix, and the process is repeated. The resulting values are placed in the next column of the output matrix. This process continues until the entire output matrix is filled. The number of filters output is set to 32. Padding is applied to ensure that the size of the output matrix matches the size of the original image. The activation function used is ReLU, which helps in faster training of the model. This configuration of layers enables the CNN to perform feature extraction on the input data and learn meaningful representations that can be used for classification tasks. The convolutional layer with 3 × 3 filters performs local feature detection, while the ReLU activation function introduces non-linearity to enhance the model's expressive power. The padding ensures that the spatial information is preserved during the convolutional operation. The resulting feature maps from this layer can be further processed by subsequent layers, such as pooling layers and fully connected layers, to extract higher-level features and make predictions for classification.

$$f(x) = \max(0, x) = \begin{cases} 0 \text{ for} x \leq 0 \\ x \text{ for} x > 0 \end{cases} \quad (1)$$

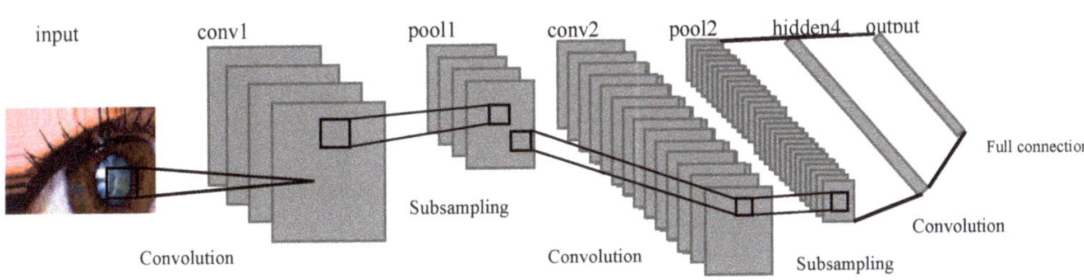

Figure 5. Illustration of the layers within LeNet-CNN.

Furthermore, a max-pooling layer with a size of 2 × 2 is incorporated into the CNN architecture. Max pooling aims to extract the most important parts of the data while increasing the efficiency of processing. It achieves this by selecting only the maximum values within a grid and storing them in the second layer. Similar to the previous layer, a 3 × 3 convolutional layer is defined with padding and ReLU activation function as shown in Equation (1). Additionally, a 2 × 2 max pooling is applied. The number of filters output remains the same at 64.

Moving on to the third layer, it consists of a fully connected layer, which is connected to every node in the previous layer. However, since the data are now in a 1-dimensional form, a process called flattening is performed to convert the data from its original 3-dimensional shape to a 1-dimensional form. The flattened data are then connected to a dense layer with a size of 512, followed by the ReLU activation function.

In the fourth layer, the results obtained from the previous layers are fed into the SoftMax classifier. This final layer enables the model to perform the subsequent classification task.

In summary, the CNN architecture involves the following steps: convolution layer (3×3) with padding and ReLU activation, max-pooling layer (2×2), another convolution layer (3×3) with padding and ReLU activation, max-pooling layer (2×2), and a fully connected layer with flattening, dense layer (512) with ReLU activation. The output of these layers is then passed through the SoftMax classifier for classification purposes.

1. Input Layer: The first step in the LeNet-CNN architecture is the input layer, which receives small-sized images with dimensions of 32×32. These images serve as the input data for the network;
2. Convolutional Layer: The second step involves using filters or kernels to perform convolutions on the input images. This process generates feature maps that capture spatial features of the images, such as detecting edges or important patterns. In LeNet-CNN, there are typically multiple convolutional layers that extract specific image features;
3. Pooling Layer: The third step employs pooling techniques to reduce the size of the feature maps and reduce the complexity of the data before passing it to the next layer. LeNet-CNN utilizes max pooling, which selects the maximum value within a small, specified region of the feature maps;
4. Fully Connected Layer: This layer is responsible for connecting to the feature maps that have undergone convolution and pooling. It consists of fully efficient cells capable of learning patterns and relationships within the data. It resembles the dense layer of an artificial neural network but with spatial limitations;
5. Output Layer: The final step is the output layer, which has the same number of cells as the desired classification classes. Each cell provides OneHot encoded output, indicating the probability of the image belonging to each class.

In summary, the LeNet-CNN architecture involves an input layer, convolutional layers for feature extraction, pooling layers for downsampling, fully connected layers for pattern learning, and an output layer for classification.

2.3. Model Training

After defining the layers of the CNN, the model can be trained by setting the following parameters:

1. Parameter setting

   We set the parameters of the model as follows:

- Batch Size: 16. It represents the number of data samples that will be propagated through the network in each training iteration;
- Epochs: 25. It indicates the number of complete passes the model will make through the training dataset during training;
- Learning Rate: 0.1. It determines the step size at which the model's weights are updated during training;
- Loss Function: binary cross-entropy. It is the average value of the cross-entropy between the actual and predicted probability distributions of two classes (Class 0 and Class 1). We also use Relu as the activation function as shown in Figure 6. A lower average value is preferable, and the binary cross-entropy loss is defined as follows:

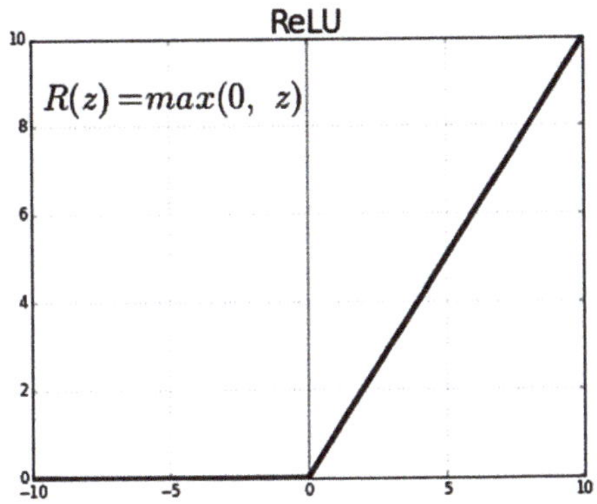

Figure 6. ReLU activation function.

In this work, we use binary cross-entropy loss function, which is often referred to as log loss or logistic loss. It is a commonly used loss function in binary classification tasks, where the goal is to predict one of two possible outcomes (usually 0 or 1). It measures the difference between predicted probabilities and actual binary labels. The formula for binary cross-entropy loss is shown in Equation (2),

$$Loss = -\frac{1}{output\ size}\sum_{i=1}^{output\ size} y_i \cdot \log \hat{y}_i + (1 - y_i) \cdot \log(1 - \hat{y}_i) \quad (2)$$

where:

- *Loss* is the binary cross-entropy loss for a single data point;
- y is the actual binary label (either 0 or 1);
- $\hat{y}$ is the predicted probability of the positive class (class 1) by the model.

2. Training the model:

During training process, we first create training and testing datasets. Then, we iterate through the specified number of training epochs. The training dataset is divided into batches according to the specified batch size. We then feed each batch of data through the CNN model and calculate the loss value using the binary cross-entropy loss function. We use the gradient descent algorithm optimizer, Stochastic Gradient Descent (SGD), to adjust the parameter values in the model, improving the model efficiency. The gradient provides information about the size and direction of parameter updates, guiding the loss value towards the minimum point on the surface. Then, we iteratively adjust the parameter values in each training iteration until the specified number of training epochs is reached. The equation for finding the lowest point of the area of SGD is shown in Figure 7.

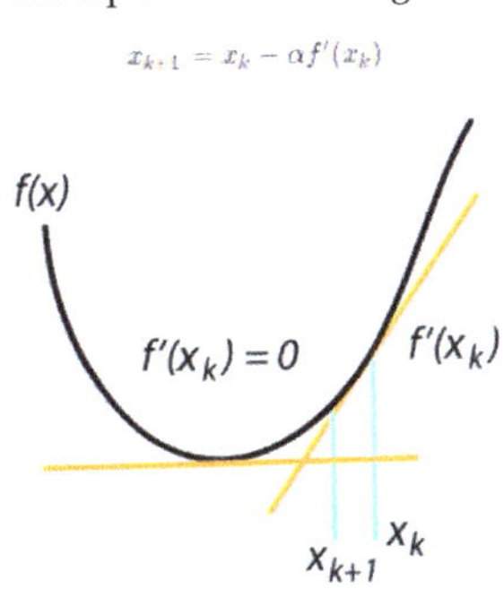

Figure 7. Equation for finding the lowest point of the area of SGD.

For the model evaluation, we utilize the test dataset to evaluate the trained model by calculating performance metrics such as accuracy, precision, recall, F1 score, or error rate. After training the model, it is important to assess its performance using the test dataset. The test dataset contains data that the model has not seen during training and serves as an unbiased evaluation of its generalization capability. Performance metrics such as accuracy, precision, recall, F1 score, or error rate are computed to measure how well the model predicts the correct class labels.

For iterative training and evaluation, we repeat the training and evaluation steps for the specified number of training iterations. In order to improve the model's performance, the training and evaluation steps are iteratively repeated for the designated number of training iterations. This process allows the model to learn from the data, update its parameters, and improve its predictions over time. By iteratively training and evaluating the model, it has the opportunity to learn and adapt to the patterns and complexities present in the dataset. After completing the training and evaluation iterations, the results are analyzed and summarized.

3. Evaluation metrics

The evaluation metrics and statistics utilized are as follows:

- True positive (TP) signifies the number of instances that are accurately predicted as positive with respect to the positive class;
- False positive (FP) indicates the number of instances that are erroneously predicted as positive when compared to the positive class;
- True negative (TN) denotes the number of instances that are correctly predicted as negative with respect to the negative class.
- False negative (FN) represents the number of instances that are incorrectly predicted as negative when compared to the negative class;
- N represents the total number of data points within the dataset.

These metrics serve as crucial measures for evaluating the model's performance and its ability to accurately classify instances into positive and negative classes. By analyzing these metrics, one can assess the model's accuracy, precision, recall, and F1 score, all of which are fundamental indicators of its efficacy in addressing the given task. The accuracy metric provides information about the percentage of instances that the model correctly predicts. It is calculated as the ratio of the sum of true positive (TP) and true negative (TN) to the total number of data points (N). The accuracy metric ranges from 0 to 1, where a value closer to 1 indicates that the model's predictions are highly accurate. The accuracy can be computed as shown in Equation (3).

$$\text{Accuracy} = (\text{TP} + \text{TN})/\text{N} \tag{3}$$

Precision is a metric that measures the accuracy of a model in predicting the relevant instances in the positive class. It is calculated by dividing the number of true positive predictions (TP) by the sum of true positive predictions and false positive predictions (TP + FP). The value of precision ranges from 0 to 1, where a higher precision value indicates a higher level of accuracy in predicting the positive class. The precision can be computed as shown in Equation (4).

$$\text{Precision} = \text{TP}/(\text{TP} + \text{FP}) \tag{4}$$

Recall, also known as sensitivity or true positive rate, is a metric that measures the accuracy of a model in predicting the occurrences of a specific event in the positive class. It is calculated by dividing the number of true positive predictions (TP) by the sum of true positive predictions and false negative predictions. The recall can be computed as Equation (5):

$$\text{Recall} = \text{TP}/(\text{TP} + \text{FN}) \tag{5}$$

F1 score is the harmonic mean of precision and recall. It is a metric that provides a balanced measure of a model's performance by taking into account both precision and recall simultaneously. The F1 score is calculated by taking the reciprocal of the arithmetic mean of the reciprocals of precision and recall. F1 score combines the precision and recall values into a single metric, allowing us to evaluate the overall effectiveness of a model in terms of both correctly predicting positive instances (precision) and capturing all actual positive instances (recall). It provides a consolidated measure that balances the trade-off between precision and recall. The F1 score ranges from 0 to 1, where a value closer to 1 indicates a better balance between precision and recall and reflects a more accurate and reliable model performance. It is particularly useful when the data are imbalanced or when both precision and recall are equally important in the evaluation of the model's effectiveness. The F1 score can be computed as shown in Equation (6).

$$F1 = 2TP/(2TP + FP + FN) \tag{6}$$

True negative rate, also known as specificity, is a metric that measures the model's ability to correctly predict negative instances or non-disease conditions. Specifically, it represents the proportion of actual negative instances that are correctly identified as negative by the model.

Specificity is an important performance measure in binary classification problems, particularly when the negative class or non-disease condition is of interest. It quantifies the model's ability to discriminate and classify negative instances accurately, minimizing the occurrence of false positives. A higher specificity value indicates that the model is proficient in identifying true negative instances and has a lower tendency to misclassify non-disease conditions as positive. It signifies the model's capability to minimize the occurrence of Type I errors, which are false positive predictions. Specificity is calculated as the ratio of true negatives (TN) to the sum of true negatives and false positives (TN + FP). It ranges from 0 to 1, where a value closer to 1 signifies a higher level of accuracy in predicting negative instances and a better ability to distinguish non-disease conditions correctly. The calculation of specificity is shown in Equation (7).

$$TNR = TN/(TN + FP) \tag{7}$$

3. Results

3.1. Datasets

In this research, publicly available datasets [58,59] were utilized. The datasets consisted of 3500 images of eyes displaying symptoms of glaucoma and 3500 images of normal eyes. The datasets were divided into a training set, comprising 80% of the data, and a testing set, comprising the remaining 20%. The dataset used in this study was obtained from public websites [58,59]. It consisted of a collection of 3500 images showing symptoms of glaucoma and an equal number of images of normal eyes. The dataset was split into an 80% training set and a 20% testing set for the purpose of model training and evaluation. The dataset employed in this investigation was sourced from publicly accessible websites [58,59]. It encompassed a total of 3500 images portraying the manifestation of glaucoma symptoms and an equivalent number of images depicting normal eyes. To facilitate the training and evaluation process, the dataset was partitioned into an 80% training subset and a 20% testing subset. For this research, a publicly available dataset [58,59] was utilized. The dataset comprised 3500 images displaying symptoms of glaucoma and an equal number of images depicting normal eyes. To train the model, 80% of the data were used as the training set, while the remaining 20% served as the testing set for evaluation purposes. The image examples of eyes with cataract symptoms and normal eye conditions are shown in Figures 8 and 9, respectively.

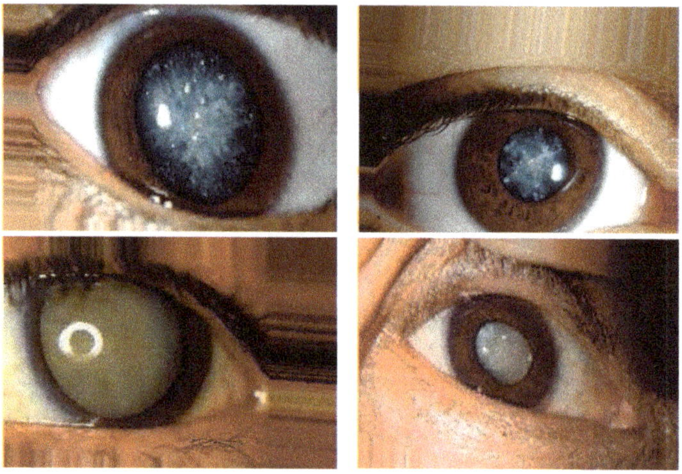

Figure 8. Image examples of eyes with cataract symptoms.

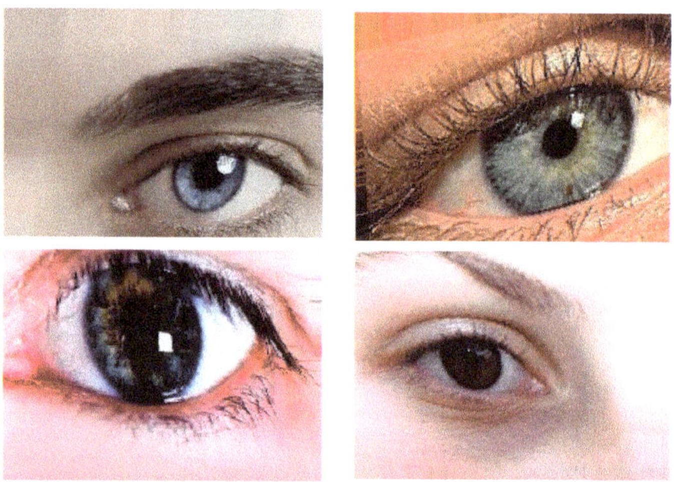

Figure 9. Image examples of normal eye conditions.

3.2. Convolutional Neural Network Testing Results

Initially, CNN was trained with parameter values set as follows: batch size of 128, 25 epochs, learning rate of 0.1, binary cross-entropy loss function, and Stochastic Gradient Descent (SGD) optimizer. The testing results showed an accuracy value of 0.94. The initial testing of the CNN was conducted with specific parameter values. The batch size was set to 128, the number of epochs was 25, the learning rate was 0.1, the loss function used was binary cross-entropy, and the optimizer employed was Stochastic Gradient Descent (SGD). The obtained result yielded an accuracy value of 0.94. The CNN was initially trained and tested with a specific set of parameters. The batch size was set to 128, the number of epochs was 25, the learning rate was 0.1, the loss function used was binary cross-entropy, and the optimizer employed was SGD. The testing results showed an accuracy of 0.94. During the initial testing phase, the CNN was trained using specific parameter configurations. The batch size was set to 128, the number of epochs was 25, the learning rate was 0.1, the loss function was binary cross-entropy, and the optimizer used was SGD. The achieved accuracy for the testing phase was 0.94 as shown in Table 1.

Table 1. Accuracy results of CNNs.

	Precision	Recall	F1 Score	Support	Accuracy
0	0.92	0.96	0.94	701	0.94
1	0.95	0.92	0.94	699	0.94

The CNN achieved a high level of performance in classifying eye images with cataract symptoms, correctly identifying 670 images. Moreover, it demonstrated proficiency in distinguishing normal eye images, accurately classifying 641 samples. However, the model exhibited some misclassifications, as it erroneously labeled 31 images with cataract symptoms as normal and 58 normal eye images as showing cataract symptoms.

Figure 10 shows that based on the provided confusion matrix, the true positive rate (sensitivity) is calculated as the ratio of correctly identified positive cases (disease) to the total number of actual positive cases. In this case, the sensitivity is determined to be 0.92, indicating a high level of accuracy in correctly identifying eye images with cataract symptoms. Similarly, the true negative rate (specificity) is computed as the ratio of correctly identified negative cases (normal) to the total number of actual negative cases. With a specificity value of 0.95, the model demonstrates a strong ability to accurately classify normal eye images. These results highlight the model's robust performance in accurately detecting cataract symptoms and distinguishing normal eye images, as evidenced by the high sensitivity and specificity values obtained from the analysis of the confusion matrix.

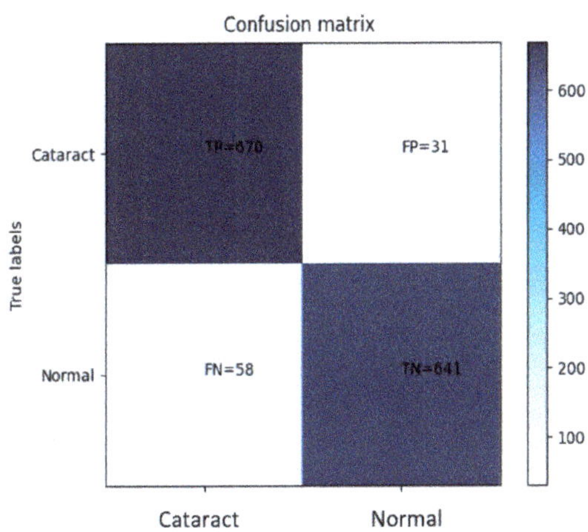

Figure 10. Plot confusion matrix of CNN.

Figure 11 shows the initial model exhibited a high accuracy score; however, upon analyzing Figure 10, it became apparent that as the training dataset size increased, the accuracy curve showed continuous improvement, while the validation accuracy curve failed to stabilize. This behavior was similarly observed in the loss curve, where the training loss continued to decrease without the validation loss reaching a steady state. Consequently, these results indicate that the model's performance is suboptimal. To address this issue, a new parameter configuration was implemented, consisting of a reduced batch size of 16, a fixed number of 25 epochs, a learning rate of 0.1, and the utilization of the binary cross-entropy loss function with the SGD optimizer. With these adjusted parameters, the updated model achieved an accuracy of 0.96 as shown in Table 2. These findings suggest that the modified parameter settings effectively enhanced the model's performance, resulting in a higher accuracy score. The alterations in batch size, learning rate, loss function, and optimizer played pivotal roles in optimizing the model's training process, leading to improved classification accuracy in distinguishing between normal and diseased eye images.

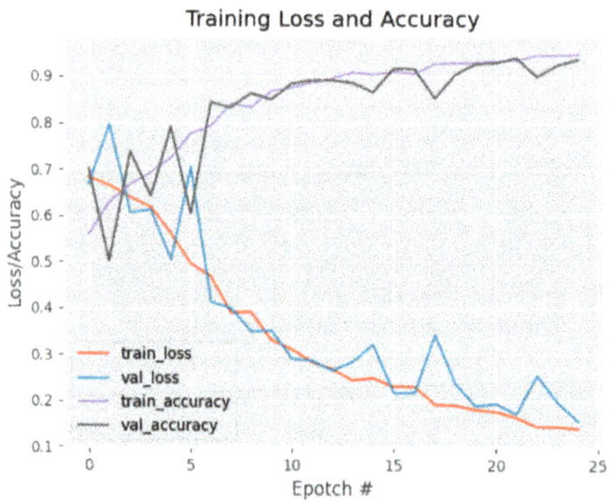

Figure 11. Graph showing accuracy and loss per epoch of the initial model.

Table 2. Results from the CNN testing.

	Precision	Recall	F1 Score	Support	Accuracy
0	0.95	0.96	0.96	701	0.96
1	0.96	0.95	0.96	699	0.96

Based on the obtained confusion matrix, the classification performance of the model can be evaluated. It correctly classified 676 eye images displaying cataract symptoms while misclassifying 25 images. Additionally, the model accurately identified 662 normal eye images but misclassified 37 images as normal when they actually exhibited cataract symptoms. These results indicate a relatively high accuracy in distinguishing between cataract-affected and normal eyes. However, the presence of misclassifications suggests the need for further improvement. Analyzing the misclassified images can provide valuable insights into the underlying patterns or features that contribute to the incorrect categorization. This analysis can guide the refinement of the model architecture, adjustment of hyperparameters, or augmentation of the dataset to mitigate misclassifications and enhance the overall performance of the model. By iteratively repeating this process, it is possible to achieve a more robust and accurate model for cataract classification.

Figure 12 shows the confusion matrix; the true positive rate, also known as sensitivity or recall, is defined as the proportion of correctly predicted positive instances out of all actual positive instances. In this scenario, the sensitivity is calculated to be 0.95, indicating that the model accurately identified 95% of the diseased cases. Similarly, the true negative rate, also referred to as specificity, measures the proportion of correctly predicted negative instances out of all actual negative instances. The specificity value is determined to be 0.96, indicating that the model correctly classified 96% of the non-diseased cases as negative. These findings highlight the model's ability to effectively capture the distinctive features associated with cataract symptoms and normal eye images. The high sensitivity and specificity values signify the model's proficiency in accurately identifying both positive (disease) and negative (non-disease) instances, underscoring its overall performance and reliability in distinguishing between the two categories.

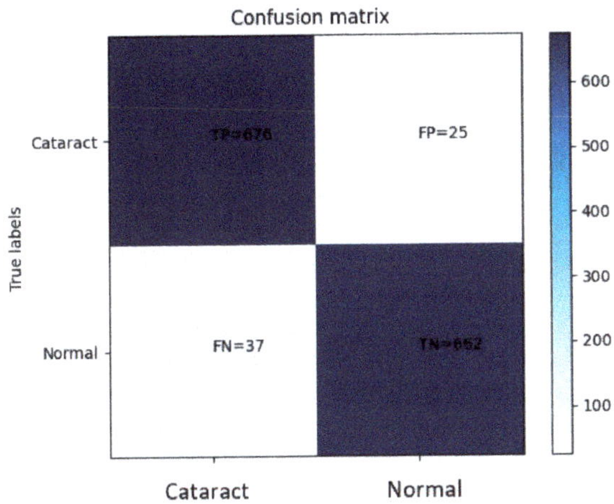

Figure 12. The graph obtained from the plot confusion matrix of CNN.

Figure 13 shows the findings from Figure 14, revealing an interesting trend in the model's performance as the training dataset size increases. The accuracy graph demonstrates a positive correlation, showing a consistent improvement in the model's ability to correctly classify the data. This indicates that the model becomes more accurate with the inclusion of more training instances. Similarly, the loss graph exhibits a negative correlation, indicating a reduction in the model's error as more data are utilized for training. Notably, the validation accuracy and validation loss graphs demonstrate a more stable behavior compared to the previous parameter settings. This stability suggests that the model's performance is not overly influenced by the specific training instances and can generalize well to unseen data. Consequently, the model's configuration appears to be more suitable, as it achieves higher accuracy and lower loss while maintaining a balanced performance on the validation set. Overall, these observations highlight the effectiveness of the current model configuration in capturing the underlying patterns of the data and producing reliable predictions. The increasing accuracy, decreasing loss, and stable validation metrics collectively indicate that the model is well-optimized and capable of making accurate predictions on both training and unseen data.

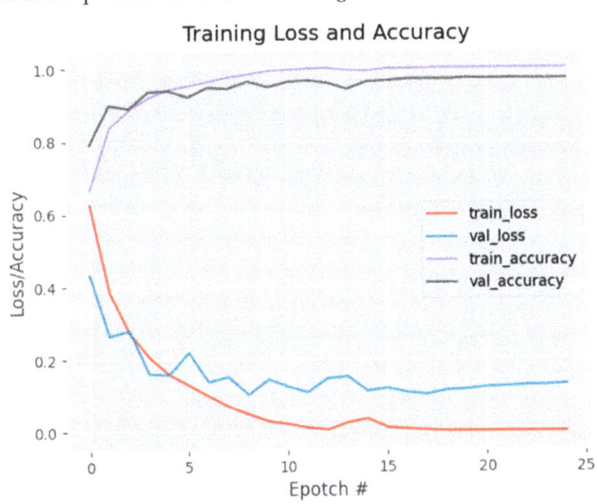

Figure 13. Graph showing accuracy and loss per epoch of the proposed model.

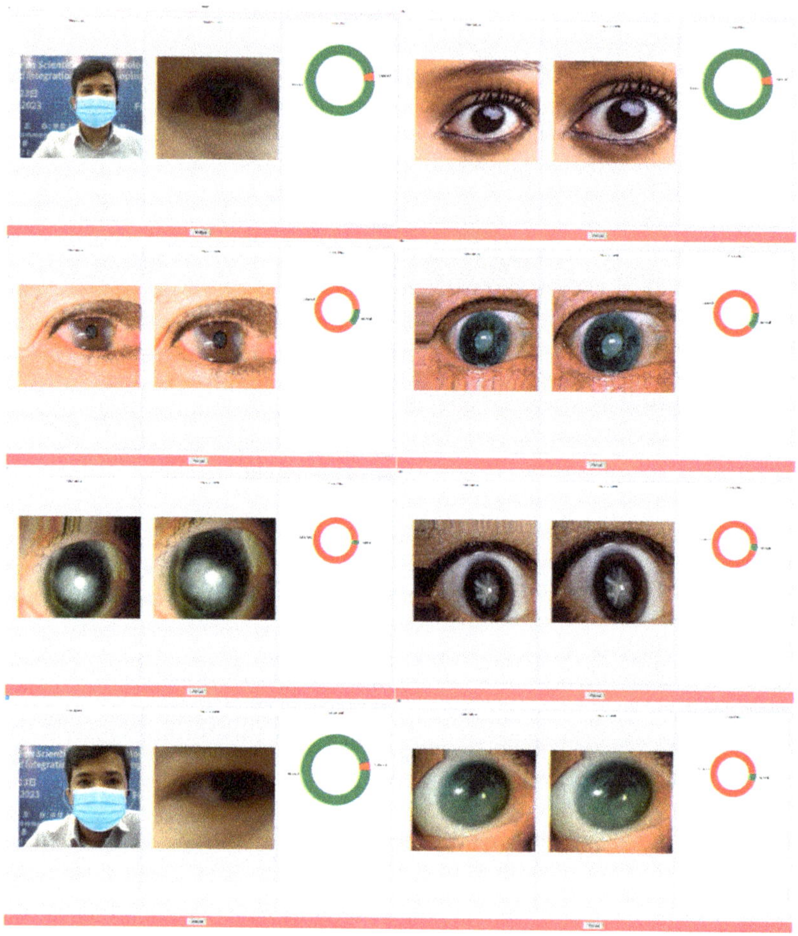

Figure 14. Results of the cataract detection experiment.

Figure 14 presents the experimental results of cataract detection, showcasing the graphical user interface utilized in the experiment. The graphical interface serves as the input medium for the image processing system to perform cataract detection. The system promptly returns the detection results to the user through a user-friendly and convenient cataract detection device. Notably, this system exhibits a high level of accuracy in cataract detection, surpassing current state-of-the-art studies.

3.3. Comparison of Analysis Results

The empirical findings presented in Table 3 elucidate the conspicuously elevated efficacy of the LeNet-CNN model in relation to other methodologies within the ambit of ocular disease classification. The LeNet-CNN model attains an impressive 96% accuracy rate, markedly outperforming its counterparts such as the SVM at 92%, decision tree at 86%, random forest at 90%, Naive Bayes at 79%, and neural network at 87%. Moreover, scrutinizing the true positive rate (sensitivity) reveals LeNet-CNN's remarkable attainment of 95%, while SVM registers at 91%, decision tree at 89%, random forest at 91%, Naive Bayes at 77%, and neural network at 87%. Parallelly, the evaluation of the true negative rate (specificity) underscores LeNet-CNN's substantial achievement at 96%, while SVM reaches 94%, decision tree at 84%, random forest at 89%, Naive Bayes at 81%, and neural network at 86%. These comprehensive metrics collectively underscore the LeNet-CNN model's discernibly heightened precision and effectiveness in the correct classification of both positive and negative instances. This salient performance renders the LeNet-CNN model the preeminent choice for discerning pathological deviations in ocular disorders. The amalgamation of superior accuracy, augmented sensitivity, and enhanced specificity consolidates its standing as an optimal tool for the meticulous identification and categorization of aberrant ocular conditions.

Table 3. The performance comparison between the LeNet-CNN and other machine learning techniques.

	SVM	Decision Tree	Random Forest	Naive Bayes	Neural Network	LeNet-CNN
Accuracy	0.92	0.86	0.90	0.79	0.87	0.96
Sensitivity	0.91	0.89	0.91	0.77	0.87	0.95
Specificity	0.94	0.84	0.89	0.81	0.86	0.96

Based on the comparative analysis presented in Table 4, the research employing image processing technology for corneal abnormality detection exhibits notable clarity and accuracy in its results. The process of segregating corneal abnormalities using the proposed technique demonstrates a higher level of accuracy compared to other existing research endeavors. This can be attributed to the utilization of a more extensive and comprehensive dataset, surpassing the scale employed in previous studies, consequently resulting in improved accuracy levels. Our proposed model outperforms other works regarding the accuracy at 96% except only [60] in which it achieved 96.8% highlighted in bold.

Table 4. Numerical results of related studies using digital camera images.

References	Method	Accuracy	Training Data	Testing Data	Cross-Validation
[29]	Single Layer Perceptron	85%	30	20	No
[59]	Support Vector Machine	88.39%	125	49	No
[61]	K-nearest Neighbor	94.5%	80	80	No
	Support Vector Machine	69.4%			
[60]	Support Vector Machine	**96.8%**	58	36	No
	Artificial Neural Network	92.3%			
[62]	K-nearest Neighbor	83%	78	42	No
	Support Vector Machine	75.2%			
	Naïve Bayes	76.6%			
[63]	Convolutional Neural Network	78%	100	30	No
Our model	LeNet-CNN	**96%**	80	20	No

In the context of research focusing on the design and development of software for corneal abnormality detection through image processing technology, the LeNet-CNN model was chosen due to its inherent efficiency and exceptional accuracy. By employing the LeNet-CNN model, the research achieved a heightened level of accuracy, surpassing the benchmarks set by other studies. Therefore, the proposed technique, which includes the incorporation of the LeNet-CNN model, convincingly demonstrates its efficacy and precision in the corneal abnormality detection process. This substantiates the efficiency and accuracy of the proposed technique, accentuating its capability to accurately identify and classify corneal abnormalities within the tested samples. The study's findings underscore the potential of the proposed technique, particularly with the utilization of the LeNet-CNN model, in significantly enhancing the detection of corneal abnormalities.

4. Discussions

In this work, we present an investigation and the development of a system aimed at detecting and classifying corneal abnormalities using advanced image processing technology. The primary objectives of this research endeavor encompass the following key points: (1) understanding the process of corneal abnormality discrimination utilizing image processing techniques; (2) designing, constructing, and advancing a sophisticated software program for detecting corneal abnormalities; and (3) evaluating the performance and efficacy of the developed program in detecting corneal abnormalities. A thorough review and analysis of pertinent theories and previous research works were conducted to assimilate and adapt the relevant knowledge into the design and development processes. A comprehensive dataset comprising 3500 images, including both normal and corneal abnormality cases, was meticulously curated for training and testing purposes. Prior to training the dataset, preprocessing methodologies were applied to optimize the image data, ensuring its suitability for subsequent machine learning procedures. The LeNet model, a type of CNN, was selected as the primary model for training the prepared dataset. The performance and accuracy of the LeNet model were systematically benchmarked against alternative techniques proposed in the literature. The results unequivocally demonstrated the superior accuracy and effectiveness of the LeNet-CNN model over other approaches. Furthermore, the LeNet model was deployed in the design and implementation of an intuitive graphical user interface (GUI) that seamlessly interfaces with the dedicated LeNet database, purposefully constructed for this research initiative. This integrated system facilitates user-friendly interactions and efficient retrieval of corneal abnormality detection results. The culmination of this research underscores the immense potential and efficacy of the proposed methodology in accurately identifying and discerning corneal abnormalities. By leveraging the power of the LeNet-CNN model and the bespoke software program developed, this research significantly advances the state-of-the-art in corneal abnormality detection, thus holding tremendous promise for clinical applications and enhancing diagnostic precision in the field of ophthalmology.

Remarkable findings have emerged based on the comprehensive investigation into the process of discerning anomalies associated with corneal diseases, employing image processing technology coupled with the application of machine learning techniques, specifically the LeNet-CNN model. Notably, the LeNet-CNN model has exhibited superior performance compared to other relevant research endeavors, boasting an accuracy rate of 96%, a sensitivity rate of 95%, and a specificity rate of 96%. This outcome is in alignment with the stipulated research objectives. Furthermore, a detailed exposition elucidating the intricacies of the LeNet-CNN model underscores its multifaceted computational framework, drawing inspiration from the intricate analysis of human neurological data during the transmission of neural signals. This computational paradigm capitalizes on the extraction and integration of sub-features, contributing to a holistic analysis of image data. Consequently, the selection of the LeNet-CNN model for designing a corneal disease anomaly detection methodology utilizing image processing technology serves as an apt choice, with the obtained results attesting to the concordance with the research objectives.

The LeNet-CNN model boasts several salient advantages that render it highly suitable for the discernment of corneal disease anomalies, as follows:

1. Efficacious foundational architecture: The LeNet-CNN model has evolved from seminal investigations and empirical experiments dating back to 1998. Its basic architecture rests on the CNN, a highly refined structure renowned for its computational prowess in image processing. Consequently, leveraging the LeNet-CNN model for corneal disease anomaly discernment engenders a credible and efficient approach;
2. Substantiation of salient sub-features: The LeNet-CNN model adopts an intricate analytical approach to image analysis, heavily influenced by the examination of human data in the context of neural information transmission. By virtue of employing methodologies that enable the identification and aggregation of pertinent sub-features,

the LeNet-CNN model possesses the capacity to pinpoint critical and indispensable attributes pertaining to corneal disease anomalies at an elevated level of precision;

3. Discerning class categorization with accuracy: The LeNet-CNN model exhibits remarkable accuracy in the classification of image classes, as evidenced by an analysis of the dataset utilized in the research, with accuracy, sensitivity, and specificity rates reaching 96%, 95%, and 96%, respectively. These compelling values underscore the LeNet-CNN model's proficiency and efficiency in categorizing corneal disease anomalies;
4. Versatility in diverse research domains: Beyond its utility in the classification of corneal disease anomalies, the LeNet-CNN model exhibits remarkable flexibility, rendering it adaptable for employment in a range of related research endeavors that necessitate robust image processing capabilities.

In conclusion, the LeNet-CNN model demonstrates exceptional accuracy in discerning corneal disease anomalies, complemented by an intricate analytical process adept at identifying critical attributes. Consequently, its implementation in research for the classification and diagnosis of corneal diseases in real-world settings proves highly conducive to the achievement of research objectives. Future research can explore more comprehensive data, including detailed severity levels and protein deposition locations, to advance the system into a multiclassification framework. Additionally, efforts can be directed towards refining the training process and image enhancement techniques by acquiring high-quality, noise-free images. Environmental factors, such as lighting conditions and image positions, impacted the system's performance, suggesting the potential for a notification mechanism to guide users on optimal image capture.

Author Contributions: Conceptualization, T.G. and P.P.; methodology, T.G.; software, T.G.; validation, T.G., P.P. and M.K.; formal analysis, T.G. and M.K.; investigation, M.K.; resources, P.P. and M.K.; data curation, P.P. and M.K.; writing—original draft preparation, T.G.; writing—review and editing, M.K.; visualization, T.G.; supervision, M.K.; project administration, T.G.; funding acquisition, T.G. All authors have read and agreed to the published version of the manuscript.

Funding: This research project is supported by the Thailand Science Research and Innovation (TSRI) Basic Research Fund: Fiscal year 2023 under project number FRB660073/0164 (Program Digital Transformation).

Institutional Review Board Statement: Not applicable.

Informed Consent Statement: Not applicable.

Data Availability Statement: The Cataract Image dataset is available at https://www.kaggle.com/alexandramohammed/cataract-image (accessed on 5 January 2023). The Cataract Detection App—Using CNN Neural Network is available at https://github.com/KrishnaRauniyar/CataractDetectionApp-UsingCNN-NeuralNetwork/tree/master/Cataract(accessed on 5 January 2023).

Conflicts of Interest: The authors declare no conflict of interest.

References

1. Rana, J.; Galib, S.M. Cataract detection using smartphone. In Proceedings of the 3rd International Conference on Electrical Information and Communication Technology (EICT), Khulna, Bangladesh, 7–9 December 2017; pp. 1–4.
2. Pascolini, D.; Mariotti, S. Global estimates of visual impairment: 2010. *Brit. J. Ophthalmol.* **2012**, *96*, 614–618. [CrossRef] [PubMed]
3. Allen, D.; Vasavada, A. Cataract and surgery for cataract. *BMJ* **2006**, *333*, 128–132. [CrossRef]
4. Yang, J.-J.; Li, J.; Shen, R.; Zeng, Y.; He, J.; Bi, J.; Li, Y.; Zhang, Q.; Peng, L.; Wang, Q. Exploiting ensemble learning for automatic cataract detection and grading. *Comput. Methods Programs Biomed.* **2016**, *124*, 45–57. [CrossRef]
5. Flaxman, S.R.; Bourne, R.R.A.; Resnikoff, S.; Ackland, P.; Braithwaite, T.; Das, A.; Jonas, J.B.; Keeffe, J.; Kempen, J.H.; Leasher, J.; et al. Global causes of blindness and distance vision impairment 1990–2020: A systematic review and metaanalysis. *Lancet Glob. Health* **2017**, *5*, e1221–e1234. [CrossRef]
6. Cao, L.; Li, H.; Zhang, Y.; Zhang, L.; Xu, L. Hierarchical method for cataract grading based on retinal images using improved Haar wavelet. *Inf. Fusion* **2020**, *53*, 196–208. [CrossRef]
7. Afolabi, O.J.; Mabuza-Hocquet, G.P.; Nelwamondo, F.V.; Paul, B.S. The use of U-Net lite and extreme gradient boost (XGB) for glaucoma detection. *IEEE Access* **2021**, *9*, 47411–47424. [CrossRef]

8. Qiao, L.; Zhu, Y.; Zhou, H. Diabetic retinopathy detection using prognosis of microaneurysm and early diagnosis system for non-proliferative diabetic retinopathy based on deep learning algorithms. *IEEE Access* **2020**, *8*, 104292–104302. [CrossRef]
9. Hu, S.; Wang, X.; Wu, H.; Luan, X.; Qi, P.; Lin, Y.; He, X.; He, W. Unified diagnosis framework for automated nuclear cataract grading based onsmartphone slit-lamp images. *IEEE Access* **2020**, *8*, 174169–174178. [CrossRef]
10. Gao, X.; Wong, D.W.K.; Ng, T.-T.; Cheung, C.Y.L.; Cheng, C.-Y.; Wong, T.Y. Automatic grading of cortical and PSC cataracts using retroillumination lens images. In Proceedings of the Asian Conference on Computer Vision, Daejeon, Korea, 5–9 November 2012; Springer: Berlin, Germany, 2012; pp. 256–267.
11. Niya, C.P.; Jayakumar, T.V. Analysis of different automatic cataract detection and classification methods. In Proceedings of the IEEE International Advance Computing Conference (IACC), Banglore, India, 12–13 June 2015; pp. 696–700.
12. Raju, B.; Raju, N.S.D.; Akkara, J.; Pathengay, A. Do it yourself smartphone fundus camera—DIYretCAM. *Indian J. Ophthalmol.* **2016**, *64*, 663. [CrossRef]
13. Fan, W.; Shen, R.; Zhang, Q.; Yang, J.-J.; Li, J. Principal component analysis based cataract grading and classification. In Proceedings of the 17th International Conference on E-health Networking, Application & Services (HealthCom), Boston, MA, USA, 14–17 October 2015; pp. 459–462.
14. Xiong, L.; Li, H.; Xu, L. An approach to evaluate blurriness in retinal images with vitreous opacity for cataract diagnosis. *J. Healthcare Eng.* **2017**, *2017*, 5645498. [CrossRef]
15. Budai, A.; Bock, R.; Maier, A.; Hornegger, J.; Michelson, G. Robust vessel segmentation in fundus images. *Int. J. Biomed. Imag.* **2013**, *2013*, 154860. [CrossRef] [PubMed]
16. Hernandez-Matas, C.; Zabulis, X.; Triantafyllou, A.; Anyfanti, P.; Douma, S.; Argyros, A.A. FIRE: Fundus image registration dataset. *Model. Artif. Intell. Ophthalmol.* **2017**, *1*, 16–28. [CrossRef]
17. Zhang, Z.; Yin, F.S.; Liu, J.; Wong, W.K.; Tan, N.M.; Lee, B.H.; Cheng, J.; Wong, T.Y. ORIGA-light: An online retinal fundus image database for glaucoma analysis and research. In Proceedings of the Annual International Conference of the IEEE Engineering in Medicine and Biology Society (EMBC), Buenos Aires, Argentina, 31 August–4 September 2010; pp. 3065–3068.
18. Porwal, P.; Pachade, S.; Kamble, R.; Kokare, M.; Deshmukh, G.; Sahasrabuddhe, V.; Meriaudeau, F. Indian diabetic retinopathy image dataset (IDRiD): A database for diabetic retinopathy screening research. *Data* **2018**, *3*, 25. [CrossRef]
19. Decencière, E.; Cazuguel, G.; Zhang, X.; Thibault, G.; Klein, J.-C.; Meyer, F.; Marcotegui, B.; Quellec, G.; Lamard, M.; Danno, R.; et al. TeleOphta: Machine learning and image processing methods for teleophthalmology. *IRBM* **2013**, *34*, 196–203. [CrossRef]
20. Staal, J.; Abràmoff, M.D.; Niemeijer, M.; Viergever, M.A.; Van Ginneken, B. Ridge-based vessel segmentation in color images of the retina. *IEEE Trans. Med. Imag.* **2004**, *23*, 501–509. [CrossRef]
21. Morales-Lopez, H.; Cruz-Vega, I.; Rangel-Magdaleno, J. Cataract detection and classification systems using computational intelligence: A survey. *Arch. Comput. Methods Eng.* **2020**, *28*, 1761–1774. [CrossRef]
22. Gali, H.E.; Sella, R.; Afshari, N.A. Cataract grading systems: A review of past and present. *Curr. Opin. Ophthalmol.* **2019**, *30*, 13–18. [CrossRef]
23. Shaheen, I.; Tariq, A. Survey analysis of automatic detection and grading of cataract using different imaging modalities. In *Applications of Intelligent Technologies in Healthcare*; Springer: Cham, Switzerland, 2019; pp. 35–45.
24. Zhang, X.; Hu, Y.; Fang, J.; Xiao, Z.; Higashita, R.; Liu, J. Machine learning for cataract classification and grading on ophthalmic imaging modalities: A survey. *arXiv* **2020**, arXiv:2012.04830. [CrossRef]
25. Gao, X.; Li, H.; Lim, J.H.; Wong, T.Y. Computer-aided cataract detection using enhanced texture features on retro-illumination lens images. In Proceedings of the 18th IEEE International Conference on Image Processing, Brussels, Belgium, 11–14 September 2011; pp. 1565–1568.
26. Yang, M.; Yang, J.-J.; Zhang, Q.; Niu, Y.; Li, J. Classification of retinal image for automatic cataract detection. In Proceedings of the 15th International Conference on e-Health Networking, Applications and Services (Healthcom 2013), Lisbon, Portugal, 9–12 October 2013; pp. 674–679.
27. Guo, L.; Yang, J.-J.; Peng, L.; Li, J.; Liang, Q. A computeraided healthcare system for cataract classification and grading based on fundus image analysis. *Comput. Ind.* **2015**, *69*, 72–80. [CrossRef]
28. Fuadah, Y.N.; Setiawan, A.W.; Mengko, T.L.R.; Budiman. Mobile cataract detection using optimal combination of statistical texture analysis. In Proceedings of the 4th International Conference on Instrumentation, Communications, Information Technology, and Bio-medical Engineering (ICICI-BME), Bandung, Indonesia,, 2–3 November 2015; pp. 232–236.
29. Fuadah, Y.N.; Setiawan, A.W.; Mengko, T.L.R. Performing high accuracy of the system for cataract detection using statistical texture analysis and K-nearest neighbor. In Proceedings of the International Seminar on Intelligent Technology and Its Applications (ISITIA), Surabaya, Indonesia, 20–21 May 2015; pp. 85–88.
30. Caixinha, M.; Amaro, J.; Santos, M.; Perdigão, F.; Gomes, M.; Santos, J. In-vivo automatic nuclear cataract detection and classification in an animal model by ultrasounds. *IEEE Trans. Biomed. Eng.* **2016**, *63*, 2326–2335. [CrossRef]
31. Harini, V.; Bhanumathi, V. Automatic cataract classification system. In Proceedings of the International Conference on Communication and Signal Processing (ICCSP), Melmaruvathur, India, 6–8 April 2016; pp. 0815–0819.
32. Jagadale, A.B.; Sonavane, S.S.; Jadav, D.V. Computer aided system for early detection of nuclear cataract using circle Hough transform. In Proceedings of the 3rd International Conference on Trends in Electronics and Informatics (ICOEI), Tirunelveli, India, 23–25 April 2019; pp. 1009–1012.

33. Sigit, R.; Triyana, E.; Rochmad, M. Cataract detection using single layer perceptron based on smartphone. In Proceedings of the 3rd International Conference on Informatics and Computational Sciences (ICICoS), Semarang, Indonesia, 29–30 October 2019; pp. 1–6.
34. Uchino, E.; Suzuki, K.; Sato, N.; Kojima, R.; Tamada, Y.; Hiragi, S.; Yokoi, H.; Yugami, N.; Minamiguchi, S.; Haga, H.; et al. Classification of glomerular pathological findings using deep learning and nephrologist–AI collective intelligence approach. *Int. J. Med. Inform.* **2020**, *141*, 104231. [CrossRef] [PubMed]
35. Junayed, M.S.; Jeny, A.A.; Atik, S.T.; Neehal, N.; Karim, A.; Azam, S.; Shanmugam, B. AcneNet—A deep CNN based classification approach for acne classes. In Proceedings of the 12th International Conference on Information & Communication Technology and System (ICTS), Surabaya, Indonesia, 18 July 2019; pp. 203–208.
36. Gao, X.; Lin, S.; Wong, T.Y. Automatic feature learning to grade nuclear cataracts based on deep learning. *IEEE Trans. Biomed. Eng.* **2015**, *62*, 2693–2701. [CrossRef] [PubMed]
37. Zhang, L.; Li, J.; Zhang, I.; Han, H.; Liu, B.; Yang, J.; Wang, Q. Automatic cataract detection and grading using deep convolutional neural network. In Proceedings of the 14th International Conference on Networking, Sensing and Control (ICNSC), Calabria, Italy, 16–18 May 2017; pp. 60–65.
38. Ran, J.; Niu, K.; He, Z.; Zhang, H.; Song, H. Cataract detection and grading based on combination of deep convolutional neural network and random forests. In Proceedings of the International Conference on Network Infrastructure and Digital Content (IC-NIDC), Guiyang, China, 22–24 August 2018; pp. 155–159.
39. Pratap, T.; Kokil, P. Computer-aided diagnosis of cataract using deep transfer learning. *Biomed. Signal Process. Control.* **2019**, *53*, 101533. [CrossRef]
40. Kim, D.; Jun, T.J.; Eom, Y.; Kim, C.; Kim, D. Tournament based ranking CNN for the cataract grading. In Proceedings of the 41st Annual International Conference of the IEEE Engineering in Medicine and Biology Society (EMBC), Berlin, Germany, 23–27 July 2019; pp. 1630–1636.
41. Hossain, M.R.; Afroze, S.; Siddique, N.; Hoque, M.M. Automatic detection of eye cataract using deep convolution neural networks (DCNNs). In Proceedings of the IEEE Region 10 Symposium (TENSYMP), Dhaka, Bangladesh, 5–7 June 2020; pp. 1333–1338.
42. Zhang, X.; Lv, J.; Zheng, H.; Sang, Y. Attention-based multi-model ensemble for automatic cataract detection in B-scan eye ultrasound images. In Proceedings of the International Joint Conference on Neural Networks (IJCNN), Glasgow, UK, 19–24 July 2020; pp. 1–10.
43. Khan, M.S.M.; Ahmed, M.; Rasel, R.Z.; Khan, M.M. Cataract detection using convolutional neural network with VGG-19 model. In Proceedings of the IEEE World AI IoT Congress (AIIoT), Seattle, WA, USA, 10–13 May 2021; pp. 0209–0212.
44. Kaggle. Ocular Disease Recognition. 2021. Available online: https://www.kaggle.com/andrewmvd/ocular-diseaserecognitionodir5k (accessed on 11 February 2021).
45. Pratap, T.; Kokil, P. Efficient network selection for computer-aided cataract diagnosis under noisy environment. *Comput. Methods Programs Biomed.* **2021**, *200*, 105927. [CrossRef]
46. Junayed, M.S.; Sakib, A.N.M.; Anjum, N.; Islam, M.B.; Jeny, A.A. EczemaNet: A deep CNN-based eczema diseases classification. In Proceedings of the IEEE 4th International Conference on Image Processing, Applications and Systems (IPAS), Genova, Italy, 9–11 December 2020; pp. 174–179.
47. Hung, J.-Y.; Perera, C.; Chen, K.-W.; Myung, D.; Chiu, H.-K.; Fuh, C.-S.; Hsu, C.-R.; Liao, S.-L.; Kossler, A.L. A deep learning approach to identify blepharoptosis by convolutional neural networks. *Int. J. Med. Inform.* **2021**, *148*, 104402. [CrossRef]
48. Gonzalez, C.R.; Woods, R.E. *Digital Image Processing*; Pearson: London, UK, 2007; Volume 85.
49. Heidari, M.; Mirniaharikandehei, S.; Khuzani, A.Z.; Danala, G.; Qiu, Y.; Zheng, B. Improving the performance of CNN to predict the likelihood of COVID-19 using chest X-ray images with preprocessing algorithms. *Int. J. Med. Inform.* **2020**, *144*, 104284. [CrossRef]
50. Szegedy, C.; Vanhoucke, V.; Ioffe, S.; Shlens, J.; Wojna, Z. Rethinking the inception architecture for computer vision. In Proceedings of the 2016 IEEE Conference on Computer Vision and Pattern Recognition (CVPR), Las Vegas, NV, USA, 27–30 June 2016; pp. 2818–2826.
51. Howard, A.G.; Zhu, M.; Chen, B.; Kalenichenko, D.; Wang, W.; Weyand, T.; Andreetto, M.; Adam, H. MobileNets: Efficient convolutional neural networks for mobile vision applications. *arXiv* **2017**, arXiv:1704.04861.
52. Simonyan, K.; Zisserman, A. Very deep convolutional networks for large-scale image recognition. *arXiv* **2014**, arXiv:1409.1556.
53. Simon, M.; Rodner, E.; Denzler, J. ImageNet pre-trained models with batch normalization. *arXiv* **2016**, arXiv:1612.01452.
54. He, K.; Zhang, X.; Ren, S.; Sun, J. Deep residual learning for image recognition. In Proceedings of the IEEE Conference on Computer Vision and Pattern Recognition (CVPR), Las Vegas, NV, USA, 27–30 June 2016; pp. 770–778.
55. Jeny, A.A.; Sakib, A.N.M.; Junayed, M.S.; Lima, K.A.; Ahmed, I.; Islam, M.B. SkNet: A convolutional neural networks based classification approach for skin cancer classes. In Proceedings of the 23rd International Conference on Computer and Information Technology (ICCIT), Dhaka, Bangladesh, 19–21 December 2020; pp. 1–6.
56. Santra, T. Automatic Retinal Disease Classification Using Machine Learning and AI. *Vixra.org* **2019**. Available online: https://vixra.org/abs/1910.0445 (accessed on 5 January 2023).
57. Chandhakanond, P.; Aimmanee, P. Hemorrhage segmentation in mobile-phone retinal images using multiregion contrast enhancement and iterative NICK thresholding region growing. *Sci. Rep.* **2022**, *12*, 21513. [CrossRef] [PubMed]

58. Mohammed, A. Cataract Image 2020. Available online: https://www.kaggle.com/alexandramohammed/cataract-image (accessed on 28 August 2020).
59. Rauniyar, K. Cataract Detection App—Using CNN Neural Network 2019. Available online: https://github.com/KrishnaRauniyar/CataractDetectionApp-UsingCNN-NeuralNetwork/tree/master/Cataract (accessed on 23 January 2021).
60. Khan, A.A.; Akram, M.U.; Tariq, A.; Tahir, F.; Wazir, K. Automated Computer Aided Detection of Cataract. In Proceedings of the International Afro-European Conference for Industrial Advancement, Marrakesh, Morocco, 21–23 November 2016.
61. Nayak, J. Automated classification of normal, cataract and post cataract optical eye images using SVM classifier. In Proceedings of the World Congress on Engineering and Computer Science, San Francisco, CA, USA, 23–25 October 2013.
62. Agarwal, V.; Gupta, V.; Vashisht, V.M.; Sharma, K.; Sharma, N. Mobile application based cataract detection system. In Proceedings of the 2019 3rd International Conference on Trends in Electronics and Informatics (ICOEI), Tirunelveli, India, 23–25 April 2019.
63. Yusuf, M.; Theophilous, S.; Adejoke, J.; Hassan, A.B. Web-Based Cataract Detection System Using Deep Convolutional Neural Network. In Proceedings of the 2019 2nd International Conference of the IEEE Nigeria Computer Chapter (NigeriaComputConf), Zaria, Nigeria, 14–17 October 2019.

Journal of
Imaging

MDPI

Systematic Review

Denoising of Optical Coherence Tomography Images in Ophthalmology Using Deep Learning: A Systematic Review

Hanya Ahmed [1,*,†], Qianni Zhang [1,†], Robert Donnan [2,†] and Akram Alomainy [1,†]

1 Department of Electronic Engineering and Computer Science, Queen Mary University of London, London E1 4NS, UK
2 Department of Engineering and Materials Science, Queen Mary University of London, London E1 4NS, UK
* Correspondence: h.t.ahmed@qmul.ac.uk
† These authors contributed equally to this work.

Abstract: Imaging from optical coherence tomography (OCT) is widely used for detecting retinal diseases, localization of intra-retinal boundaries, etc. It is, however, degraded by speckle noise. Deep learning models can aid with denoising, allowing clinicians to clearly diagnose retinal diseases. Deep learning models can be considered as an end-to-end framework. We selected denoising studies that used deep learning models with retinal OCT imagery. Each study was quality-assessed through image quality metrics (including the peak signal-to-noise ratio—*PSNR*, contrast-to-noise ratio—*CNR*, and structural similarity index metric—*SSIM*). Meta-analysis could not be performed due to heterogeneity in the methods of the studies and measurements of their performance. Multiple databases (including Medline via PubMed, Google Scholar, Scopus, Embase) and a repository (ArXiv) were screened for publications published after 2010, without any limitation on language. From the 95 potential studies identified, a total of 41 were evaluated thoroughly. Fifty-four of these studies were excluded after full text assessment depending on whether deep learning (DL) was utilized or the dataset and results were not effectively explained. Numerous types of OCT images are mentioned in this review consisting of public retinal image datasets utilized purposefully for denoising OCT images ($n = 37$) and the Optic Nerve Head (ONH) ($n = 4$). A wide range of image quality metrics was used; *PSNR* and *SNR* that ranged between 8 and 156 dB. The minority of studies ($n = 8$) showed a low risk of bias in all domains. Studies utilizing ONH images produced either a *PSNR* or *SNR* value varying from 8.1 to 25.7 dB, and that of public retinal datasets was 26.4 to 158.6 dB. Further analysis on denoising models was not possible due to discrepancies in reporting that did not allow useful pooling. An increasing number of studies have investigated denoising retinal OCT images using deep learning, with a range of architectures being implemented. The reported increase in image quality metrics seems promising, while study and reporting quality are currently low.

Keywords: deep learning; ophthalmology; image processing; optical coherence tomography

1. Introduction

Optical coherence tomography (OCT) stands at the forefront of modern medical imaging techniques, harnessing the power of low-coherence infrared light to delve deep into biological structures with unprecedented clarity and precision [1]. This revolutionary technology affords longer exposure times due to its inherent biological safety, presenting a stark departure from the ionizing radiation associated with conventional X-rays. Moreover, when juxtaposed with Magnetic Resonance Imaging (MRI) and Computerized Tomography (CT), OCT emerges as a cost-effective alternative, democratizing access to high-quality diagnostic imaging. However, amidst the brilliance of OCT lies a challenge inherent to all imaging modalities—noise. Inevitably introduced during the imaging process, noise mingles with the signal emanating from the object under scrutiny, influencing the resultant intensity observed by the detecting pixel [2]. Of particular concern is speckle noise, a byproduct of low coherence in irradiance, which casts a shadow over the signal-to-noise ratio, obscuring critical details within the imagery. Within the realm of ophthalmology, OCT serves as a versatile tool, facilitating the acquisition of cross-sectional and volumetric images that illuminate the intricate landscape of biological tissues and retinal structures. These images serve as invaluable aids in diagnosing a myriad of ocular diseases, ranging from diabetic retinopathy (DR) [3] to age-related macular degeneration (AMD) [4], guiding clinicians toward tailored treatment strategies.

Meanwhile, in the realm of computer science and Artificial Intelligence (AI), Machine Learning (ML) emerges as a beacon of innovation. By mining insights from past data, ML algorithms prognosticate future trends, obviating the need for explicit programming or human intervention [5]. At its core, ML epitomizes the essence of pattern recognition, endowing computers with the capacity to glean insights from vast datasets with unparalleled efficiency.

Within the ML landscape, deep learning (DL) commands center stage, propelled by leaps and bounds in computational prowess and the proliferation of "big data". Convolutional Neural Networks (CNNs) epitomize this evolution, revolutionizing DL with their ability to extract features, classify images, and recognize patterns at breakneck speeds. By emulating the intricate workings of the human brain through the deployment of filters and intricate layers, CNNs herald a new era of computational efficiency.

The intersection of DL and OCT heralds a realm of boundless possibilities, marked by advancements in volumetric data handling, heightened sensitivity, and specificity in detecting structural alterations and the tantalizing prospect of denoising retinal images to unprecedented levels of clarity. This synthesis prompts a critical examination, as we delve into recent studies illuminating the applications of DL to OCT imagery, evaluating their impact on image quality assessment.

Moreover, as we stand on the precipice of a new era in medical imaging and computational innovation, it behooves us to explore the clinical ramifications of integrating these cutting-edge computational techniques into the fabric of healthcare delivery. Through a judicious examination of recent computational innovations and their potential clinical applications, we chart a course toward enhanced diagnostic accuracy, streamlined treatment pathways, and, ultimately, improved patient outcomes. Thus, this review not only serves as a testament to the symbiotic relationship between technology and healthcare but also as a compass guiding future research endeavors and clinical initiatives.

2. Overview of Optical Coherence Tomography

Optical coherence tomography (OCT) stands as a maturing imaging technology, offering resolution ranging from millimeters to sub-millimeters and boasting a penetration depth comparable to that achieved in human skin [6]. This innovative technique predominantly employs low-coherence infrared light to safely delve into biological tissues, affording longer exposure times in contrast to X-rays. Figure 1 illustrates the conceptual system configuration of Michelson interferometry, the foundation upon which OCT operates. Within this setup, the interferometric probe beam, formed by recombined reflected beams at the beam splitter, is directed toward the surface under examination, with a detector poised to capture the backscatter emanating from this surface [7].

Central to the functioning of OCT is the notion of coherence, a defining characteristic of light wherein all rays maintain a consistent and calculable phase over a defined period. However, the utilization of low coherence, while advantageous in probing biological tissues, introduces an unintended consequence—the introduction of noise altering pixel intensity and distorting the resulting image. This phenomenon manifests as artifacts, leading to a compromised signal-to-noise ratio within OCT images [7]; while various forms of noise may afflict the imaging process, speckle noise emerges as the predominant type encountered in OCT imagery [8].

At the heart of OCT lies the Michelson interferometer, serving as its primary setup. Here, an optical probe directs low-coherent light toward the sample, penetrating its surface and awaiting the rebound of reflected light. Subsequently, this reflected light is channeled to the interferometer via an optical fiber for meticulous analysis, as depicted in Figure 1 [9].

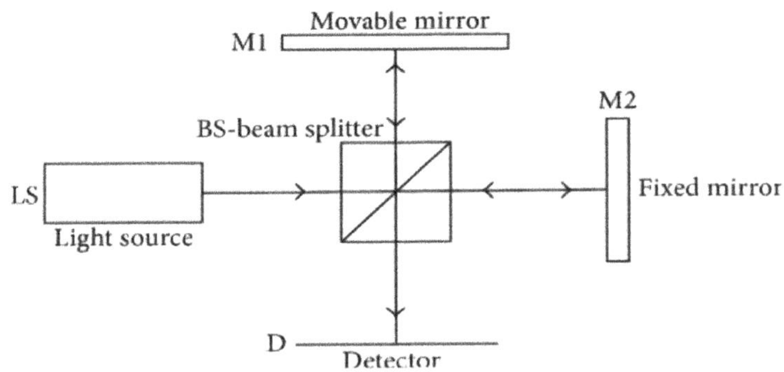

Figure 1. A system diagram of the principle of Michelson interferometry used in OCT [9]. The signal from a collimated light source (LS) is partitioned into two orthogonal beam paths by a beam splitter (BS); one collinear and the other normal to the ls pointing. The normal beam is reflected off a movable mirror (M1), the collinear off a fixed mirror (M2). The reflected beams are recombined at the bs and the co-propagating superposition is recorded as an interferogram at the detector (D). Image [9].

3. Material and Methods

This paper presents a systematic review focusing on studies concerning diagnostic accuracy. The reporting of this study adheres to the PRISMA-DTA guidelines [10].

3.1. Eligibility Criteria

This systematic review is designed to address specific queries utilizing the PICO framework, which encompasses Population, Intervention, Control, and Outcomes. Our investigation centers on several key questions. Firstly, we aim to explore the various implementations and accuracy outcomes associated with employing deep learning (DL) techniques for denoising retinal optical coherence tomography (OCT) imagery. Secondly, we seek to evaluate the effectiveness of these DL-based denoising models in mitigating noise and enhancing the quality of retinal OCT images. Thirdly, we will assess the performance of DL-based denoising models against other denoising methods or untreated OCT images used as benchmarks. Finally, our primary objective is to evaluate the effectiveness of DL-based denoising models by analyzing improvements in image quality through the application of widely used image metrics. Inclusion criteria for studies are as follows:

- Population (P): Studies focusing on the utilization of DL models with retinal imagery obtained from either clinical or research settings.
- Intervention and Control (I, C): Studies employing DL-based models for tasks such as image denoising, speckle reduction, or super-resolution, compared with a reference test.
- Outcomes (O): Studies reporting any estimate of image quality metrics (such as *PSNR*, *CNR*, *SSIM*) applied at either the image or pixel level.

Our exclusion criteria encompass studies that do not meet specific standards, including a lack of clear explanation regarding the utilized DL model, absence of an effective comparative analysis with other state-of-the-art denoising methods, and classification as reviews rather than original research contributions. By meticulously delineating these inclusion and exclusion criteria, our goal is to ensure the robustness and reliability of the systematic review findings. This approach facilitates comprehensive insights into the utility and efficacy of DL in enhancing the quality of retinal OCT imagery.

3.2. Search Methods for Identifying Studies

An electronic search was conducted within the following electronic databases: Google Scholar, Scopus, Medline (via PubMed), Embase, and ArXiv, covering entries up to 16 November 2023. Results were filtered to include publications from 2010 onward, as deep learning (DL) for computer vision and image analysis gained prominence following its development in 2012 by Krizhevsky et al. [11]. Language restrictions were not applied. Customized keywords were utilized for each database. Refer to Table 1 for the specific search queries employed. Moreover, additional studies were identified through screening conference proceedings and journal articles. Furthermore, manual cross-referencing of the bibliographies of included papers was performed to ensure comprehensive coverage.

Table 1. The results of the electronic search in multiple databases.

Database	Keywords	Results	Date
Google Scholar	Image denoising AND deep learning AND "optical coherence tomography"	4220	16 November 2023
Medline	Image denoising AND deep learning AND "optical coherence tomography"	32	16 November 2023
Scopus	TITLE-ABS-KEY (image AND denoising AND deep AND learning AND "optical coherence tomography")	68	16 November 2023
Embase	("image denoising"/exp OR "image denoising" OR (("image"/exp OR image) AND ("denoising"/exp OR denoising))) AND ("deep learning"/exp OR "deep learning" OR (deep AND ("learning"/exp OR learning))) AND ("optical coherence tomography"/exp OR "optical coherence tomography")	37	16 November 2023
ArXiv	Image denoising AND deep learning AND "optical coherence tomography"	9	16 November 2023

3.3. Study Selection

To efficiently manage citations, BibTeX was employed as a tool. Initial screening involved the removal of duplicate entries based on titles and abstracts. Subsequently, a thorough evaluation of articles was conducted to identify eligible studies in accordance with the predetermined inclusion and exclusion criteria. This meticulous process ensured the selection of relevant and appropriate studies for the systematic review. Additionally, any discrepancies or uncertainties during the screening process were resolved through discussion among the research team, ensuring consistency and accuracy in study selection.

3.4. Data Collection and Extraction

Data collection was conducted independently from the included studies and meticulously revised to address any discrepancies or disagreements. Comprehensive information was extracted, encompassing various data items vital for analysis. These included bibliographic details such as authors' names and publication years, details regarding the data modality and type of dataset utilized, hardware specifications, and dataset size (including train/validation/test sets, if provided). Moreover, inclusion and exclusion criteria at the image level, if available, were noted, along with the specified objective of the study (e.g., image denoising, speckle reduction, super resolution). Information regarding pre-processing techniques, data augmentation strategies, and the deep learning (DL) approach employed, including the neural network (NN) architecture utilized, was also recorded. Additionally, details regarding the loss function employed and the image quality metrics used for evaluation were documented. The resulting findings from each study were thoroughly examined. In cases where an article compared multiple NN architectures, the most accurate one was reported to ensure clarity and consistency in the analysis. This comprehensive approach to data extraction facilitated a robust and thorough examination of the included studies, contributing to the reliability and validity of the systematic review findings.

3.5. Risk of Bias and Applicability

This review focuses specifically on the denoising aspect, and to assess the risk of bias in the included studies, we adapted and employed the QUADAS-2 tool. This modified tool encompasses four main domains addressing the risk of bias: the data selection, index test, reference standard, and flow-and-timing. Additionally, it evaluates three domains regarding the applicability of the study to patient selection, the index test, and the reference standard.

Within the "data selection" domain, we scrutinized papers with vague data-split strategies and limited information on the dataset, which could potentially lead to data leakage, indicating a high risk of bias. Moving on to the "index test" domain, we assessed indicators such as the lack of description of the model and the absence of details regarding test recreation and reproducibility. The "flow-and-timing" domain was evaluated based on indicators such as the implementation of multiple reference standards (i.e., state-of-the-art denoisers) within the same article and the appropriateness of intervals between the index test and reference standard. Finally, within the "reference standard" domain, we considered indicators like inadequate information on reference standard definition and the utilization of only one reference test.

In cases where concerns arose regarding the relevance of the studies, certain factors were meticulously reviewed. These included the dataset used, the procedure employed for creating clean data, the specific deep learning (DL) model utilized, and its performance concerning image quality metrics. Table 2 outlines the key questions utilized in our assessment process, providing a structured framework for evaluating the risk of bias across the included studies. Through this comprehensive approach, we aimed to ensure the rigor and reliability of our review findings.

Table 2. Modified leading questions of QUADAS-2 for critical appraisal.

Domain	Leading Questions
Data Selection	1—Are any data imbalances addressed in the article? 2—Was the dataset split explained correctly for training, validation, and testing? 3—Did the study collect sufficient noisy–clean image pairs?
Index test	1—Was the methodology sufficiently explained for reproducibility? 2—Were the results of deep learning models explained without knowledge of state-of-the-art denoisers? 3—Did the study apply any image quality metrics, sensitivity or robustness analysis on their model?
Flow and Timing	1—Was the full dataset utilized in the analysis? 2—Did each image have a reference clean image? 3—Were the reference clean images produced similarly? 4—Did the model show a sufficient interval between the reference and index test?
Reference Standard	1—Were state-of-the-art results of denoisers mentioned and utilized for interpretation? 2—Did the study describe the noisy–clean image procedure and minimize bias? 3—Were limitations, biases, and generalization issues reported sufficiently?

3.6. Data Synthesis and Analysis

Due to the diverse array of study designs and image quality measures utilized, our quantitative synthesis was primarily confined to examining the outcomes related to image denoising and speckle reduction. Given the substantial variability in the image quality metrics employed for quantifying denoising and super resolution, the scope for direct comparison was somewhat constrained, while a minority of studies reported metrics such as contrast-to-noise ratio (*CNR*), equivalent number of looks (*ENL*), and the structural similarity index measure (*SSIM*), the predominant metrics presented were peak signal-to-noise ratio (*PSNR*) or signal-to-noise ratio (*SNR*). Consequently, conducting a comparative quantitative analysis across studies was limited to reporting any or all evaluation metrics mentioned, calculated by:

$$PSNR = 10\log(\frac{L^2}{MSE}), \tag{1}$$

$$SNR = 10\log(\frac{P_{signal}}{P_{noise}}), \tag{2}$$

L denotes the maximum possible pixel value, and MSE is the mean squared error of the image. P_{signal} and P_{noise} are the mean and standard deviation of pixel values, respectively. Next, the structural similarity index (*SSIM*) is a well-known image quality metric that focuses on perceived similarity. The *SSIM* focuses on texture, quality degradation and visible structures. The *SSIM* is defined as

$$SSIM = \frac{(2\sigma_{nc} + c_2)(2\mu_n\mu_c + c_1)}{(\mu_n^2 + \mu_c^2 + c_1)(\sigma_n^2 + \sigma_c^2 + c_2)} \tag{3}$$

μ_n, μ_c, and σ_n, σ_c are the mean value and standard variation in noisy (n)–clean (c) image pairs, respectively. Lastly, contrast-to-noise ratio (*CNR*) utilizes ROIs of background and signal areas for speckle repression with respect to both areas. Equivalent number of looks (*ENL*) is a metric assessing the smoothing of the predicted image. It does not require a reference image since it utilizes selected ROIs of background and signal. *ENL* is defined as

$$ENL = \frac{\mu_b^2}{\sigma_s^2} \tag{4}$$

σ_s is the standard deviation of the signal representation, and μ_b is the mean value for background representation. *CNR* is calculated through

$$CNR = 10\log(\frac{\mu_s - \mu_b}{\sqrt{\sigma_b^2 + \sigma_s^2}}) \tag{5}$$

μ_s and σ_s are the mean value and standard deviation of the signal representation, respectively. For background representation, μ_b and σ_b are the mean value and standard deviation.

4. Results

4.1. Study Selection and Study Characteristics

Out of the initial pool of 4399 studies identified, a rigorous evaluation was conducted for 41 studies based on the criteria outlined in Table 2, utilizing their full texts. Subsequently, 54 studies were excluded following a thorough assessment of their full texts. The reasons for exclusion were carefully documented and categorized, with detailed explanations provided in Table S1. Ultimately, after meticulous manual screening, a total of 41 studies were deemed eligible for inclusion in our review. Notably, the number of studies included per year exhibited an upward trend over the observation period, as depicted in Figure 2. This trend underscores the increasing interest and attention devoted to the subject matter over time, highlighting the evolving landscape of research in this field.

The studies examined in this review were categorized into two main groups based on the type of optical coherence tomography (OCT) images utilized within the field of ophthalmology: Optic Nerve Head (ONH) images and retinal images. A summary of these studies is presented in Tables 3 and 4, respectively, highlighting the utilization of various deep learning (DL) models for tasks such as image denoising, speckle reduction, and super resolution. Specifically, retinal image datasets utilized in this review were predominantly public datasets purposely employed for denoising OCT images (n = 37). Notable datasets included

DUKE [12], Topcon [13], OPTIMA [14], Cirrcus [15], and Heindberg [16]. Conversely, ONH images ($n = 4$) primarily comprised private datasets created by researchers.

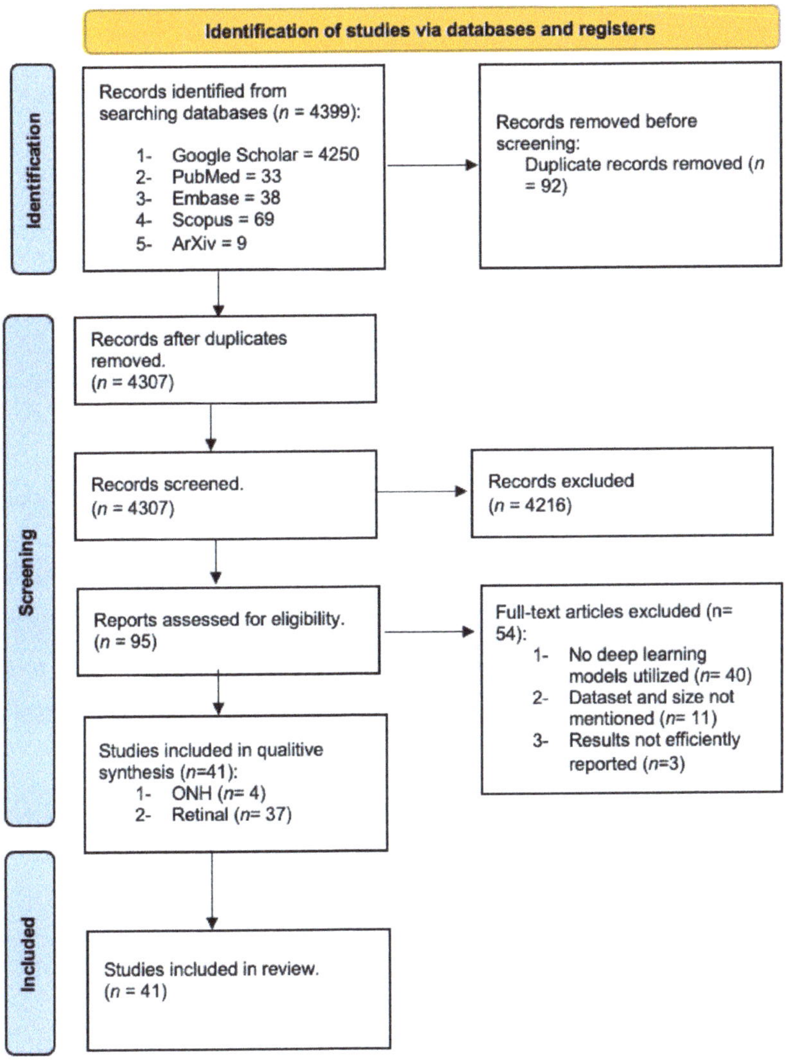

Figure 2. Flowchart of the search following the PRISMA guidelines.

The majority of studies ($n = 39$) incorporated multiple reference tests to evaluate their proposed methods against previous state-of-the-art denoisers. These reference tests encompassed both traditional programming and DL denoising models. Specifically, 26 studies implemented DL models and traditional programming as reference tests, while 9 studies utilized traditional programming-based state-of-the-art denoisers such as BM3D and NLM. However, five studies did not specify any established reference test.

Regarding the choice of dataset for denoising, the DUKE dataset ($n = 22$) was the most frequently utilized, followed by Topcon ($n = 6$), with ONH images being the least employed ($n = 4$). Various DL models were deployed and integrated into hybrid frameworks, as depicted in Figure 3, which illustrates the distribution of studies implementing each DL model. Notably, 44% of the studies implemented a hybrid generative adversarial network (GAN), with the super-resolution GAN (SR-GAN) being the most utilized ($n = 3$) alongside the conditional GAN (cGAN) ($n = 3$). Additionally, 13% of the studies employed a traditional U-Net model ($n = 6$) for denoising, with only one hybrid framework reported.

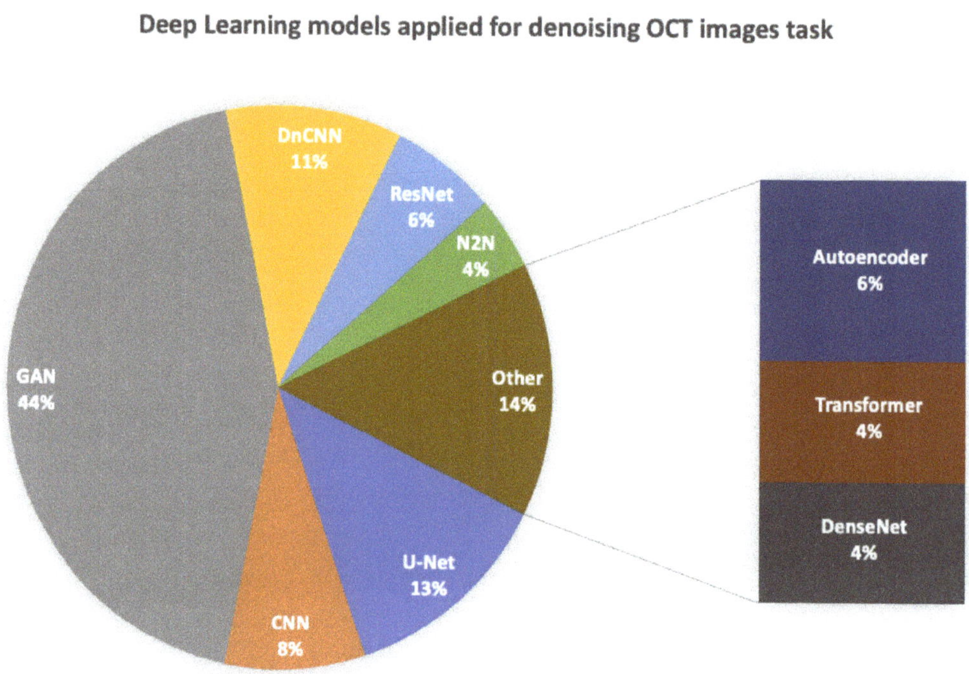

Figure 3. Deep learning models approaches involved for denoising OCT images (n = studies).

Evaluation of denoising, speckle reduction, and studies on super resolution primarily relied on metrics such as *PSNR* and *SNR* as the major image quality indicators. Furthermore, other widely used image quality metrics included the *SSIM* ($n = 27$), *CNR* ($n = 24$), and *ENL* ($n = 16$). These metrics are comprehensively displayed in Tables 3 and 4, providing insight into the methodologies and outcomes of the reviewed studies.

Table 3. Summary of findings in the selected studies utilizing Optic Nerve Head (ONH) datasets for image denoising, speckle reduction, and super-resolution.

Paper	Data Size	Pre-Processing and Augmentation	Model	Compared to	Loss Function	Image Quality Metrics
Devalla, 2019 [17]	24,832 B-Scans (23,280/1552)	Augmentation: Rotation, flip, elastic deformation Pre-Processing: NA	U-Net with residual blocks	NA	MAE	SNR = 8.14 dB CNR = 7.63 dB MSSIM = 0.65
Cheong, 2021 [18]	2628 B-Scans (2328/300)	Augmentation: rotation, translation, flip, and scaling Pre-Processing: NA	Hybrid model (framework containing U-Net followed by a choice of ResNet, ResXNet, or EfficientNet)	NA	Shadow + content + style	PSNR = 11.1% AGM = 57.2% CNR = 154% SSIM = 187%
Tian, 2020 [19]	350 HQ scans (315/35)	Augmentation: flip, rotation, cropping	RMA-GAN	SRCNN, EDSR, ESRGAN	Content + perceptual + adversarial + MS-SSIM + TV	PSNR = 25.7 dB SSIM = 0.77
Hu, 2020 [20]	2500 B-Scans	Augmentation: NA	MSUN with self-fusion	NA	L1 + MSE	PSNR = 10.1 dB SSIM = 0.57
Akter, 2020 [21]	157 B-Scans (100/57)	Augmentation: NA Pre-processing: contrast adjustment, sharpening filter, manually removed noise using Fiji: ImageJ	U-Net	WIN5-RB, Autoencoder, DnCNN, Dense-UNet	MSE	PSNR = 29.8 dB SSIM = 0.90 MSE = 0.005 MAE = 0.03
Halupka, 2018 [22]	55,080 B-Scans (40,711/5587/8780)	Augmentation: flip, rotation, cropping	GAN	WGAN, BM3D, DD-CDWT	MSE + VGG + adversarial	PSNR = 32.3 dB SSIM = 0.78 MSE = 40.3 MS-SSIM = 0.92

Keywords: MSE—Mean squared error; MAE—Mean absolute error; SNR—Signal-to-noise ratio; PSNR—Peak signal-to-noise ratio; CNR—Contrast-to-noise ratio; SSIM—Structural similarity index measure; MS-SSIM—Multi-scale structural similarity index measure; RMA-GAN—Realistic mixed attention GAN; SRCNN—Super-resolution CNN; EDSR—Enhanced deep super-resolution network; ESRGAN—Enhanced super-resolution GAN; MSUN—Multi-scale U-Net; WGAN—Wasserstein GAN.

Table 4. Summary of findings in the selected studies utilizing retinal datasets for image denoising, speckle reduction, and super-resolution.

Paper	Data Size	Pre-Processing and Augmentation	Model	Compared to	Loss Function	Image Quality Metrics
Wei, 2018 [23]	Duke [12] 26 B-Scans (22/4)	Pre-processing: KNN for clustering Augmentation: scaling	DnCNN	BM3D, NLM, BM3D-SAPCA, LPG-PCA, Low Rank, FFDnet	NA	PSNR = 28.2 dB CNR = 3.9 dB MSR = 6.2
Chen, 2020 [24]	36 B-Scans (25/11)	Pre-processing: aligning, averaging, thresholding, adding speckle noise	DN-GAN	MSBTD, SBSDI, BM3D, K-SVD, Tikkhonov, SRResNet, GAN-SRResNet, DCSRN, GAN-U-Net	L1 + perceptual	PSNR = 27.9 dB SSIM = 0.9 FBE = 3.6
Gour, 2020 [25]	Duke [12] and Topcon [13], 23 B-Scans	NA DnCNN	Adaptive Median Filtering, wavelet thresholding, Tikhonov, BM3D, K-SVD, MSBTD, Anisotropic diffusion, STAT, Bayesian, Isotropic diffusion, SE-CNN	MSE	PSNR = 27.5 dB SSIM = 0.68	
Hassan, 2021 [26]	10,000 B-Scans (8000/2000)	Pre-processing: added speckle noise	D-GAN	Wavelet, Bilateral, NLM, BM3D	Euclidean + perceptual + adversarial	PSNR = 35.4 dB MSE = 0.19
Ma, 2018 [15]	Duke [12] and Topcon [13] 521 B-Scans (512/9)	Pre-processing: registration, alignment and enhancing contrast Augmentation: flip, scaling, rotation, non-rigid transformation	cGAN	NLM, BM3D, STROLLR, K-SVD, MAP, DnCNN, ResNet	MSE + L1 + edge	SNR = 60.1 dB CNR = 10.0 dB ENL = 126.9 dB EPI = 1.0
Guo, 2020 [27]	Duke [12], A2A SD-OCT, 90 B-Scans (10/80)	NA	Nonlocal GAN	NLM, BM3D, K-SVD, BM4D, GCBD, GAN-MSE, DnCNN, GAN-WDP, DeGAN	Binary cross-entropy	SNR = 40.1 dB ENL = 981.3 dB CNR = 7.4 dB
Qiu, 2020 [28]	47 B-scans (37/10)	Pre-processing: averaged and registered the B-scans to create denoised image pairs	DnCNN	NLM, BM3D	Perceptually sensitive (SSIM loss)	PSNR = 26.4 dB SSIM = 0.71 MSE = 89.6 MS-SSIM = 0.91

Table 4. *Cont.*

Paper	Data Size	Pre-Processing and Augmentation	Model	Compared to	Loss Function	Image Quality Metrics
Huang, 2021 [13]	OCT2017 [29], 84,500 B-scans (83,416/32/968)	Augmentation: crop	AC-SRResNet	BM3D, U-Net, SRResNet	L1	SNR = 41.8 dB CNR = 44.6 dB EPI = 0.72
Halupka, 2018 [22]	69 OCT volumes (51/7/11)	Pre-processing: averaged and registered the B-scans to create denoised image pairs	GAN	BM3D, DD-CDWT, CNN-WGAN	Adversarial + MSE + perceptual	PSNR = 32.3 dB SSIM = 0.78 MS-SSIM = 0.92 MSE = 40.3
Qiu, 2021 [30]	Duke [12], 52 groups of 50 B-scans each (37/15)	NA	P2PGAN-N2N	Median, NLM, BM3D	Adversarial + L1	SNR = 35.5 dB SSIM = 0.81 CNR = 4.0 dB ENL = 260.3 dB R = 0.94
Abassi, 2019 [31]	Duke [12], 28 B-Scans (10, 18)	Augmentation: flip, rotate, crop	MIFCN	KSVD, BM3D, SAIST, PG-GMM, BM4D, SSR	MSE	PSNR = 27.4 dB CNR = 3.8 dB ENL = 2750.8dB
Shi, 2019 [32]	Topcon [13] and Cirrus [15], 11 groups of 256 B-Scans (2/9)	NA	DeSpec-Net	NLM, BM3D, STROLLR, K-SVD, MAP, Intra-volume compounding, DnCNN	L1	SNR = 40.2 dB CNR = 9.7 dB ENL = 166.2 dB EPI = 0.91
Huang, 2020 [33]	Duke [12], 26 B-scans (10/16)	Pre-processing: registering and averaging images, removing any over smoothed images	DRGAN	Median, Bilateral, NLM, Wavelet, BM3D, SNR-GAN, NWSR, edge-sensitive cGAN, HDCycleGAN, Nonlocal GAN, SiameseGAN	Adversarial + reconstruction + cycle-consistency + novel noise	PSNR = 24.4 dB SSIM = 0.58 CNR = 3.2 dB EPI = 0.98 MSR = 4.8 ENL = 317.4 dB
Yu, 2018 [34]	Duke [12], 15 B-scans (8/3/4)	Pre-processing: crop, removing unaligned images	DN-GAN	BM3D, BM3DPCA, LPGPCA, FFDNET	MSE + adversarial	PSNR = 31.0 dB CNR = 3.3 dB MSR = 3.7

Table 4. *Cont.*

Paper	Data Size	Pre-Processing and Augmentation	Model	Compared to	Loss Function	Image Quality Metrics
Tajmirr-iahi, 2021 [35]	Topcon [13], 240 B-Scans (200/40)	Augmentation: rotation, shift, flip, and crop	Autoencoder	GT-SC-GMM, BM3D, MSBTD, Tikhonov	MSE	SNR = 108.8 dB CNR = 82.2 dB ENL = 58.4 dB TP = 0.79 EP = 0.98 CT = 4.68
Sengupta, 2021 [36]	Duke [12], 1600 B-Scans (1400/200)	Pre-processing: crop	EdgeWaveNet	NLM, DeBlur-GAN, RDNSR-GAN, RED-GAN	L1 + adversarial + Sobel edge	PSNR = 22.8 dB SSIM = 0.61
Mehdi-zadeh, 2021 [37]	71 B-scans (51/20)	Augmentation: created patches	DnCNN	NA	L2 + L1 + perceptual + VGG	PSNR = 33.6 dB PSI = 0.23 JNB = 13.9 S3 = 0.26
Cai, 2018 [38]	Topcon [13], 256 B-scans (246/10)	Pre-processing: averaged and registered the B-scans to create denoised image pairs	ResNet	Median, NLM, BM3D	MSE	PSNR = 34.8 dB SSIM = 0.52
Zhou, 2022 [39]	5000 B-scans (4500/480/20)	Pre-processing: crop	Transformer-IP2	BM3D, PNLM, NCDF, OBNLM, DnCNN, CNN-NLM, Neighbor2Neighbor	Neighbor-2Neighnor + PNLM	SNR = 154.6 dB CNR = 7.9 dB ENL = 13,160.3 dB
Anoop, 2021 [40]	Duke [12] and Optima [14], 2720 B-scans (2176/544)	Pre-processing: noise distribution is found for each image, patches and denoised image pairs were created	DenseNet121	CAD, OBNLM, TVG, Wavelet, K-SVD, DnCNN	Cross-entropy	PSNR = 31.0 dB SSIM = 0.91
Fu, 2021 [41]	Duke [12], 21 B-scans (16/5)	Pre-processing: registering and averaging images, removed any over smoothed images	ADGAN	Wavelet, NLM, BM3D, NWSR, HDCycleGAN	Adversarial + cycle-consistency	PSNR = 27.6 dB SSIM = 0.62 CNR = 3.1 dB ENL = 530.8 dB
Wang, 2021 [42]	Topcon [13] and Cirrus [15], 1920 B-scans (512/1408)	Pre-processing: creating denoised image pairs from [23]	Capsule cGAN	BM3D, K-SVD, NLM, MAP, STROLLR, DnCNN, ResNet, Cycle-GAN, cGAN	L1 + Adversarial + SSIM	SNR = 59.0 dB CNR = 11.4 dB ENL = 417.2 dB EPI = 1.0

Table 4. *Cont.*

Paper	Data Size	Pre-Processing and Augmentation	Model	Compared to	Loss Function	Image Quality Metrics
Zhou, 2022 [43]	Topcon [13] and Cirrus [15], 1920 B-scans (512/1408)	Pre-processing: registering and averaging images	Cycle-GAN with mini-cGAN	NLM, BM3D, STROLLR, K-SVD, MAP, DnCNN, DPDNN, NAGAN with mini-cGAN	L1 + MSE	SNR = 20.9 dB CNR = 12.5 dB SSI = 0.09 EPI = 0.99
Wu, 2021 [44]	3737 B-scans, (3537/200)	Pre-processing: crop and contrast enhancement	cGAN	Cycle-GAN, DnCNN, BM3D, DCWT, NLM, MPE, cGAN, EGAN, SR	Adversarial + cycle-consistency + structural consistency + regularization	SNR = 35.0 dB CNR = 7.2 dB EPI = 0.92 CRSB = 0.14
Das, 2020 [45]	Duke [12] 45 B-scans and 384 OCT volumes, (2000/17)	Pre-processing: crop	SRGAN	SBSDI, SSR, NWSR, SRGAN	Adversarial + cycle-consistency + identity mapping	PSNR = 39.2 dB CNR = 4.7 dB
Huang, 2019 [46]	Duke [12], 26 B-scans (10/16)	Pre-processing: crop	SDSR-OCT	BM3D + Bicubic, NWSR, SRCNN	Pixel + perceptual + GAN	PSNR = 28.1 dB CNR = 4.6 dB ENL = 537.5 dB EPI = 0.95
Ge, 2022 [47]	Duke [12], 10 B-scans	Pre-processing: clear images are obtained by registering and averaging and crop	Self2Self-OCT	BM3D, NWSR, DnCNN, DIP, TSI	Background noise attenuation + self-prediction	PSNR = 24.8 dB SSIM = 0.99
Ma, 2022 [48]	Duke [12], 26 B-scans (10/16)	NA	DSGAN	MIFCN, Edge-sensitive cGAN, SDSR-OCT	Adversarial + SSIM + MSE	PSNR = 28.1 dB SSIM = 0.95 CNR = 3.7 dB
Xie, 2022 [49]	Duke [12], 26 B-scans (22/4)	NA	GAN	K-SVD, BM3D, wGAN, cGAN, SDSR, HDcycleGAN, DRGAN	Adversarial + cycle-consistency + perceptual	PSNR = 27.6 dB EPI = 1.0 CNR = 3.1 dB ENL = 73.8 dB MSR = 5.1 SSIM = 0.65
Xie, 2023 [50]	Duke [12], 26 B-scans (10/16)	NA	MGAN	NLM, BM3D, DnCNN, MIFCN, SDSR-OCT	Adversarial + pixel-level error + BCE+ SSIM	PSNR = 28.1 dB SSIM = 0.95 EPI = 0.99 CNR = 3.6 dB

Table 4. *Cont.*

Paper	Data Size	Pre-Processing and Augmentation	Model	Compared to	Loss Function	Image Quality Metrics
Ahmed, 2022 [51]	Duke [12], 18 B-scans (10/8)	Pre-processing: clean images are obtained by BM3D, BM3DDEB, Weiner and HWT	DenseNet with AG	BM3D, NLM	MSE + pixel difference	PSNR = 23.5 dB CNR = 7.7 dB ENL = 585.5 dB
Ahmed, 2022 [52]	Duke [12] and dentistry, 28 B-scans (18/12)	NA	Autoencoder with MFSK and AG	BM3D, NLM, DnCNN, GAN	MSE + pixel difference	PSNR = 26.9 dB CNR = 7.0 dB ENL = 213.7 dB SSIM = 0.68
Zhou, 2023 [53]	OCT2017 [29] and OCTID [54], 5620 B-scans (5000/ 600/20)	Pre-processing: crop	Transformer-based NLM	N2N, DRGAN, Den-mimic-net, Contourlet, BM3D, INLSM, NLM, OBNLM, PNLM	MSE + gradient	CNR = 15.7 dB SNR = 51.1 dB ENL = 23,787 dB
Kande, 2020 [55]	Duke [12], 28 B-Scans (10/18)	NA	SiameseGAN	MSBTD, MIFCN, Shared Encoder, WGAN U-Net, WGAN ResNet	MS-SSIM + perceptual	PSNR = 28.3 dB SSIM = 0.83 MSR = 4.2 CNR = 2.6 dB TP = 0.68 EP = 0.66
Qiu, 2020 [56]	Duke [12], 52 groups of 50 B-scans each (37/15)	Pre-processing: crop	DBPN	BM3D, Bicubic, NWSR, U-Net	MSE	PSNR = 31.3 dB RMSE = 0.027 MS-SSIM = 0.92
Zhou, 2021 [57]	Topcon [13] and Cirrus [15], 521 B-scans (512/9)	Pre-processing: registering and averaging images Augmentation: flip, scaling, rotation, non-rigid transformation	GAN with HRNet	NLM, STROLLR, DnCNN, DPDNN, Edge-cGAN, mini-cGAN	L1 + MSE + Adversarial	SNR = 40.4 dB CNR = 11.2 dB SSI = 0.09 EPI = 0.96

Table 4. *Cont.*

Paper	Data Size	Pre-Processing and Augmentation	Model	Compared to	Loss Function	Image Quality Metrics
Ahmed, 2022 [58]	Duke [12], 18 B-scans (10/8)	NA	DnCNN	BM3D, Weiner, NLM	CNR + pixel difference	PSNR = 29.6 dB CNR = 11.5 dB ENL = 1196.6 dB

Keywords: MSE—Mean squared error; MAE—Mean absolute error; SNR—Signal-to-noise ratio; PSNR—Peak signal-to-noise ratio; CNR—Contrast-to-noise ratio; SSIM—Structural similarity index measure; MS-SSIM—Multi-scale structural similarity index measure; ENL—Equivalent number of looks; EPI—Edge preservation index; RMA-GAN—Realistic mixed attention GAN; SRGAN—Super-resolution GAN; EGAN—Enhanced GAN; SRCNN—Super-resolution CNN; SSR—Self Super-resolution; SRResNet—Super-resolution ResNet; WGAN—Wasserstein GAN; DeGAN—Denoising GAN; BCE—Binary Cross Entropy; MGAN—Multi-task generative adversarial network; HWT—Hyperanalytic Wavelet Transform; AG—Attention Gate; MKSF—Multi-kernel speckle filtering block; N2N—Neighbor2Neighbor; DRGAN—Disentangled representation generative adversarial network; NLM—Nonlocal means.

4.2. Risk of Bias and Applicability

Every study included in the review underwent a thorough assessment of risk of bias, with the results meticulously documented in Table S2. Out of the studies evaluated, 10 (27%) were identified as having a low risk of bias across all four domains. Notably, the domain presenting the most challenges was the "index test", with only 22 studies (59.5%) categorized as having a low risk of bias in this area. This finding underscores the importance of critically evaluating the methodology and execution of the index test within each study to ensure the reliability and validity of the findings. Through this rigorous risk-of-bias assessment, we aimed to provide a comprehensive evaluation of the methodological robustness of the included studies, thereby enhancing the credibility and trustworthiness of our review outcomes.

4.3. Findings of the Studies

When focusing on peak signal-to-noise ratio (*PSNR*) and signal-to-noise ratio (*SNR*) as metrics for denoising images, a significant variation in deep learning (DL) techniques across different types of optical coherence tomography (OCT) images was observed. Specifically, for studies utilizing Optic Nerve Head (ONH) images, *PSNR* or *SNR* values ranged from 8.1 to 25.7 dB, while, for retinal datasets, these metrics ranged from 26.4 to 158.6 dB.

The majority of studies predominantly applied public retinal datasets such as Duke [12], Topcon [13], and Cirrus [15], primarily due to their extensive availability and widespread use. However, it is noteworthy that these datasets typically do not provide "clean" versus "noisy" image pairs required for denoising tasks. Consequently, researchers resorted to generating clean images through alternative means. For instance, clean images were generated by averaging multiple B-scans or employing traditional programming techniques. In some cases, unsupervised DL techniques were also utilized to generate clean images. These approaches aimed to provide a reliable basis for assessing the effectiveness of DL-based denoising methods, despite the absence of explicitly labeled clean and noisy image pairs within the datasets.

5. Discussion

Numerous DL tasks consisting of detection, segmentation and classification are challenging in ophthalmology since OCT is the main imaging technique in that specific field. OCT introduces speckle noise that adds artifacts, thereby impairing image quality and confounding correct clinical interpretation [59]. Thus, DL has provided solutions for denoising and speckle reduction in OCT images to overcome this problem. In this systematic review, many studies were compiled and analyzed to assess the application of DL for image denoising of OCT images in ophthalmology. Over the past ten years, multiple researchers have shown a developing and promising body of evidence supporting DL for this task. Even so, there was a limited supply of quality studies for comparison across traditional programming and DL. A number of findings require more detailed discussion. The challenge of addressing numerous deep learning (DL) tasks in ophthalmology, including detection, segmentation, and classification, is compounded by the prevalent use of optical coherence tomography (OCT) as the primary imaging technique in this field. OCT introduces speckle noise, which can adversely affect image quality, thereby complicating accurate clinical interpretation. DL techniques have emerged as promising solutions for denoising and reducing speckle artifacts in OCT images, with this systematic review analyzing a multitude of studies to assess the application of DL for image denoising in ophthalmology. Over the past decade, researchers have generated a growing body of evidence supporting the efficacy of DL for this purpose. As shown in Figure 3, the main DL models implemented were the GAN, DnCNN, Autoencoder, U-Net, Noise2Noise, and transformer, etc.

Generative adversarial networks (GANs) exhibit promise in synthesizing synthetic OCT images with reduced noise levels. By training a generator network to generate realistic images and a discriminator network to differentiate between real and generated images, GANs have demonstrated their capability to denoise OCT scans effectively while preserving clinically relevant features. However, GANs can be challenging to train and prone to mode collapse, where the generator produces limited variations in images, potentially limiting their diversity and generalizability in clinical settings. In contrast, Denoising Convolutional Neural Networks (DnCNNs) utilize deep convolutional layers to understand the underlying structure of noisy OCT images and produce clean counterparts. Through iterative training on paired noisy–clean image datasets, DnCNNs efficiently enhance image clarity. However, DnCNNs may struggle with complex noise patterns and require large amounts of labeled data for training, which may be resource-intensive and time-consuming to acquire. The U-Net architecture, characterized by its symmetric encoder–decoder structure with skip connections, has showcased remarkable performance in semantic segmentation tasks, including the denoising of OCT images. By integrating contextual information from various spatial scales, U-Net effectively preserves anatomical structures. Nonetheless, U-Net architectures may suffer from memory inefficiency and computational overhead, particularly when handling high-resolution OCT images. Autoencoders, consisting of an encoder and decoder network, learn to reconstruct input data from a compressed representation. Trained on noisy OCT images, autoencoders can encode essential features suitable for clinical interpretation. However, autoencoders may struggle with capturing complex image structures and may require careful tuning of hyperparameters to achieve optimal performance. Transformers, initially designed for natural language processing tasks, have recently been applied in image processing, including denoising. Leveraging self-attention mechanisms to capture long-range dependencies, transformers effectively preserve spatial information in OCT images. However, transformers may suffer from scalability issues when applied to large-scale image datasets due to their computational complexity and memory requirements. The Noise2Noise approach involves training deep learning models directly on pairs of noisy images, eliminating the need for clean reference images during training. By exploiting inherent redundancies in noisy data, Noise2Noise denoises OCT images without access to ground truth clean images, making it suitable for real-world clinical applications where clean reference images may not be readily available. Nevertheless, Noise2Noise may struggle with highly variable noise patterns and may require careful regularization techniques to prevent overfitting to noise artifacts.

These innovative DL techniques have the potential to revolutionize clinical workflows in ophthalmology by automating the denoising process and improving the efficiency and accuracy of OCT image interpretation. By providing clinicians with high-quality, denoised images, these techniques can facilitate more confident diagnoses and treatment decisions, ultimately leading to improved patient outcomes. However, the scarcity of high-quality studies for comparison across traditional programming and DL remains limited, necessitating a closer examination of key findings.

Firstly, a significant proportion of studies (62%) conducted their DL model training and testing on public datasets, validating their reported image quality metrics and comparing them with other state-of-the-art DL models (reference tests). These studies demonstrated relatively high *PSNR* or *SNR* values for denoised images, indicating substantial improvement in image quality attributable to DL models. Given the considerable noise inherent in OCT images, DL techniques hold significant promise for enhancing diagnosis, segmentation, and detection tasks.

Secondly, only studies utilizing the Duke [12] public dataset had access to clean versus noisy image pairs for training purposes. Other studies, utilizing datasets like Topcon [12] and Cirrus [15], lacked clean images and thus resorted to creating their own algorithms or applying traditional programming techniques to obtain "ground truths" for their B-scans. This variance in ground truth generation methods complicates cross-study comparisons and underscores the importance of clearly outlining and validating strategies for creating clean images to ensure the production of robust models.

Thirdly, there was considerable variability in both the reporting and conduct of the studies, posing challenges for comprehensive quantitative and qualitative analyses across the board. Given the wide array of available image quality metrics beyond *PSNR* or *SNR* (such as *CNR*, *ENL*, and *SSIM*), it was observed that studies typically reported only two or three metrics. This lack of uniformity in reporting was further exacerbated by the implementation of multiple and varied reference tests. For instance, several studies utilized a diverse range of reference tests, resulting in a vague impact assessment on either *PSNR* or *SNR*. Consequently, there is a pressing need for a standardized structure for image denoising studies, encompassing specific reference tests (including the appropriate balance of traditional programming and deep learning models) and a predetermined set of specified image quality metrics (such as *PSNR*, *SSIM*, *CNR*, and *ENL*). Consequently, the feasibility of conducting a meta-analysis was limited, as only a small subset of studies provided similar image quality metrics beyond *PSNR* or *SNR*. This situation underscores the significant consequences stemming from the lack of minimum standards and tools available in the field of image denoising, where diagnostic studies are currently limited to utilizing only two tools (STRAD-AI and QUADS-2). Since the studies included in this review solely reported either *PSNR* or *SNR* of the denoised images in comparison to multiple reference tests, it remains uncertain whether DL provides significant assistance in image denoising and speckle reduction. Ideally, the impact of DL for denoising OCT images in ophthalmology should be demonstrated in practice-based settings and validated by its ability to improve further objectives such as detection [60], classification [61], and segmentation [62], which the majority of included studies did not consider.

Fourthly, a significant proportion (74%) of papers were excluded from the review due to the absence of DL methodologies, instead focusing on improving traditional programming methods such as wavelets and shearlets [63], NLM [64], and BM3D [65]. This suggests that DL has not been extensively investigated for OCT image denoising. The remaining papers were excluded due to inadequate descriptions of datasets and their utilization, as well as a lack of reference testing to demonstrate their impact on *PSNR* and *SNR*.

Fifthly, and perhaps most importantly, none of the studies incorporated input from clinicians regarding the produced results. This is particularly concerning considering that the primary purpose of denoising OCT images is to assist clinicians in diagnosing various retinal diseases. Ideally, clinicians should provide feedback on the reported denoised images, assessing whether useful data has been removed or added that could significantly impact diagnosis accuracy.

Lastly, it is essential to emphasize that OCT is the primary instrument in ophthalmology for capturing multiple types of images (including ONH and retinal images) crucial for detecting retinal diseases. This systematic review has revealed both strengths and limitations, with a systematic and extensive assessment of studies conducted to compare DL for image denoising, speckle reduction, and super resolution for OCT images in ophthalmology; while there was a restricted timeline justified by the focus on DL, a substantial and diverse body of evidence has been presented, validating the necessity for such limitations. Additionally, the absence of a meta-analysis was attributed to the mixed nature of reporting and the lack of quality in comparative result analysis.

Therefore, there has been a growing number of studies investigating the denoising of OCT images in ophthalmology using deep learning, with various computational architectures being explored. Among the reported metrics of image quality, the peak signal-to-noise ratio (*PSNR*) has emerged as a reliable metric for intercomparison, with values spanning from 8.1 to 25.7 dB for ONH images and 26.4 to 158.6 dB for retinal datasets. Moving forward, it is imperative for future studies to clearly outline reference tests and datasets, relying on a common, extensive, and clinically meaningful outcome basis to drive progress in the field.

6. Conclusions

In summary, the landscape of research in denoising Optical Coherence Tomography (OCT) images within ophthalmology has seen a notable surge, with a diverse array of studies delving into the application of deep learning techniques. This exploration has encompassed various computational architectures, reflecting a dynamic and evolving field seeking optimal solutions for image enhancement. Among the array of metrics used to evaluate image quality, the peak signal-to-noise ratio (*PSNR*) has emerged as a consistent benchmark for intercomparison, offering insights into the effectiveness of denoising methodologies. Notably, *PSNR* values have exhibited substantial ranges, underscoring the complexity and variability inherent in OCT image denoising efforts. Looking ahead, it is essential for future investigations to prioritize transparency and standardization in methodologies, particularly in outlining reference tests and datasets. By establishing a common foundation grounded in clinically meaningful outcomes, researchers can foster more robust advancements and ensure the translation of findings into tangible benefits for clinical practice and patient care.

Supplementary Materials: The following supporting information can be downloaded at: https://www.mdpi.com/article/10.3390/jimaging10040086/s1, Table S1: List of excluded studies and the reasons; Table S2: Risk-of-bias assesment of papers reviewed.

Author Contributions: H.A. and R.D. wrote the main manuscript text, and H.A. prepared Figures 1–3. All authors reviewed the manuscript. All authors have read and agreed to the published version of the manuscript.

Funding: This research received no external funding.

Informed Consent Statement: Not applicable.

Conflicts of Interest: The authors declare no conflicts of interest.

Abbreviations

MSE—Mean squared error; SNR—Signal-to-noise ratio; PSNR—Peak signal-to-noise ratio; CNR—Contrast-to-noise ratio; SSIM—Structural similarity index measure; ENL—Equivalent number of looks; BCE—Binary Cross Entropy; MGAN—Multi-task generative adversarial network; HWT—Hyperanalytic Wavelet Transform; AG—Attention Gate; MKSF—Multi-kernel speckle filtering block; N2N—Neighbor2Neighbor; DRGAN—Disentangled representation generative adversarial network; NLM—Nonlocal means; EPI—Edge preservation index; RMA-GAN—Realistic mixed attention GAN; SRGAN—Super-resolution GAN; EGAN—Enhanced GAN; SRCNN—Super-resolution CNN; SSR—Self Super-resolution; SRResNet—Super-resolution ResNet; WGAN—Wasserstein GAN; DeGAN—Denoising GAN; BCE—Binary Cross Entropy; MGAN—Multi-task generative adversarial network. MSUN—Multi-scale U-Net.

References

1. Aumann, S.; Donner, S.; Fischer, J.; Müller, F. Optical Coherence Tomography (OCT): Principle and Technical Realization. In *High Resolution Imaging in Microscopy and Ophthalmology*; Springer International Publishing: Cham, Switzerland, 2019; pp. 59–85.
2. Huang, S.; Tang, C.; Xu, M.; Qiu, Y.; Lei, Z. BM3D-based total variation algorithm for speckle removal with structure-preserving in OCT images. *Appl. Opt.* **2019**, *58*, 62336243. [CrossRef] [PubMed]
3. Tey, K.Y.; Teo, K.; Tan, A.C.S.; Devarajan, K.; Tan, B.; Tan, J.; Schmetterer, L.; Ang, M. Optical coherence tomography angiography in diabetic retinopathy: A review of current applications. *Eye Vis.* **2019**, *6*, 110. [CrossRef] [PubMed]
4. Müller, P.L.; Liefers, B.; Treis, T.; Rodrigues, F.G.; Olvera-Barrios, A.; Paul, B.; Dhingra, N.; Lotery, A.; Bailey, C.; Taylor, P.; et al. Reliability of Retinal Pathology Quantification in Age-Related Macular Degeneration: Implications for Clinical Trials and Machine Learning Applications. *Transl. Vis. Sci. Technol.* **2021**, *10*, 44 [CrossRef] [PubMed]
5. Schmidhuber, J. Deep learning in neural networks: An overview. *Neural Netw.* **2015**, *61*, 85117.
6. Colston, B.W.; Sathyam, U.S.; DaSilva, L.B.; Everett, M.J.; Stroeve, P.; Otis, L.L. Dental OCT. *Opt. Express* **1998**, *3*, 230238. [CrossRef] [PubMed]
7. Lai, Y.-C.; Lin, J.-Y.; Yao, C.-Y.; Lyu, D.-Y.; Lee, S.-Y. Interactive OCT-Based Tooth Scan and Reconstruction. *Sensors* **2019**, *19*, 4234. [CrossRef] [PubMed]
8. Wong, A.; Mishra, A.; Bizheva, K.; Clausi, D.A. General Bayesian estimation for speckle noise reduction in optical coherence tomography retinal imagery. *Opt. Express* **2010**, *18*, 8338–8352. [CrossRef]
9. Novacam Technologies, Inc. How Low-Coherence Interferometry (LCI) Works. 2021. Available online: https://www.novacam.com/technology/how-lci-works/ (accessed on 17 November 2023).
10. McInnes, M.D.F.; Moher, D.; Thombs, B.D.; McGrath, T.A.; Bossuyt, P.M.; Clifford, T.; Cohen, J.F.; Deeks, J.J.; Gatsonis, C.; Hooft, L.; et al. Preferred reporting items for a systematic review and meta-analysis of diagnostic test accuracy studies: The PRISMA-DTA statement. *J. Am. Med. Assoc.* **2018**, *319*, 388396. [CrossRef]

11. Krizhevsky, A.; Sutskever, I.; Hinton, G.E. Imagenet classification with deep convolutional neural networks. *Adv. Neural Inf. Process. Syst. (NIPS)* **2012**, *25*, 10971105. [CrossRef]
12. Fang, L.; Li, S.; Nie, Q.; Izatt, J.A.; Toth, C.A.; Farsiu, S. Sparsity based denoising of spectral domain optical coherence tomography images. *Biomed. Opt. Express* **2012**, *3*, 927942. [CrossRef]
13. Huang, Y.; Zhang, N.; Hao, Q. Real-time noise reduction based on ground truth free deep learning for optical coherence tomography. *Biomed. Opt. Express* **2021**, *12*, 20272040. [CrossRef] [PubMed]
14. Optima—Cyst Segmentation Challenge. 2015. Available online: https://optima.meduniwien.ac.at/optima-segmentation-challenge-1/ (accessed on 17 November 2023).
15. Ma, Y.; Chen, X.; Zhu, W.; Cheng, X.; Xiang, D.; Shi, F. Speckle noise reduction in optical coherence tomography images based on edge-sensitive cgan.*Biomed. Opt. Exp.* **2018**, *9*, 5129–5146. [CrossRef] [PubMed]
16. Mokhtari, M.; Kamasi, Z.G.; Rabbani, H. Automatic detection of hyperreflective foci in optical coherence tomography B-scans using morphological component analysis. In Proceedings of the 39th Annual International Conference of the IEEE Engineering in Medicine and Biology Society (EMBC), Jeju, Republic of Korea, 11–15 July 2017; pp. 1497–1500.
17. Devalla, S.K.; Subramanian, G.; Pham, T.H.; Wang, X.; Perera, S.; Tun, T.A.; Aung, T.; Schmetterer, L.; Thiery, A.H.; Girard, M.J.A. A Deep Learning Approach to Denoise Optical Coherence Tomography Images of the Optic Nerve Head. *Sci. Rep.* **2019**, *9*, 14454. [CrossRef] [PubMed]
18. Cheong, H.; Devalla, S.K.; Chuangsuwanich, T.; Tun, T.A.; Wang, X.; Aung, T.; Schmetterer, L.; Buist, M.L.; Boote, C.; Thiery, A.H.; et al. OCT-GAN: Single step shadow and noise removal from optical coherence tomography images of the human optic nerve head. *Biomed. Opt. Express* **2021**, *12*, 14821498. [CrossRef] [PubMed]
19. Tian, C.; Yang, J.; Li, P.; Zhang, S.; Mi, S. Retinal fundus image superresolution generated by optical coherence tomography based on a realistic mixed attention GAN. *Med. Phys.* **2022**, *49*, 318531998. [CrossRef] [PubMed]
20. Hu, D.; Malone, J.D.; Atay, Y.; Tao, Y.K.; Oguz, I. Retinal OCT Denoising with Pseudo-Multimodal Fusion Network. In *OMIA 2020: Ophthalmic Medical Image Analysis*; Springer: Cham, Switzerland, 2020; pp. 125–135.
21. Akter, N.; Perry, S.; Fletcher, J.; Simunovic, M.; Roy, M. Automated Artifacts and Noise Removal from Optical Coherence Tomography Images Using Deep Learning Technique. In Proceedings of the 2020 IEEE Symposium Series on Computational Intelligence (SSCI), Canberra, ACT, Australia, 1–4 December 2020; pp. 2536–2542.
22. Halupka, K.; Lee, B.A.M.; Lucy, K.; Rai, R.; Ishikawa, H.; Wollstein, G.; Schuman, J.; Garnavi, R. Retinal optical coherence tomography image enhancement via deep learning. *Biomed. Opt. Express* **2018**, *9*, 62056221. [CrossRef] [PubMed]
23. Wei, X.; Liu, X.; Yu, A.; Fu, T.; Liu, D. Clustering-Oriented Multiple Convolutional Neural Networks for Optical Coherence Tomography Image Denoising. In Proceedings of the 11th International Congress on Image and Signal Processing, BioMedical Engineering and Informatics (CISP-BMEI), Beijing, China, 13–15 October 2018; pp. 1–5.
24. Chen, Z.; Zheng, Z.; Shen, H.; Zheng, Z.; Dai, P.; Ouyang, P. DN-GAN: Denoising generative adversarial networks for speckle noise reduction in optical coherence tomography images. *Biomed. Signal Process. Control* **2020**, *55*, 101632. [CrossRef]
25. Gour, N.; Khanna, P. Speckle denoising in optical coherence tomography images using residual deep convolutional neural network. *Multimed. Tools Appl.* **2020**, *79*, 1567915695. [CrossRef]
26. Hasan, M.J.; Alom, M.S.; Fatema, U.; Wahid, M.F. Deep Learning Based Retinal OCT Image Denoising using Generative Adversarial Network. In Proceedings of the International Conference on Automation, Control and Mechatronics for Industry 4.0 (ACMI), Rajshahi, Bangladesh, 8–9 July 2021; pp. 1–6.
27. Guo, A.; Fang, L.; Qi, M.; Li, S. Unsupervised Denoising of Optical Coherence Tomography Images with Nonlocal-Generative Adversarial Network. *IEEE Trans. Instrum. Meas.* **2020**, *70*, 112. [CrossRef]
28. Qiu, B.; Huang, Z.; Liu, X.; Meng, X.; You, Y.; Liu, G.; Yang, K.; Maier, A.; Ren, Q.; Lu, Y. Noise reduction in optical coherence tomography images using a deep neural network with perceptually sensitive loss function. *Biomed. Opt. Express* **2020**, *11*, 817830. [CrossRef]
29. Kermany, D.S.; Goldbaum, M.; Cai, W.; Valentim, C.C.; Liang, H.; Baxter, S.L.; McKeown, A.; Yang, G.; Wu, X.; Yan, F. Identifying medical diagnoses and treatable diseases by image-based deep learning. *Cell* **2018**, *172*, 1122–1131. [CrossRef]
30. Qiu, B.; Zeng, S.; Meng, X.; Jiang, Z.; You, Y.; Geng, M.; Li, Z.; Hu, Y.; Huang, Z.; Zhou, C.; et al. Comparative study of deep neural networks with unsupervised Noise2Noise strategy for noise reduction of optical coherence tomography images. *J. Biophotonics* **2021**, *14*, e202100151. [CrossRef] [PubMed]
31. Abbasi, A.; Monadjemi, A.; Fang, L.; Rabbanni, H.; Zhang, Y. Three-dimensional optical coherence tomography image denoising through multi-input fully convolutional networks. *Comput. Biol. Med.* **2019**, *108*, 18. [CrossRef]
32. Shi, F.; Cai, N.; Gu, Y.; Hu, D.; Ma, Y.; Chen, Y.; Chen, X. DeSpecNet: A CNN-based method for speckle reduction in retinal optical coherence tomography images. *Phys. Med. Biol.* **2019**, *64*, 175010. [CrossRef]
33. Huang, Y.; Xia, W.; Lu, Z.; Liu, Y.; Chen, H.; Zhou, J.; Fang, L.; Zhang, Y. Noise-Powered Disentangled Representation for Unsupervised Speckle Reduction of Optical Coherence Tomography Images. *IEEE Trans. Med. Imaging* **2021**, *40*, 26002614. [CrossRef] [PubMed]
34. Yu, A.; Liu, X.; Wei, X.; Fu, T.; Liu, D. Generative Adversarial Networks with Dense Connection for Optical Coherence Tomography Images Denoising. In Proceedings of the 11th International Congress on Image and Signal Processing, BioMedical Engineering and Informatics (CISP-BMEI), Beijing, China, 13–15 October 2018; pp. 1–5.

35. Tajmirriahi, M.; Kafieh, R.; Amini, Z.; Rabbani, H. A Lightweight Mimic Convolutional Auto-Encoder for Denoising Retinal Optical Coherence Tomography Images. *IEEE Trans. Instrum. Meas.* **2021**, *70*, 18. [CrossRef]
36. Sengupta, S.; Singh, A.; Lakshminarayanan, V. EdgeWaveNet: Edge aware residual wavelet GAN for OCT image denoising. In Proceedings of the Medical Imaging 2021: Imaging Informatics for Healthcare, Research, and Applications, Online, 15–20 February 2021; Volume 11601, pp. 110–115.
37. Mehdizadeh, M.; MacNish, C.; Xiao, D.; Alonso-Caneiro, D.; Kugelman, J.; Bennamoun, M. Deep feature loss to denoise OCT images using deep neural networks. *Biomed. Opt.* **2021**, *26*, 046003. [CrossRef]
38. Cai, N.; Shi, F.; Hu, D.; Chen, Y. A ResNet-based universal method for speckle reduction in optical coherence tomography images. In Proceedings of the IEEE International Symposium on Biomedical Imaging (ISBI), Washington, DC, USA, 4–7 April 2018.
39. Zhou, Q.; Wen, M.; Ding, M.; Zhang, X. Unsupervised despeckling of optical coherence tomography images by combining cross-scale CNN with an intra-patch and inter-patch-based transformer. *Opt. Express* **2022**, *30*, 1880018820. [CrossRef]
40. Anoop, B.N.; Kalmady, K.S.; Udathu, A.; Siddharth, V.; Girish, G.N.; Kothari, A.R.; Rajan, J. A cascaded convolutional neural network architecture for despeckling OCT images. *Biomed. Signal Process. Control* **2021**, *66*, 102463. [CrossRef]
41. Fu, Z.; Yu, X.; Ge, C.; Aziz, M.Z.; Liu, L. ADGAN: An Asymmetric Despeckling Generative Adversarial Network for Unpaired OCT Image Speckle Noise Reduction. In Proceedings of the IEEE 6th Optoelectronics Global Conference (OGC), Shenzhen, China, 15–18 September 2021; pp. 212–216.
42. Wang, M.; Zhu, W.; Yu, K.; Chen, Z.; Shi, F.; Chen, X. Semi-Supervised Capsule cGAN for Speckle Noise Reduction in Retinal OCT Images. *IEEE Trans. Med. Imaging* **2021**, *40*, 11681183. [CrossRef]
43. Zhou, Y.; Yu, K.; Wang, M.; Ma, Y.; Peng, Y.; Chen, Z.; Zhu, W.; Shi, F.; Chen, X. Speckle Noise Reduction for OCT Images Based on Image Style Transfer and Conditional GAN. *IEEE J. Biomed. Health Inform.* **2022**, *26*, 139150. [CrossRef] [PubMed]
44. Wu, M.; Chen, W.; Chen, Q.; Park, H. Noise Reduction for SD-OCT Using a Structure-Preserving Domain Transfer Approach. *IEEE J. Biomed. Health Inform.* **2021**, *25*, 34603472. [CrossRef] [PubMed]
45. Das, V.; Dandapat, S.; Bora, P.K. Unsupervised Super-Resolution of OCT Images Using Generative Adversarial Network for Improved Age-Related Macular Degeneration Diagnosis. *IEEE Sens. J.* **2020**, *20*, 87468756. [CrossRef]
46. Huang, Y.; Lu, Z.; Shao, Z.; Ran, M.; Zhou, J.; Fang, L.; Zhang, Y. Simultaneous denoising and super-resolution of optical coherence tomography images based on generative adversarial network. *Opt. Express* **2019**, *27*, 1228912307. [CrossRef] [PubMed]
47. Ge, C.; Yu, X.; Li, M.; Mo, J. Self-Supervised Denoising of single OCT image with Self2Self-OCT Network. In Proceedings of the IEEE 7th Optoelectronics Global Conference (OGC), Shenzhen, China, 6–11 December 2022; pp. 200–204.
48. Ma, Z.; Xie, Q.; Fan, F.; Zhu, J. DSGAN: A generative model for speckle noise reduction in retinal optical coherence tomography images. In Proceedings of the SPIE 12320, Optics in Health Care and Biomedical Optics XII, Online, 5–12 December 2022; p. 123201H.
49. Xie, K.; Luo, M.; Chen, H.; Yang, M.; He, Y.; Liao, P.; Zhang, Y. Speckle denoising of optical coherence tomography image using residual encoder–decoder CycleGAN. *Signal Image Video Process.* **2023**, *17*, 1521–1533. [CrossRef]
50. Xie, Q.; Ma, Z.; Zhu, L.; Fan, F.; Meng, X.; Gao, X.; Zhu, J. Multi-task generative adversarial network for retinal optical coherence tomography image denoising. *Phys. Med. Biol.* **2022**, *68*, 045002. [CrossRef] [PubMed]
51. Ahmed, H.; Zhang, Q.; Donnan, R.; Alomainy, A. Framework of Unsupervised based Denoising for Optical Coherence Tomography. In Proceedings of the International Conference on Biomedical Signal and Image Processing (ICBIP), Suzhou, China, 19–21 August 2022; pp. 19–24.
52. Ahmed, H.; Zhang, Q.; Donnan, R.; Alomainy, A. Attention Based Speckle Reduction for Optical Coherence Tomography in Ophthalmology and Dentistry. In Proceedings of the International Congress on Image and Signal Processing, BioMedical Engineering and Informatics (CISP-BMEI), Beijing, China, 5–7 November 2022; pp. 1–6.
53. Zhou, Q.; Wen, M.; Yu, B.; Lou, C.; Ding, M.; Zhang, X. Self-supervised transformer based non-local means despeckling of optical coherence tomography images. *Biomed. Signal Process. Control* **2023**, *80*, 104348. [CrossRef]
54. Gholami, P.; Roy, P.; Parthasarathy, M.K.; Lakshminarayanan, V. OCTID: Optical coherence tomography image database. *Comput. Electr. Eng.* **2020**, *81*, 106532. [CrossRef]
55. Kande, N.A.; Dakhane, R.; Dukkipati, A.; Yalavarthy, P.K. SiameseGAN: A Generative Model for Denoising of Spectral Domain Optical Coherence Tomography Images. *IEEE Trans. Med. Imaging* **2021**, *40*, 180192. [CrossRef]
56. Qiu, B.; You, Y.; Huang, Z.; Meng, X.; Jiang, Z.; Zhou, C.; Liu, G.; Yang, K.; Ren, Q.; Lu, Y. N2NSR-OCT: Simultaneous denoising and super-resolution in optical coherence tomography images using semisupervised deep learning. *J. Biophotonics* **2020**, *14*, e202000282. [CrossRef]
57. Zhou, Y.; Li, J.; Wang, M.; Zhu, W.; Peng, Y.; Chen, Z.; Wang, L.; Wang, T.; Yao, C.; Wang, T.; et al. High-Resolution Hierarchical Adversarial Learning for OCT Speckle Noise Reduction. In *MICCAI 2021: Medical Image Computing and Computer Assisted Intervention*; Springer: Cham, Switzerland, 2021; pp. 372–381.
58. Ahmed, H.; Zhang, Q.; Donnan, R.; Alomainy, A. Unsupervised Region-Based Denoising for Optical Coherence Tomography Framework. In Proceedings of the International Conference on Computational Intelligence and Applications (ICCIA), Nanjing, China, 24–26 June 2022; pp. 267–273.
59. Hao, S.; Hao, G. Research on OCT Image Processing Based on Deep Learning. In Proceedings of the International Conference on Electronics Information and Emergency Communication (ICEIEC), Beijing, China, 17–19 July 2020; pp. 208–212.

60. Abbas, Q.; Qureshi, I.; Yan, J.; Shaheed, K. Machine Learning Methods for Diagnosis of Eye-Related Diseases: A Systematic Review Study Based on Ophthalmic Imaging Modalities. *Arch. Comput. Methods Eng.* **2022**, *29*, 3861–3918. [CrossRef]
61. Syed, A.M.; Faizan, M.; Akbar, M.U.; Fatima, J. Automated Techniques for Detection and Classification of Diabetic Macular Edema: A Review. *Asian J. Eng. Sci. Technol.* **2016**, 39–43.
62. Stankiewicz, A.; Marciniak, T.; Dabrowski, A.; Stopa, M.; Marciniak, E.; Obara, B. Segmentation of Preretinal Space in Optical Coherence Tomography Images Using Deep Neural Networks. *Sensors* **2021**, *21*, 7521. [CrossRef] [PubMed]
63. Jian, Z.; Yu, L.; Rao, B.; Tromberg, B.J.; Chen, Z. Three-dimensional speckle suppression in optical coherence tomography based on the curvelet transform. *Opt. Express* **2010**, *18*, 1024103. [CrossRef]
64. Buades, A.; Coll, B.; Morel, J.M. A non-local algorithm for image denoising. In Proceedings of the 2005 IEEE Computer Society Conference on Computer Vision and Pattern Recognition (CVPR'05), San Diego, CA, USA, 20–25 June 2005; pp. 60–65.
65. Dabov, K.; Foi, A.; Katkovnik, V.; Egiazarian, K. Image Denoising by Sparse 3-D Transform-Domain Collaborative Filtering. *IEEE Trans. Image Process.* **2007**, *16*, 20802095. [CrossRef]

MDPI AG
Grosspeteranlage 5
4052 Basel
Switzerland
Tel.: +41 61 683 77 34

Journal of Imaging Editorial Office
E-mail: jimaging@mdpi.com
www.mdpi.com/journal/jimaging

www.ingramcontent.com/pod-product-compliance
Lightning Source LLC
Chambersburg PA
CBHW061513210926
51850CB00008B/1905
9783725861897